BROADBAND SIGNALLING EXPLAINED

Edited by
Dick Knight
BT, UK

JOHN WILEY & SONS, LTD
Chichester • New York • Weinheim • Brisbane • Singapore • Toronto

Copyright © 1995, 2000 by John Wiley & Sons Ltd,
Baffins Lane, Chichester,
West Sussex PO19 1UD, England

National 01243 779777
International (+44) 1234 779777
e-mail (for orders and customer service enquiries): cs-books@wiley.co.uk
Visit our Home Page on http://www.wiley.co.uk or http://www.wiley.com

All Rights Reserved. No part of this publication may be reproduced, stored in a retrieval system, or transmitted, in any form or by any means, electronic, mechanical, photocopying, recording, scanning or otherwise, except under the terms of the Copyright Designs and Patents Act 1988 or under the terms of a licence issued by the Copyright Licensing Agency, 90 Tottenham Court Road, London, UK W1P 9HE, UK, without the permission in writing of the Publisher, with the exception of any material supplied specifically for the purpose of being entered and executed on a computer system, for exclusive use by the purchaser of the publication.

Neither the authors nor John Wiley & Sons Ltd accept any responsibility or liability for loss or damage occasioned to any person or property through using the material, instructions, methods or ideas contained herein, or acting or refraining from acting as a result of such use. The authors and Publisher expressly disclaim all implied warranties, including merchantability of fitness for any particular purpose.

Designations used by companies to distinguish their products are often claimed as trademarks. Readers, however, should contact the appropriate companies for more complete information regarding trademarks and registration.

Other Wiley Editorial Offices

John Wiley & Sons, Inc., 605 Third Avenue,
New York, NY 10158-0012, USA

WILEY-VCH Verlag GmbH
Pappelallee 3, D-69469 Weinheim, Germany

Jacaranda Wiley Ltd, 33 Park Road, Milton,
Queensland 4064, Australia

John Wiley & Sons (Asia) Pte Ltd, 2 Clementi Loop #02-01,
Jin Xing Distripark, Singapore 129809

John Wiley & Sons (Canada) Ltd, 22 Worcester Road
Rexdale, Ontario, M9W 1L1, Canada

British Library Cataloguing in Publication Data

Broadband signalling explained / edited by Dick Knight.
   p.   cm.
   'Wiley–BT series.'
   Includes bibliographical references.
   ISBN 0–471–97846–9 (alk. paper)
   1. Broadband communication systems. 2. Signal processing.
I. Knight, Dick.
TK5103.4.B7678 2000
621.382'16–dc21   99–38882
                                                                          CIP

Library of Congress Cataloging-in-Publication Data

A catalogue record for this book is available from the British Library

ISBN 0 471 97846 9

Typeset in 10/12pt Palatino by Vision Typesetting, Manchester
Printed and bound in Great Britain by Antony Rowe Ltd, Chippenham, Wiltshire
This book is printed on acid-free paper responsibly manufactured from sustainable forestry, in which at least two trees are planted for each one used for paper production.

Dedication

I would like to dedicate this book to my wife, Gill.
Without her patience, support, love and understanding
this book would never have been published.

Dedication

BROADBAND SIGNALLING EXPLAINED

About the Wiley — BT Series

The titles in the Wiley-BT Series are designed to provide clear, practical analysis of voice, image and data transmission technologies and systems, for telecommunications engineers working in the industry. New and forthcoming works in the series also cover the Internet software systems, solutions, engineering and design.

Other titles in the Wiley-BT Series:

Software Engineering Explained
Mark Norris; Peter Rigby and Malcolm Payne, 1992, 210pp, ISBN 0-471-92950-6

Teleworking Explained
Mike Gray, Noel Hodson and Gil Gordon, 1993, 310pp, ISBN 0-471-93975-7

The Healthy Software Project
Mark Norris, Peter Rigby and Malcolm Payne, 1993, 198pp, 0-471-94042-9

High Capacity Optical Transmission Explained
Dave Spirit and Mike O'Mahony, 1995, 268pp, 0-471-95117-X

Exploiting the Internet
Andy Frost and Mark Norris, 1997, 262pp, 0-471-97113-8

Media Engineering
Steve West and Mark Norris, 1997, 250pp, 0-471-97287-8

ISDN Explained, Third Edition
John Griffiths, 1992, 306pp, 0-471-93480-1

Total Area Networking, Second Edition
John Atkins and Mark Norris, 1999, 326pp, ISBN 0-471-98464-7

Designing the Total Area Network
Mark Norris and Steve Pretty, 2000, 352pp, ISBN 0-471-85195-7

Broadband Signalling Explained
Dick Knight, 2000, 448pp, ISBN 0-471-97846-9

Contents

Forward   ix
   John Griffiths

About the Editor   xi
About the Contributors   xiii
Acknowledgments   xxi

Introduction   1
   Dick Knight

1 ATM Switching   7
      Dick Knight

2 Introduction to Signalling   23
      Dick Knight and Bryan Law

3 Signalling Standards   43
      Bryan Law and Dick Knight

4 Requirements for Signalling—Broadband Services   61
      Roger Morley and Dick Knight

5 Access Signalling   79
      Bryan Law and Dick Knight

6 The ATM Forum Signalling Protocols and their Interworking   107
      Nick Cooper

7 The ATM Forum's Private Network Network Interface   125
      Jennifer Scott and Ian Jones

8 B-ISUP, ITU-T's Internodal Broadband Signalling Protocol   147
      Ian Jones

9 The VB5 Interface   169
      Mick Hale, Alex Gillespie and Keith James

10 The use of Session Control in DAVIC to Provide Interactive Multimedia Services   191
      Richard Miles, Paul Reece and Laurent Boon

11 Design for Performance of Broadband Signalling and Services   213
      Dave Morris and Ann Elvidge

12 Broadband VPN Signalling   227
    Frank Allard

13 UMTS: The Mobile Part of Broadband Communications for the Next Century   241
    Alan Clapton, Nigel Lobley, Steve Dutnall, Mark Dando and Pedro Serna

14 Signalling with Objects   261
    Paul McDonald

15 The Call Control Protocol in a Separated Call and Bearer Environment   277
    Dick Knight and Bryan Law

16 Supporting Applications with Network Intelligence and B-ISDN   301
    Gary Bruce and Jon Clark

Appendix A: Answers to Questions   327

Appendix B: The Development of Broadband Signalling Platforms   353
    Chris Shephard, Peter Hovell, Steve Boswell and John King

Appendix C: The Broadband Call Control Demonstrator—A Demonstrator Platform for ITU-T, DAVIC and TINA-C Implementations   379
    Paul Reece, Richard Macey and Peter Clarke

Appendix D: Index of International Standards   403
    Anselm Martin and Dick Knight

Appendix E: Abbreviations   415

Index   423

Foreword

For the last 30 years telecommunications users have thought that they were looking for more bandwidth so that their information transfer could be faster or with better quality. The heroic development of Optical Fibre Transmission Systems has largely met this need. I have in my possession a small reel of 6 km of optical fibre which, under the right circumstances, could provide me with 50 Gbit/s bandwidth but still I am not happy. The reason is that bandwidth is not enough; what I need is connectivity.

The joy with which the Internet was embraced demonstrates the truth of this. It was not the large bandwidths that drove the Internet. It was the fact that you appeared to be connected to all the computers all over the world at the same time. Now we all know that appearances can be deceptive but, on the other hand, it is perception that counts for the user. The Internet-working Protocol was the magic which unlocked the door to connectivity. However, now we want this associated with enough bandwidth to browse moving pictures and high quality sound. A move to broadband networks is necessary and, as this must retain the connectivity that users have come to expect, the requirements for the signalling system are very demanding. It must handle all the connectivity demands of the user whilst making reasonably efficient use of the network infrastructure. Telecommunications engineers of the future will therefore need to have a thorough grounding in the principles and practice of the signalling system used to manage and control the broadband network. Dick Knight's book is thus very timely and an excellent addition to the BT 'something' Explained series.

I am very pleased to be allowed by Dick to write this foreword. Not only is he an excellent engineer as you will realise from the contents of this book, but also he is a wonderful person to know. His expedition to deliver aid to the people of Romania in 1993 has been beautifully described in Gill and Dick Knight's book 'My Helicopter is Full of Pasta'.

I am sure that you will find some of the excitement and achievement of that book has endured into this one.

Professor John Griffiths
Queen Mary and Westfield College
University of London

About the Editor

Dick Knight joined the Post Office Telecommunications Circuit Labs as an apprentice in 1971 and moved to the Post Office Research Centre (now BT Labs) in 1979. He worked on System X development in the areas of the Test Network Subsystem and customer line card design acceptance testing. In 1986 he was made responsible for the software to implement a programmable ISDN signalling design acceptance tester. Dick designed and implemented a protocol to collect network management information for the 'Jetphone' network in 1992. From 1993 to 1994 he moved onto Broadband Signalling design by managing the collaborative RACE MAGIC project on behalf of fourteen international companies—a project whose results have impacted broadband signalling design in the world. With a short break in 1996 to design and patent an application protocol to collect information such as burglar alarms over the ISDN D channel packet access, he has continued to work on various aspects of Broadband Signalling, both on internal projects and representing BT at ETSI SPS5 and ITU-T Study Group 11 where he has contributed some of the MAGIC results. Dick continues to co-ordinate some of BT's standards representation in the technical area of signalling.

This is Dick's second book—his first, co-authored with his wife, told the story of twelve ordinary people's adventures during a Romanian national disaster when flooding devastated an area the size of Wales. It is called 'My Helicopter is Full of Pasta', and, like this book, the author's royalties are donated to charity.

About the Contributors

Frank Allard, the author of Chapter 12, graduated from De Montfort University in Leicester in 1995 with a BEng Honours in Electrical & Electronic Engineering and from Aston University in 1996 with a MSc in Telecommunications before he joined BT as a student to provide a solution for the interworking between QSIG and ISUP. He lead the Feature Transparency project for BT, Concert, and the Joint Venture (JV) partners, which aims to provide narrowband Global Virtual Private Network Supplementary Services (for QSIG and DPNSS) transparently over a common platform, independent of the switch manufacturers, and was involved in Global Signalling network design for BT European JV partners. He has also been involved in Broadband QSIG signalling protocols, representing BT at ECMA TG15 meetings. He now works at the technical direction of the BT JV in France.

Laurent Boon, who jointly wrote Chapter 10, graduated from Durham University in 1995 with an Honours degree in Applied Physics and Computer Science. He joined BT in the same year where he has been working on Broadband network design and prototyping.

Steve Boswell, who contributed to Appendix B, joined BT Laboratories in 1994 after graduating from Swansea University with a B.Sc. in Computer Science with Electronics in 1993, followed by an M.Sc in Communication Systems. Since then he has worked in the Network Intelligence Centre and primarily been involved in the design and development of software for a range of systems, including network protocols, billing interfaces and broadband signalling. He is currently a member of the Intelligent Solutions Unit investigating telephony over IP and Network Centric Computing.

Gary Bruce (joint author of Chapter 16) received a BSc Honours degree in Information Technology from Leicester Polytechnic in 1988. After graduating, he joined the digital services division at BT Laboratories and has been involved in the production of signalling test equipment and the formal specification of signalling systems. In 1993, he joined the network intelligence engineering centre to develop broadband signalling systems and to standardise broadband signalling requirements within the ITU-T. He is an Associate Member of the IEE.

Alan Clapton, the joint author of Chapter 13, has worked for BT since 1967; his career has included aeronautical and maritime radio developments as well as GSM service and network developments, Recent work has included supporting Cellnet on GSM data and value added services as well as intelligent networks. In parallel he has lead the BT labs. team developing UMTS technical aspects.

Jonathan Clark (joint author of Chapter 16) received a BSc Honours degree in Electronic Engineering from Lancaster University in 1988. After graduating, he joined the digital services division at BT Laboratories and was involved in the testing of signalling test equipment and then built the Jetphone Network Management System communications agent. He has been involved in various IN projects and represented BT in Eurescom projects concerned with a Pan European IN and a Harmonised IN/B-ISDN Networks. He obtained the BT MSc in Telecommunication Engineering in 1996 and is an Associate Member of the IEE.

Peter Clarke, who contributed to Appendix C, graduated from Leeds University with a PhD in Polymer Physics in 1982. He joined BT in 1984 working initially on X.25 services and Narrowband ISDN. Subsequently he became involved in studies of broadband services. He developed an implementation of the signalling protocol Q.2931 and worked in a team considering the suitability of DAVIC protocols to offer broadband services. Currently he is supporting BTs SVC work within the JAMES project.

Nick Cooper, who wrote Chapter 6, joined BT Laboratories in 1973 after graduating from Cambridge University. His work has included X.25 packet switch design and the support of packet switching within the narrowband ISDN environment. He currently leads a team exploring signalling and call control issues for ATM and has represented BT on signalling matters at the ATM Forum.

Mark Dando, who contributed to Chapter 13, has been working at BT Laboratories since 1995. He currently co-ordinates the radio aspects and advises on the industry structure of UMTS for the BT Group. His background also involves investigation of different radio planning strategies for Cellnet's live GSM network.

Steve Dutnall, who contributed to Chapter 13, joined BT Laboratories in October 1996, after completing a 6 month project placement at BT as part of his MSc. He has worked on the impact of IN onto GSM (CAMEL) and

UMTS. He is currently contributing to the BT Groups projects on UMTS, focusing on the transport mechanisms and network evolution for this third generation system.

Anne Elvidge, joint author of Chapter 11, joined BT Laboratories following her PhD at the University of Surrey in 1985, and initially worked on voice network design and performance. Following a secondment to Syncordia, she moved into modeling SS No. 7 and broadband signalling networks. She currently represents BT at the ITU-T on network operations, and following completion of her MBA at Henley Management College in 1996, she is active in business modeling, providing support for business cases for new services.

Alex Gillespie, who was one of the authors of Chapter 9, gained a masters degree from Cambridge University and a doctorate from Durham University after his first class honours degree and prize for the year from St. Andrews, Scotland. He has since worked on both sides of the Atlantic on a number of aspects of telecommunications. Since 1988 he has been at BT and has been the editor of a number of telecommunications standards, including management standards for V5, VB5 and ATM switches handling SVCs. He is currently chairman of ETSI TMN2-3 on access and switch management and is rapporteur for the corresponding ITU-T work, Q21 of study group 4. He is the author of the book *Access Networks: Technology and V5 Interfacing*, and is currently part of the access OSS design team at BT Laboratories.

Mick Hale, who was one of the authors of Chapter 9, joined BT as an apprentice in 1973. He initially worked on transmission, maintenance and private services installation in Colchester before transferring to London to work on the installation of a corporate data network for CSS. He moved to BT Laboratories in 1988, initially working on embedded control systems, before moving to what is now the fibre systems unit. There he worked on the development and deployment of TPON, and later PON systems for the BT interactive TV trial. He is now working on access network ATM standards and solution design for IP dial access.

Peter Hovell, one of the authors of Appendix B, graduated from the Royal Military College of Science in 1978 with a BSc degree in Electrical Engineering. After a number of years working for the Civil Aviation Authority on the development of Radar processing equipment Peter joining BT. Initially he worked on the system specification and design of novel local access mechanisms before becoming involved with two RACE projects dealing with broadband ATM demonstrators. During the last few years Peter has been associated with a number of projects constructing broadband demonstrators at BT labs and is currently leading an initiative to investigate the core network design requirements for very large IP networks.

Keith James, who was one of the authors of Chapter 9, joined BT as an apprentice. After receiving a first class honours degree in electrical and electronic engineering in 1982, he became involved in access network signalling. Recent activities have been centred around ETSI and ITU-T standardisation work, where he is recognised as a leading expert in V5, NMDS and VB5 interfaces. He is a chartered engineer and a member of the IEE.

Ian Jones (who wrote Chapter 8 and co-authored chapter 7) joined BT in September 1994 after graduating from Aston University, with a MSc in Telecommunications Technology. After joining, Ian worked on producing software models to demonstrate the behaviour of broadband internodal and access signalling protocols. He now works on defining the future signalling capabilities related to BTs broadband IP programmes. Ian has represented BT at the ITU-T B-ISUP standardisation meetings and also at the UK PNO-IG ATM task force committee. He is also an associate member of the IEE.

John King, contributing to Appendix B, graduated in 1989 with a BSc(Hons) in Electronics and Microcomputer Systems from the University of Dundee. He joined the Signalling and Protocols unit where he was involved in investigating methods for formally specifying and testing signalling protocols. Latterly, his move to the Intelligence and Internet Engineering unit has seen him gain new skills in the broadband and internet arena. John has an MSc in Telecommunications for BT from the University of London.

Bryan Law (who wrote Chapters 2, 3, 5 and 15) joined BT (then General Post Office Telephones) as an apprentice in 1958. After working on subscriber apparatus and Strowger exchange maintenance he was then involved in transmission repeater station duties. In 1967 he joined the Engineer-in-Chief's Office, being responsible for trunk cable installation standards and then commissioning of FDM and digital transmission equipment. He returned to his former field of switching in 1980 when joining the Telecommunications Strategic Studies department undertaking information flow studies for System X and the specification of CCITT SS No. 7 for use in the BT network. During this period, he represented BT at the international standards bodies responsible for CCITT SS No. 7 and became the UK co-ordinator for the studies in this area. In 1989 he joined the RACE project studying access signalling for the support of B-ISDN. He then represented BT in ETSI and ITU-T studies of B-ISDN signalling being responsible for co-ordinating BT activity in the area of access signalling. He was the UK co-ordinator for B-ISDN signalling studies, until retiring from BT in March 1998.

ABOUT THE CONTRIBUTORS

Nigel Lobley, the joint author of Chapter 13, has been working at BT Laboratories since 1990 and has worked on Network aspects for GSM, IN and UMTS. He is currently providing the technical lead for the BT Groups Network Architecture project on UMTS, has participated in EC projects on Mobile and IN Networks and has developed IN and GSM Network aspects for Cellnets live GSM network in the UK.

Richard Macey (who helped to write Appendix C) graduated from the University of Newcastle Upon Tyne with a BEng (Hons) in Electronic Engineering in 1989 and joined a team investigating broadband networks. He has worked on the control and signalling aspects of Mobile and Intelligent networks, including the development of a number of demonstrator platforms. He is currently leading a task investigating the control requirements of integrated IP and ATM networks, and working in a team developing longer-term network intelligence architectures.

Anselm Martin, who co-authored Appendix D graduated from Brunel University with a Beng (Hons) in electrical engineering. He joined BT in 1994 and worked on Signalling System No. 7 management before moving to ISDN Signalling in 1996. He currently works on ATM standardisation.

Paul McDonald, the author of Chapter 14, graduated with a BEng (Hons.) in Electronic & Microprocessor Engineering from Strathclyde University. He joined BTs research laboratories to undertake studies into ISDN/mobile interworking and B-ISDN signalling concepts. This was followed by investigations into broadband services and simulation for the RACE MAGIC project. For his last two years at BT he looked at various architecture issues for B-ISDN and developed a WWW based interface for the BT Magic service. Since 1998 Paul has been working for the GSM Research group at Motorola where he is currently studying third generation mobile networks and services.

Richard Miles (who co-authored Chapter 10) joined BT in 1986 as a Trainee Technician Apprentice. He then worked on optical access network technology until 1990, when he gained sponsorship to attend the University of Kent. He graduated in 1993 with a BEng(Hons) in Electronic Engineering and joined the Intelligence Design and Performance Unit, where he has been designing and prototyping systems for the delivery of broadband services.

Roger Morley (the joint author of Chapter 4) started with BT in 1980 as a Trainee Technician Apprentice (TTA) in the Nottingham Telephone Area. Upon completing his apprenticeship in 1983, he undertook a variety of Technician level work involving customer apparatus installation, exchange jumpering and finally Strowger telephone exchange maintenance. In 1985, he was successful in gaining a BT minor award to read for a B.Eng. degree in Telecommunications Engineering at Queen Mary College, London. Upon graduation, he then worked at BT Laboratories on a PC based X.21 communications card for use over BTs pilot ISDN service. His interest in ISDN and ISDN applications grew from this point and was eventually moved into BTs ISDN Services and Applications Laboratory where he developed notable expertise in ISDN signalling systems, Group3/4 facsimile and the end-end testing of international ISDN links. In 1993, he gained an M.Sc. degree in Telecommunication and Information systems from the University of Essex which started the development of his B-ISDN and ATM networking skills. In 1994, he moved over to work on B-ISDN service definition within RACE MAGIC, which was specifying signalling systems for advanced, multi-media, multi-party services for release 2/3 of B-ISDN. Following the completion of RACE MAGIC in 1994, he was assigned to BT internal projects which are now concerned with introducing broadband signalling and intelligence into BTs future broadband networks. Mr. Morley is also an Associate Member of the IEE.

Dave Morris, joint author of Chapter 11, received an MSc in Opto-electronics and Information Theory from Queen's University of Belfast in 1988. Prior to this he obtained a BSc in Physical Electronics in 1986. Since 1988 he has been working at BT Laboratories, initially on the design of high speed optical modulators and then on third generation mobile systems. He currently works on the performance analysis and dimensioning of narrowband and broadband networks.

Paul Reece (who co-authored Chapter 10 and Appendix C) graduated from Coventry Polytechnic with a BSc(Hons) in Mathematics in 1991 and joined the broadband and Data networks Unit. He was involved in a wide range of activities related to broadband signalling and network intelligence for six years. He moved to the Intelligence Design and Performance Unit where he now leads a workpackage on network intelligence architectures and design.

Jennifer Scott (joint author of Chapter 7) joined the Submarine Systems Section of BT Labs in 1988. In 1994 she joined the Broadband and Data Networks Unit, working on Call Control and Signalling for ATM where she became a member of the Cellstream project, responsible for definition of SVC service requirements for BT's Cellstream ATM Service and the Business Broadband Platform. She is now solution designer for two of BT's dial IP products.

Pedro Serna, who contributed to Chapter 13, joined BT Laboratories in November 1996 and has been assessing and analysing radio technologies for UMTS. He graduated from Cantabria University, Spain, in 1994 and gained a diploma thesis from Ulm University Germany on the investigation of CDMA systems in 1995.

Chris Shephard (joint author of Appendix B) read Theoretical Physics at the Universities of East Anglia and then Birmingham. He joined BT in 1980 and worked for a number of years on the development of the System X switching systems. On the basis of this experience he accepted a job with Siemens in Florida within a large project designed to adapt their range of digital public switches (EWSD) to the U.S. market and regulatory requirements. He returned to BT after two years to work on local network access switches and later on Broadband ISDN systems (during which time he was the BT project manager for TRIBUNE). Chris is now part of a development group established to exploit the capabilities of Network Computers.

Acknowledgements

Many people have helped and assisted in the mechanics of writing individual chapters and the help that all of them provided is gratefully acknowledged, however there are a few that deserve special mention.

I am indebted to all of the contributors of each chapter for their original work and personal commitment to this book. Anselm Martin provided me with valuable assistance in producing Chapters 1 and 2, as well as updating the tables of standards in Appendix D. Ian Jones and Danny Hernandez stepped in to assist with proof reading, as well as providing questions and answers for Chapter 6.

I would also like to thank Kevin Woollard and John Griffiths for their encouragement and support, as well as Andrew Jell, editor of the *BT Technology Journal*, for technical assistance.

Ian Jones expresses his thanks to all the colleagues that helped in providing input to Chapter 8, and especially Sirak Bahlbi of BT.

Frank Allard has expressed his thanks to all his colleagues, for numerous discussions, with special thanks to Colin Bates who gave him the opportunity to be involved in Broadband, Dick Knight for his support, and Mel Bale for his valuable comments on Chapter 12.

Gary Bruce and Jon Clark wish to thank many colleagues around the world for their contributions to standardisation work. Also, they would like to thank the colleagues closer to home for their time and effort in proof reading Chapter 16.

Introduction

The concepts and technology of Asynchronous Transfer Mode (ATM) switching is explained in Chapter 1. Reference is also made to narrowband (circuit switching) to highlight the differences and assist the reader to understand the evolution from circuit switched to ATM switching. The chapter describes the ATM cell header, ATM connection types and the ATM Adaptation Layer (AAL).

Chapter 2 provides an overview of the concepts of protocols and signalling. The different types of signalling are discussed, and the chapter introduces access signalling and Signalling System number 7 (SS7). The link layer (signalling ATM adaptation layer at the User-Network Interface (UNI) and the Network Node Interface (NNI) services and primitives are described, with the principles of the SS7 broadband message transfer part 3 (MTP-3B).

Chapter 3 identifies what is meant by standards and their value in the Broadband Integrated Services Digital Network (B-ISDN) signalling environment. International bodies responsible for producing these standards are identified, together with an overview of their working methods and documents. The various drivers behind the bodies (commercial, technological, regulatory, etc.) are discussed, and the chapter concludes by examining collaborative activities and its value in limiting the potentially vast number of standards and duplication.

The B-ISDN is being tailored to become the universal (standardised) future network, and will be capable of supporting a wide range of multimedia, multiparty applications. It is based upon the same principles as its narrowband predecessor, and hence can be regarded as a natural extension of Narrowband Integrated Services Digital Network (N-ISDN). However, the move away from constant bit rate circuit switching towards the use of ATM technology has provided much needed flexibility, especially in terms of the connection bandwidths and Quality of Service (QoS) available to the user.

Chapter 4 briefly describes the network capabilities that B-ISDN should support, and how they have been derived from a representative sample of user applications principally proposed by the International Telecommunications Union—telecommunications standardisation sector (ITU-T), the

European Telecommunications Standards Institute (ETSI) and the Digital Audio-Visual Council (DAVIC). The identification of the required network capabilities is the first step towards the specification of signalling protocols for the B-ISDN which must be flexible enough to support the wide range of current and future advanced applications and services. One such potential future B-ISDN application, which demonstrates the range of signalling functionality required, is the 'Travel Agent Service', which is examined in detail.

The access signalling protocols that provide the release 1 signalling mechanism, known as Digital Signalling System No. 2 (DSS 2), are described in Chapter 5. In addition to identifying the functions that are required to be incorporated in a signalling protocol to support call/connection establishment, in-call control and clear down for a basic point to point call, the extensions required for support of type 2 connections are also described. The chapter then considers further extensions that have been produced to enable the negotiation, modification and look ahead capabilities in relation to the connection in order to provide the Capability Set 2 Step 1 DSS 2 signalling mechanism.

The ATM Forum is another key body which has been developing specifications for ATM technology over the past six years. It has produced a number of signalling specifications covering the user network interface, the inter-carrier interface and private networking environments. While work continues to refine these agreements, there is still discussion about how they are used in large scale networks, and in particular on the interworking of public and private network environments.

Chapter 6 reviews the current status of the relevant protocols and their scope, and explores the issues related to their interworking. The role of the ATM Inter-Network Interface (AINI), on which development started during 1996, is discussed as an alternative inter-network or inter-carrier interface to the Broadband Inter-Carrier Interface (B-ICI).

The ATM Forum completed the first version of its Private Network Network Interface (P-NNI) in March 1996, and many ATM switch manufacturers are now offering early implementations. P-NNI offers a different type of inter-network or inter-nodal interface from the traditional SS7-based approach favoured to-date by 'public' network operators. In spite of its name, however, P-NNI may find its place in network service provider networks as well as in 'private' or customer networks. Some of the perceived limitations of the current specification for such an environment are currently being addressed in subsequent addenda to the P-NNI version 1.0 specification.

The P-NNI really consists of two parts: a signalling protocol based around the ATM Forum's UNI signalling specification; and a dynamic source routing protocol. Chapter 7 provides an overview of the functionality and mechanics of P-NNI, and compares it with the functionality offered by the ITU-T's Broadband Integrated Services User Part (B-ISUP).

Chapter 8 describes B-ISUP, the inter-nodal signalling protocol defined by the ITU-T. Its purpose is to establish, maintain and release ATM connections in a public network, via the use of signalling messages. The initial B-ISUP recommendation was approved in February 1995, and is known as Capability Set 1 (CS-1). Since then, progress has been made to extend the capabilities of B-ISUP with the development of B-ISUP 2000. The chapter provides an overview of the B-ISUP signalling protocol capabilities, and describes its impact on the traditional SS7 architecture.

The VB5 interface is an ATM-based broadband interface at the 'V' reference point, the interface between a user access and a Service Node (SN), also known as the SNI (Service Node Interface) within the ITU-T. The B-ISDN architectural model allows for a number of different user accesses, such as straightforward digital sections and simple multiplexors, but it is anticipated that the most frequent implementation, particularly for mass deployment, will be via a broadband (Access Network) with VB5 interfaces.

The VB5 interface is specified similarly to its narrowband cousin V5 in two variants; VB5.1, which supports allocation of resources via the management system; and VB5.2, which adds a mechanism for allocating AN resources under the control of the SN, enabling call-by-call concentration in the AN. Chapter 9 describes the background to the development of VB5, and explains the two variants VB5.1 and VB5.2.

Chapter 10 introduces DAVIC, its objectives and plans. It then considers the split of the functional model into the three roles; service consumer system, delivery system and service provider system. Once the vertical split of the functional model has been considered, the horizontal split into S1–S4 levels is covered. The chapter expands on the concepts of session and connection control through the use of the S2, S3 and S4 levels in the set-up, transfer and release of a user to network session, and the end-to-end control of the user-to-user session. Finally, the future plans/directions for DAVIC and the adoption of Internet technologies within the DAVIC specifications are considered.

As the telecommunications industry evolves, and broadband services start to arrive, customers are increasingly coming to expect the perception of instantaneous access to service providers, together with transparency to network failures. System performance dictates that response times need to be minimised, sufficient redundant capacity be installed in case of failure, and controls be embedded within the design to manage the exceptional situations (such as media stimulated events) that continually threaten network integrity. A vital part of this system is the broadband signalling network, which underpins the dialogue with the customer, and which enables the delivery of the service. Network design based on a 'top-down', 'end-to-end' methodology plays a fundamental role in delivering solutions that meet customers' performance needs. Simple, approximate performance models at an early stage in this lifecycle are valuable to uncover major performance problems which affect the design of the architecture. Performance issues identified can then be fed back into the design process in an iterative way, to ensure that the design solution will conform to performance requirements.

Chapter 11 outlines performance studies of a number of design scenarios for providing broadband services. These studies consider interactive multimedia services, looking at service demand, physical network topology, signalling message flows, the mapping of functional entities to physical components, and routing, as part of the network design process. The most significant performance issues identified relate to bottlenecks, capacity requirements, and load-dependent response times as perceived by the customer.

Chapter 12 describes the broadband Virtual Private Network (VPN) Signalling. In corporate voice networks, the VPN configuration is particularly useful where a shared network infrastructure can be more cost effective and flexible than dedicated leased lines. Current standards for VPN use the international narrowband private signalling system, QSIG. Private network signalling standards for broadband networks are also developing rapidly, and signalling systems such as ECMA's BQSIG and ATM-Forum's PNNI will play a leading role in the Broadband VPN of the future.

Chapter 13 outlines the path from developing broadband ISDN/ATM systems towards the European vision of the next generation mobile system: Universal Mobile Telecommunications System (UMTS). The capabilities of UMTS in terms of services and features compared to the second generation mobile system, the global system for mobile communications (GSM), and the network technologies behind N-ISDN and B-ISDN are discussed with proposals for technical developments to satisfy the UMTS requirements. The UMTS radio interface aspects are described, and considerations on the network requirements to support these radio aspects are discussed in this chapter. The requirements within the network for control, switching and transport to support the mobility, service and interconnecting networks' aspects of UMTS are outlined in conjunction with the technical solutions currently under discussion, and some latest thoughts on a potential high-level architecture.

Conventional signalling systems, such as Q.931 and Q.2931, have grown from simple techniques for creating a single connection between two parties. However, with the B-ISDN, it will be desirable to be able to construct multiparty, multiconnection calls for multimedia services. Chapter 14 highlights the work of the collaborative project 'RACE MAGIC' to develop a new object-based signalling protocol to meet these requirements, and describes how this new protocol overcomes the shortcomings of conventional signalling. The concepts outlined in this chapter have been used to influence the development of the broadband signalling protocols for advanced services.

The separation of call and bearer to support future teleservices and services in broadband networks is outlined in Chapter 15. Various mechanisms to undertake the separation are considered, and the method chosen by ITU and ETS) is described in detail. A full protocol model for broadband access signalling is derived and explained through the basic steps leading to the establishment of an association between end points supporting call control information flows. The call control protocol is outlined and illustrated by information flows.

With the new millennium, many telecommunications network operators are gearing up to provide telecommunications networks that will satisfy the majority of demands which new feature-rich applications will place upon them. To some people, supporting application requirements for a telecommunications network might mean adding more intelligence to the network. Others may wish to transmit data at faster speeds, communicate more freely on the move, or be able to mix and match the appropriate network infrastructure as they, or their applications, choose to do so. These diverse application requirements are shaping the strategies being pursued by telecommunications network operators to deliver integrated telecommunications networks for the next millennium.

After looking back at some of the developments taking place within the broadband and network intelligence domains, Chapter 16 summarises some of the application and service requirements that must be addressed when building future telecommunications networks. From the plethora of technical proposals aimed at developing the necessary environment to fulfil these requirements, this concluding chapter explores the alternatives. Of course, for all of these alternative routes there will be many questions, technical or otherwise, that yet remain unanswered. Some of these issues will be discussed, and various options will be considered to provoke thought or socialise possible solutions.

It has been the intention to provide a book that is not just a definitive work on broadband signalling, but also a textbook that can be used by students to gain an understanding of the theory, concepts and technology of advanced communication networks. Appendix A provides the answers to all of the questions that have been provided for students and lecturers at the end of each of the chapters.

Mao Zedong once said 'All genuine knowledge originates in direct experience'. Appendix B describes the experiences gained through a range of development projects, all concerned with the design, development, demonstration and maintenance of real broadband signalling systems. The authors of this appendix have participated in the development of a variety of systems offering broadband capabilities, all of which support some form of broadband signalling between the service user and service provider. The process of designing and building such systems can provide invaluable insights into issues such as development methods, signalling performance, programming interface definitions, etc., none of which are apparent from purely paper-based studies or through the standards specification process.

This 'coal-face' experience covers both collaborative and BT projects, with each system being subject to a wide variety of service requirements. In addition, the period of work covered by the developments described in the paper encompasses broadband signalling systems that, on the one hand, pre-date the published ITU-T standards on Broadband ISDN systems and, on the other, includes a comprehensive testbed supporting an interworking set of broadband signalling interfaces based on the appropriate signalling standards defined by both the ATM Forum as well as the ITU-T.

Appendix C provides an introduction to the broadband call control demonstrator platform developed at BT Laboratories. It describes the basic connection level functionality used to give point-to-point and point-to-multipoint call demonstrations. It then describes the introduction of a Digital Audio Visual Council (DAVIC) conformant session control capability, and how this is used as the basis of a more comprehensive demonstration. This appendix also examines how future networking concepts, such as object orientation and distributed processing environments, are being introduced, in particular the interworking between DAVIC and the Telecommunications Information Network Architecture Collaborative project (TINA-C) approaches. Finally, some of the evolutionary aspects of the broadband call control demonstrator platform are introduced.

This book relies on standards, and would be incomplete without an index to the world's standards for broadband signalling. Appendix D provides a definitive list of standards, and where to obtain them, but will be out of date almost as soon as the book is published. There are sufficient references for the reader to find the sources of the tables and keep up to date using the internet.

1

ATM Switching

Dick Knight

1.1 SWITCHING TECHNOLOGIES

Circuit switching to provide a basic 'Plain Old Telephony System' (POTS) was developed in the 1960s and 1970s. Bringing together Pulse Code Modulation (PCM) transmission systems and time division switching enabled digital systems to provide exceptional levels of reliability in telephone networks. As integrated circuits with ever-increasing levels of integration became more available, the cost of switching also reduced. Most of the developed world's telecommunications networks had been converted to digital circuit switching by the end of the 1980s.

Narrowband digital switching is built around the 64 kb/s circuits that provides the capability to transmit digitised speech. Switching bandwidths of greater than 64 kb/s is only possible by using more circuits. Generally, for control and switching purposes, circuits are transmitted between switches in groups of either 24 (in north America) or 32 (in Europe), providing transmission bandwidths of 1.536 or 2.048 Mb/s. Transmission systems may be used with further groups into higher speed transmission links, but these trunks are usually 'invisible' to the control structures in the switches and the user. It is generally not possible to aggregate circuits together, or otherwise to correlate circuits into a controlled group that provides bandwidth greater than 1.5 or 2 Mb/s. Narrowband digital switching does, however, provide the utopian solution of a global telecommunications infrastructure supporting speech and speech-like services through 64 kb/s connection oriented circuit switching.

Switching and transmission systems for data communications also developed throughout the 1960s and 1970s, producing a number of solutions for a more diverse range of applications. In particular, the X.25 networks provided the ability for data communications between terminals (and customer networks) on a global scale. This network was based upon an underlying connectionless switching system, in which units of variable length data, termed packets, were routed according to a complete address contained within every packet. Each switch acknowledged the receipt of a packet to the preceding entity ensuring a guaranteed service. The advantage of this system, to a data user, was that only relevant information was ever transmitted over the networks, compared with the circuit switched system which transmits, switches and routes idle patterns over circuits when no user information is available.

Narrowband digital switching, whilst ideally suited to the application it was developed for, had some limitations. It did not provide support for switching circuits at any bandwidth other than 64 kb/s, and was arranged to support applications requiring the interchange of information at a constant rate. The X.25 network provided the flexibility of a variable bit rate, but with a rather high overhead, in terms of both a larger header field and slower overall end-to-end transfer time.

As more diverse services began to be combined together to form ever more complex communications requirements, it became clear that a combination of the flexibility of packet switched networks, and the speed and efficiency of connection oriented-based circuit switching would be needed. The added requirement of an ever increasing demand for more and more bandwidth resulted in the development of a new switching technology, Asynchronous Transfer Mode (ATM). The goal was to be able to allocate the available bandwidth from small amounts (64 kb/s) through the artificial barrier of 2 Mb/s smoothly upwards to an almost ridiculous figure. To achieve this, the fixed 64 kb/s time division was replaced by an asynchronous system of small cells. Cells can then be allocated to provide bandwidth from 64 kb/s through to the maximum available on a link in very small incremental steps.

Both ATM and the old circuit switched techniques provide Time Division Multiplexing (TDM). In narrowband digital switching, the stream of bits contains a marker to indicate the start of a frame. Each synchronous frame consists of either 32 or 24 channels, each containing eight bits. An application is allocated a channel number, and every time that channel appears, it is allowed to fill the channel by setting the eight bits.

1.2 ATM SWITCHING

ATM does not contain a frame marker, instead it identifies only the start of a cell, and it does not allocate a fixed number of cells to applications, instead negotiating the rate and allowing the allocation of different bandwidths to different applications. Each cell holds up to 48 octets (eight-bit 'bytes' of data) with a measly five octets for the header, achieved by identifying the cell relative only to the other cells in a stream (Figure 1.1).

1.2 ATM SWITCHING

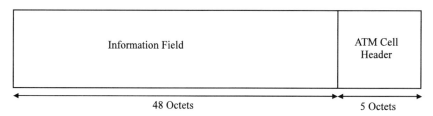

Figure 1.1 ATM Cell

Cells are routed according to the tables held in the switches, which are updated only upon a request for a connection, or by external action (management action). ATM provides switching and transmission based on the best parts of the circuit switched and packet switched networks, supporting variable length packets in a connection-oriented system whose bandwidth may be adjusted to the service requirement.

To take a trivial example, if two applications are sharing a single ATM link of 6 Mb/s, and one application requires twice as much bandwidth as the other, it will use two out of every three cells. The lower rate application fills the remaining cells.

ATM regards any transmission link as a single circuit of very high bandwidth (in excess of 2 Mb/s), that may be arbitrarily divided into a large number of theoretical circuits, Virtual Circuits (VCs). VCs may be grouped into a set of related circuits, known as a Virtual Path (VP). VCs carry a variety of information that may be regarded as having an arbitrary bandwidth at an instance in time. The VP and VC that the cell has been assigned to is indicated by the Virtual Path Identifier (VPI) and the Virtual Channel Identifier (VCI), respectively. Combined together, these fields form the VPI/VCI, which is an identifier unique to a physical path between two ATM entities (e.g. ATM switch-ATM switch, Terminal-ATM switch, etc.) and are included in the header of every ATM cell. As can be seen from Figures 1.2 and 1.3, the number of virtual channels possible is 2^{16}, or about 64 000, and the number of virtual paths is 256 at the User Network Interface (UNI) and 4096 at the Network Node Interface (NNI).

Cells entering a network at the UNI are groomed by means of a Generic Flow Control (GFC) mechanism, to ensure that bandwidth allocations are not exceeded. Once the network has accepted a cell, it will be delivered to its destination (under normal circumstances), and so at the Network Node Interface (NNI) the GFC is not required (see Figures 1.2 and 1.3).

ATM switches may be either VP or VC switches. A VC switch also performs VP switching (see Figure 1.4). It is important to note that VCs are meaningless without reference to the VP, and that the VP is meaningless without reference to the physical link. Thus, a VC only has significance within a particular VP of a particular physical link. The same VCI may occur in another VP or another physical link, or both. Similarly, the same

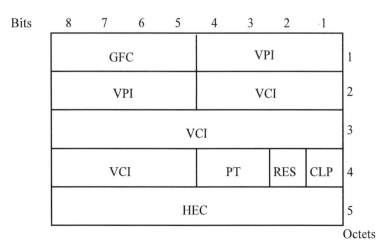

Figure 1.2 ATM Cell Header at the UNI

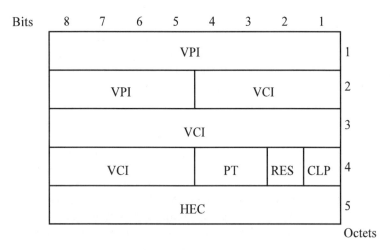

Figure 1.3 ATM Cell Header at the NNI

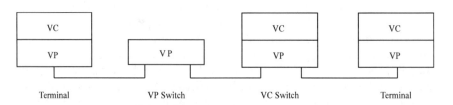

Figure 1.4 VC and VP connections in ATM

1.2 ATM SWITCHING

VPI value may be present in two different transmission links. The VPI/VCI value alone is insufficient to enable a switch to perform the routing.

An ATM switch can be regarded as a simple table consisting of the incoming physical link and VPI/VCI (VPI only for VP switches) indexing the outgoing VPI/VCI (VPI only for VP switches) and outgoing physical link to be used. Figure 1.5 shows VC L in VP a, incoming on Link 1 (top left), being switched to VC P in VP x on outgoing Link 3. The diagram also shows that a complete VP may be switched from one physical link to another, without determining the VCs it may contain (VP b on Link 1 is switched to VP z on Link 4).

ATM cell header

The fields in the ATM header are shown in Figures 1.2 and 1.3. The VPI and VCI fields have already been explained. The HEC field provides Header Error Control. If an ATM cell header is corrupted, then information will be lost from one user stream and may be erroneously inserted into another user's stream. The HEC provides a means to detect, identify and correct a corruption of a single bit in the ATM cell header. The HEC can also detect (but not correct) a corruption that affects multiple bits in the ATM cell header. The method of detection and correction involves the use of an eight bit cyclic redundancy check, and represents another field of study in protocols and networking. The HEC does not protect the information field of an ATM cell.

At the UNI, it is possible that more than one terminal may be connected

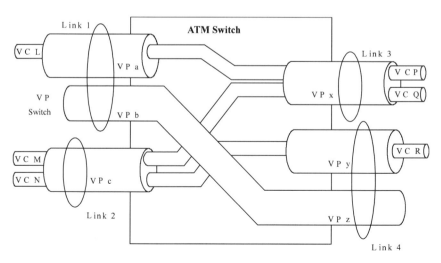

Figure 1.5 ATM Switching of VPs and VCs

(multiplexed) onto a single interface. As we shall see later, this can be quite a simple matter in ATM, using separate VPs for each terminal. However, a mechanism is still required to prevent one terminal from 'hogging' all the bandwidth. The Generic Flow Control (GFC) not only enables the network to 'turn off' a terminal, it provides a number of other functions dependant on the direction of transmission. This mechanism was designed to alleviate short-term overload conditions that may occur, and will use unassigned cells to provide flow control communication when there are no other ATM layer cells available. Up to two groups of controlled connections are supported (designating ATM cells as part of either group A or group B), which assumes a 'two queue' model. The default configuration is for a one queue model (i.e. a single group). Even if a single terminal is assumed, the transmission of cells can be turned off by the network (the 'controlling equipment'). This is termed 'uncontrolled' operation. This subject often leads to confusion, mainly due to the multiple uses of the word 'controlled'. All ATM user equipment is controlled using the GFC, but the type of control imposed depends upon the connection. ATM connections may be uncontrolled, Group A controlled or Group B controlled.

The GFC sent from the network indicates whether the terminal should stop the transmission of assigned cells towards to the network, or whether it may continue to transmit information. The GFC is further able to start a limited period of transmission on connections of a controlled group (either A or B), through the setting of a counter (the GO_CNTR). The counter is decremented when a cell is transmitted on a connection of that group, and therefore causes the group to 'auto stop'. The counter for Group A does not affect Group B, nor does it affect uncontrolled connections.

From the controlling equipment to the controlled equipment, there are three bits required. These indicate HALT/NO_HALT SET_A/NULL and SET_B/NULL. Each bit operates independently of the others, and so multiple instructions can be sent in one cell.

In the opposite direction, the controlled equipment uses the GFC field to indicate the control being applied for assigned cells (only)—unassigned cells cannot belong to Group A or Group B. This again takes three bits of the four bit field, although not all combinations are valid.

Cell loss due to congestion is prioritised by a single bit in the header, the Cell Loss Priority (CLP) bit. Cells with this bit set to 0 will be dropped by a network in preference to cells with the bit set to one.

The Payload Type (PT) field (three bits) indicates whether a cell is being used for user information or for management of the network and interfaces. Management cells, among other functions, are used for detection of link failures. If the payload type is set to user data, then the second bit in the field is used to indicate whether congestion has been experienced or not, and the third bit is left available for the user. The use of this third bit is dependant upon the particular type of user data, and is defined with the adaptation layer information.

Table 1.1 Idle Cell Header Pattern—from I432.1

	Octet 1	Octet 2	Octet 3	Octet 4	Octet 5
Header pattern	00000000	00000000	00000000	00000001	HEC = Valid code 01010010

Notes
The content of the information field is '01101010' repeated 48 times.
There is no significance to any of these individual header fields from the point of view of the ATM layer, as idle cells are not passed to the ATM layer.

Idle cells cause no action at a receiving node and are only used as fillers if there is no user data to be sent. They are also inserted and discarded for cell rate decoupling and cell delinearisation (i.e. finding the start of a cell). Idle cells are identified by the standardised pattern for the cell header shown in Table 1.1.

1.3 ATM CONNECTION TYPES

Narrowband circuit switching supports only a single type of connection. This connection is between two end points, is bidirectional (which means that both endpoints are able to transmit and receive information) and symmetrical (i.e. the two directions have to use the same characteristics, e.g. identical bandwidth). ATM offers the ability to support different types of connections (Morley and Knight 1998), and they may be asymmetrical. There are six connection types offering

- bidirectional point-to-point
- unidirectional point-to-multi-point
- unidirectional multipoint-to-point
- multipoint to multipoint connectivity
- bidirectional point-to-multi-point
- and multipoint to multipoint emulation using a network server.

Point-to-point connections

The simplest category is a point-to-point connection, which is a single switched link (i.e. an ATM Virtual Channel Connection) between two

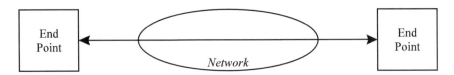

Figure 1.6 Point-to-Point Connection

endpoints. This is referred to as a Type 1 connection, and may be either bidirectional or unidirectional. Bidirectional connections are permitted to have an asymmetrical bandwidth, but must have the same ATM layer transfer capability and ATM Adaptation Layer (AAL) type in both directions (Figure 1.6).

Unidirectional point-to-multipoint connections

Unidirectional point-to-multipoint connections (Type 2) utilise the ATM switching fabrics capability to replicate ATM cells and to direct the resulting streams to different endpoints (Figure 1.7). This function is analogous to the broadcast of information, since one source may transmit information to many receivers. However, within ATM studies, this type of capability is more usually referred to as a *multicast connection*, since the number of receivers is limited by the number of connections possible within the network. An information broadcast capability implies an unlimited number of receivers, and hence unlimited network connection resources. Type 2 connections are restricted to a unidirectional capability, having only a single source and multiple sinks.

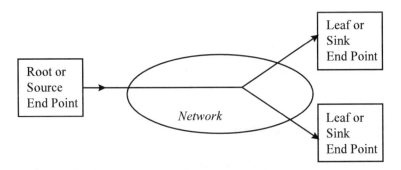

Figure 1.7 Unidirectional Point-to-Multipoint Connection (Type 2)

1.3 ATM CONNECTION TYPES 15

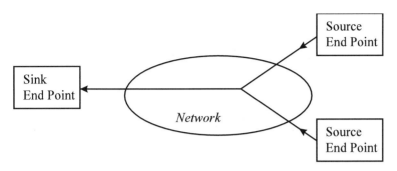

Figure 1.8 Unidirectional Multipoint-to-Point Connection (Type 3)

Unidirectional multipoint-to-point connection

Unidirectional multipoint-to-point connections (Type 3) enables multiple endpoints to send information to a single receiver through one Virtual Channel Connection. They are currently only supported in AAL type 3/4, and require an underlying transport system that is capable of multiplexing two or more connection sources into a single connection for reception by a single sink, as shown in Figure 1.8.

Multipoint-to-multipoint connections

Multipoint-to-multipoint connections (Type 4) perform conferencing functionality—a complete communications connection supporting more than two endpoints, essentially in a star configuration. Any information sent by an endpoint is received by all the other endpoints, although each endpoint only has a single bidirectional connection. There are some restrictions on the implementation of this type of connection in the AAL. AAL Type 3/4 can support this type of connection, although it may also be required in other AALs for some services. To meet this requirement, a Type 6 connection is required. The functional view of the Type 4 connection is shown in Figure 1.9, though this does not address the issue of how the connection should be implemented in the ATM switch.

Bidirectional point-to-multipoint connections

Type 2 and Type 3 connections (point-to-multipoint and multipoint to point) are unidirectional. Bidirectional point-to-multipoint connections

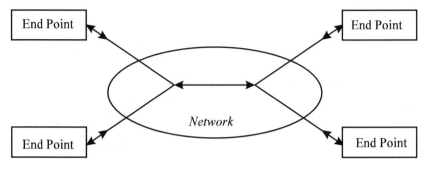

Figure 1.9 Multipoint-to-Multipoint Connection (Type 4)

(Type 5) are defined as Type 2 connections overlaid with a Type 3 to provide a bidirectional connection. The introduction of a bidirectional capability allows leaves in a distribution tree to have a 'back channel' which may be used to send information to the root. Type 5 connections can support such applications as broadcast television (TV) with a voting mechanism, since the half of the connection from the leaf to the root can be used to multiplex, on an AAL Type 3/4 level, the users' votes going back to the root. The connection topology is shown in Figure 1.10, and is always referred to as a *bidirectional point-to-multipoint connection*, although the terminology *bidirectional multipoint-to-point connection* should be equally valid.

Multipoint-to-multipoint emulation

Type 4 connections can be emulated by the addition of a server (or bridge) within a network to perform the type of functionality associated with conferencing, but utilising only point-to-point connections, as shown in Figure 1.11. The server or bridge function provides the required

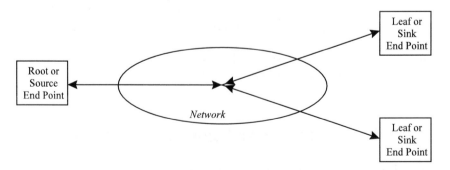

Figure 1.10 Bidirectional Point-to-Multipoint (Type 5)

1.3 ATM CONNECTION TYPES

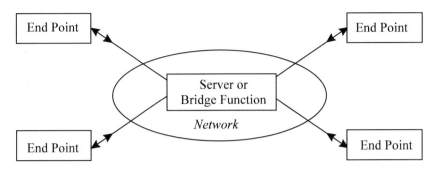

Figure 1.11 Multipoint-to-Multipoint Connection (Type 6)

combining of streams, such that each end point's incoming stream is some sort of combination of all the other end points' outgoing streams. This description does not address the issue of how the connection should be implemented, especially as the term *conference bridge* is a rather abused term, often used to refer to a piece of equipment that is providing some application related service, rather than merely multiplexing and demultiplexing ATM cells. It is also possible to provide a Type 6 connection using equipment external to the network (i.e. an end point acts as a server), but this provides requirements on the services, rather than networks and switches.

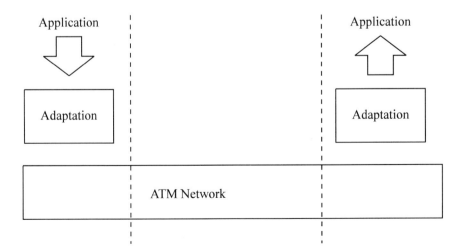

Figure 1.12 Requirements for ATM Adaptation

1.4 ATM ADAPTATION LAYER

ATM has been designed to support many types of users, each with (potentially) many types of communications needs. To provide an efficient network that enables such diverse applications as speech, file transfer, internet access, video and mobility produces a conflict of requirements. On the one hand, the network switches need to be as simple and as fast as possible—the more work the switches have to undertake, the less traffic they can support. On the other hand, there are applications requiring relatively complex levels of formatting (e.g. framing and synchronisation). ATM gets around this problem by taking the best of both worlds. The switches are kept simple by providing the cell switching based on only the header, and then a standardised approach to adapting application streams was adopted (Figure 1.12). The ATM Adaptation Layer (AAL) has significance only for the end applications.

Cells are transferred from end-to-end in the same order in which they are received, and the network delivers only those cells that are guaranteed to have a correct cell header (i.e. those first five octets). This means that the network does not provide

- time synchronisation (important for voice and video)
- framing (for data services such as file transfer)
- error checking of user information (necessary for data services)
- source information (needed to support Type 3 connections).

Constant bit rate adaptation

The types of services requiring a transfer at a constant bit rate comprise such applications as voice and video. These applications are relatively error tolerant, require timing information and are connection oriented. It is much more efficient for these applications to simply throw away the relatively low number of erroneous cells and wait for the next correct cell, rather than use a handshaking technique to enable retransmission of lost information. Sequencing information, however, is vital. The first AAL to be defined was designed to support voice and other constant bit rate services (Figure 1.13). One octet of the user information field of every cell is used to indicate clock recovery information, a sequence number and a simple error detection on the user information. AAL Type 1 leaves 47 octets for user data.

1.4 ATM ADAPTATION LAYER

Figure 1.13 ATM Switching without Time Synchronisation

Variable bit rate–data blocks

AAL Types 3 and 4 were combined together at an early stage to form AAL3/4. The basis for this was to provide a generic adaptation technique that could be provided largely independently of the application. AAL 3/4 supports block transfer of up to 64 000 octets, split into 44 octet segments (for each cell). Each segment is protected by a two octet error detection trailer and a two octet header that provides sequencing, framing and source information. It is the last item that enables AAL Type 3/4 to support Type 3 connections (multipoint-to-point connections), as shown in Figure 1.14.

Since AAL Type 3/4 protects the 44 octets of user data in every cell, a detected error could be rectified by retransmission of a single cell (Figure 1.15). The data networking community found this type of adaptation to be too heavyweight for its requirements. Given the low rate of loss in digital networks, it is generally preferred to retransmit a whole block, rather than a cell, in the event of data corruption leading to cell loss. So, whilst AAL Type 3/4 supports switched multimedia data services, with retransmission but without having to implement all of the protocols of applications, a simple but efficient adaptation layer was required to support data networks—AAL Type 5.

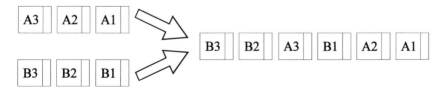

Figure 1.14 Multiplexing Information using Type 3 Connections

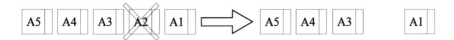

Figure 1.15 Effects of ATM Cell Header Error

Variable bit rate–large frames

AAL Type 5 takes the user data and adds a trailer and padding to take the frame size to a multiple of 48 octets. The single user bit in the ATM header is used to indicate the end of the frame, and the error checking is then applied. If any part of the frame has been lost, or the user data has been corrupted, then the error check fails and the entire frame is discarded. AAL Type 5 is very efficient for large frames and removes the arbitrary 64k limit found in AAL3/4. AAL Type 5 was selected as the underlying adaptation layer to use for signalling.

Additional support for circuit switched traffic

If this book had been written a year earlier, this chapter would have finished with the three AAL types (1, 3/4 and 5). The need to support both mobility and to carry circuit switched groups (land line trunking) resulted in the definition of another AAL—Type 2. Type 2 supports constant bit rate connection oriented traffic between two arbitrary points in an ATM network. Each point may be at either a UNI or an NNI, regardless of where the other end point is located. AAL type 2 traffic is assumed to have its own call control, and therefore it is required only to (semi-permanently) provide a bearer-only service between the two points. In both the service examples, it is assumed that the ATM switch is already set up to provide a single VC between the two points and then multiple circuit switched streams are transferred by the VC. Since circuit switched traffic transmits an idle pattern when circuits are not in use, the use of ATM is limited to pure transfer and bandwidth utilisation in an AAL2 trunk is the same whether any connections are 'live' or not. The actual adaptation requirements in AAL2 are for a virtually empty AAL, as framing and sequencing are performed by the application.

REFERENCES

Morley R J and Knight R R, Requirements for B-ISDN signalling, *BT Technology J*. Vol. 16, No. 2, 1998, pp. 20–28.

FURTHER READING

de Prycker M, *Asynchronous Transfer Mode*, Ellis Horwood, 1991.
Griffiths J M, *ISDN Explained*, 2nd ed, Wiley, 1992.

STANDARDS

I.113 *Vocabulary of terms for broadband aspects of ISDNs*, ITU-T (06/97)
I.121 *Broadband aspects of ISDN*, ITU-T (04/91)
I.150 *ATM Functional Characteristics*, ITU-T (11/95)
I.311 *B-ISDN general networking aspects*, ITU-T (03/93)
I.313 *B-ISDN networking requirements*, ITU-T (09/97)
I.321 *B-ISDN protocol reference model and its application*, ITU-T (04/91)
I.326 *Functional architecture of transport networks based on ATM*, ITU-T (11/95)
I.327 *B-ISDN functional architecture*, ITU-T (03/93)
I.356 *B-ISDN ATM layer cell transfer performance*, ITU-T (revised, 10/96)
I.357 *B-ISDN semi-permanent connection availability*, ITU-T (08/96)
I.361 *B-ISDN ATM Layer Specification*, ITU-T (11/95)
I.363 *B-ISDN ATM adaptation Layer specifications*, ITU-T (03/93)
I.363.1 *AAL type 1*, ITU-T (08/96)
I.363.2 *AAL type 2*, ITU-T (09/97)
I.363.3 *AAL type 3/4*, ITU-T (08/96)
I.363.5 *AAL type 5*, ITU-T (08/96)
I.371 *Traffic control and congestion control in B-ISDN*, ITU-T (revised, 08/96)
I.432.1 *B-ISDN user-network interface—Physical layer specification: General characteristics*, ITU-T (08/96)
I.610 *B-ISDN Operation and Maintenance Principles and Functions*, ITU-T (11/95)
Supplement 8 to Q Series Recommendations, *Signalling requirements for AAL type 2 Capability Set 1 (CS1)*, ITU-T (03/99)
X.25 *Interface between Data Terminal Equipment and Data Circuit-terminating Equipment for terminals operating in the packet mode and connected to public data networks by dedicated circuit*, ITU-T (10/96)

QUESTIONS

1. Suggest a simple method of allowing multiple terminals to share a single network connection.
2. What information is required to enable a service to use connection Type 3?
3. Why is AAL Type 5 more efficient than AAL Type 3/4 for large frames?

2
Introduction to Signalling

Dick Knight and Bryan Law

2.1 EARLY SWITCHING CONTROL

In Kansas City, USA, in the year 1806 an undertaking business started that was to change the world (Lesher 1999). The undertaker believed that his main rival had a major business advantage—a wife who was also the local switchboard operator. In the days of the manual switchboard this could have been a tremendous advantage. Anybody phoning for Almon Strowger's undertaking business could be redirected to her husband, perhaps employing a variety of excuses. Almon Strowger did not take this treatment lying down. In 1886 he patented a solution to the problem based upon the concept of automatic switching, and further patents followed, leading to the founding of the Automatic Electric Company. We are now familiar with the Strowger step-by-step analogue system of switching, employed in the developed world, based upon two motion selectors and relays to provide a system of seizing, dialling and tones to enable humans to interact with a distributed processing machine. Almon Strowger may not have become renown for being the best undertaker in the world, but he will certainly be remembered as the father of the automatic telephone system. The UK removed its last Strowger exchange on the 23rd June 1995.

It is worth looking a little closer at the principles involved in the Strowger system, as much of the terminology we use today is inherited from this early communications control system. We shall work through a simple call set up.

Strowger calls

The first step was to seize a line. Equipment was expensive and the need to concentrate traffic, especially low usage lines such as residential, was recognised early on. The concentrating equipment recognised a single command, seize, and this was indicated by the subscriber's telephone placing a low resistance loop across the two wires (termed A and B) from the exchange. The loop caused the concentrating equipment to hunt for a free first selector, and once found it performed two functions—reservation and seizure. Reservation was performed using a third wire, the P wire (present only in the exchange), preventing other concentrating traffic assuming this equipment was free. The reservation applied to both the first selector and the subscriber's line—so incoming calls on this number would now be informed that it was busy. Seizure of the first selector was achieved by extending the subscriber's pair through to the first selector, which now took over the job of reservation (backwards holding), allowing the disconnection of the concentrating equipment's seize detection relay.

The first selector was now ready to receive the first digit of the called subscriber's number. The time delay between the seizure and the ready was generally more than the time taken for the subscriber to lift the receiver off its cradle and up to their ear. To prevent the subscriber dialling before the equipment was ready, the subscriber needed an indication that the first selector had been reached, and this was achieved by providing an audible tone, a dial tone, from the first selector. The provision of a dial tone was the only difference between a first selector and subsequent selectors. The subscriber provided the digit using the mechanical dial. The dial was wound round to the finger stop and a spring then unwound and caused a stream of alternating short circuits and full disconnections at ten pulses per second. Each disconnection caused the selector to move one step (generally vertically) until the digit had been received. The end of the digit was represented by an inter-digit pause, and during this time the selector performed a similar hunt and seize mechanism for the next selector. Dial tone was also removed by the first selector, assuring the subscriber that the digit had successfully been received.

Subsequent digits stepped further selectors to reach the final selector, where there would be just two digits left, so this selector was stepped both vertically and horizontally by the subscriber's dial. The final test would be for the absence of an earth on the P wire, indicating the subscriber was free. The line was reserved by earthing the P wire, a tone was returned to the calling subscriber and ringing current was applied to the called subscriber—a large current to move the electromagnetic clapper that rang the bells. On the answer signal—a low resistance loop placed across the line when the receiver was lifted from its cradle—ringing current and ringing tone were removed and the transmission path was completed by the final selector, enabling communication by the two subscribers. Most importantly, from the point of view of the network operators, a signal was also sent to start the charging mechanism. The only other mechanism needed, release, was achieved by the receiver being placed back on the cradle by the calling subscriber disconnecting the low resistance loop—so a full disconnection (of at least the same length of time as an interdigit pause) provided the 'clear' signal.

This mechanism was used successfully from the 1920s through to the 1960s and early 1970s, and even later in some countries. The terminology used in seizing the line, setting up the call, answering and releasing were all inherited by later control systems. Generally speaking, we call this type of communication *signalling*, which is the exchange of service related information in a distributed control system of a communications network.

2.2 TYPES OF SIGNALLING

In band signalling

The Strowger system worked well for early systems for short distance calls, but any long distances introduced the need for amplification, and then it became impossible to transfer the short circuits and disconnections indicating seizure, release and digits. Several alternatives for amplified systems, many based on audible tones, were used relatively successfully. They all had one disadvantage—either a portion of the commercial speech band (300 Hz–3.1 kHz) had to be suppressed, or there was the possibility of a subscriber fraudulently introducing the signalling tones to redirect the call. This system of providing audible tones is the first type of signalling, 'in-band signalling', and, more generally, this is any system where the signalling is included in the same circuit as the speech. For technologies based on analogue and amplifiers, there were few alternatives to this type of mechanism for exchanging information about the call.

Channel associated signalling

The first digital systems were deployed as transmission links in the 1970s, and enabled a new type of out-of-band signalling to be used. Pulse Code Modulation systems encoded the analogue speech into digital representations at instances in time, and then multiplexed several channels together into a single stream. One channel was reserved for an indication of the current state of the line (looped or disconnected) of the other (speech) channels. Since each channel contained eight bits in a frame, each bit of the signalling channel could be used to indicate the state of one speech channel. Frames needed to be combined into multi-frames, so that all 22 speech channels (out of the 24) could be indicated. The remaining channel (channel 0) was used for synchronisation, multi-frame indications and transmission link management information. This type of signalling, *channel associated signalling*, had advantages over the in-band signalling, since it prevented fraud and provided the full commercial speech band for subscriber communication. It was, however, quite inefficient. Throughout an entire call, one bit in the signalling channel was held at 'one' to indicate that the line was still looped. A better system was soon found, based on experiences from the data world.

Packetised signalling

In the data world, the International Standards Organisation (ISO) produced an Open Systems Interconnect (OSI) model (see Figure 2.1) for communications based upon a system of layering to provide message-based communications. Each of the seven layers of the OSI model performed a different set of functions, termed a *service*, for the layer above. The first three layers were targeted at the communication needed between users and the network and the communication within a network to enable the messages containing the higher layers (four to seven) to be delivered to the destination user.

The seven layers are as follows:

1. The Physical Layer provides the mechanical, electrical, functional and procedural means to activate, maintain and deactivate physical connec-

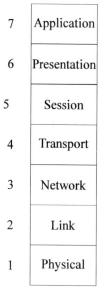

Figure 2.1 The OSI Seven-layer Model

tions for bit transmission between data link entities. The service that it provides to the next layer is to sequentially transfer bits.
2. The Data Link Layer provides functional and procedural means for the establishment, maintenance and release of data link connections among network entities, and for the transfer of data link service data units. A data link connection is built upon one or several physical connections. The Data Link Layer detects and possibly corrects errors which may occur in the Physical Layer. The service that the Data Link Layer provides to the Network Layer is the ability to control the interconnection of data circuits within the Physical Layer.
3. The Network Layer provides the means to establish, maintain and terminate network connections and the functional and procedural means to exchange network service data units between transport entities over network connections. The Network Layer provides the independence of routing and relay considerations for transport entities to communicate. The basic service of the Network Layer is to provide the transparent transfer of data between transport entities. This service allows the structure and detailed content of submitted data to be determined exclusively by layers above the Network Layer.
4. The transport layer provides transparent transfer of data between entities communicating using the session layer, and relieves them from any concern with the detailed way in which reliable and cost effective transfer of data is achieved. This means that the transport layer must understand any restrictions that the network layer provides, such as maximum size of network layer user data (network service data units). The transport layer is therefore responsible for the segmentation and reassembly of the session data into portions that can be transferred by the network layer to the destination end. All protocols defined in the transport layer have end-to-end significance, where the ends are defined as transport entities having transport associations. Therefore, the transport layer is OSI end open system oriented, and transport protocols operate only between OSI end open systems.
5. The Session Layer provides the means for the presentation layer to communicate. It includes session connection between two presentation entities, the support of orderly data exchange interactions, and the release of the connection in an orderly manner. A session connection is created when requested by a presentation entity at a session service access point. The session connection exists until it is released by either the presentation entities or the session entities. The only function of the Session Layer for connectionless-mode communication is to provide a mapping of transport addresses to session addresses.
6. The Presentation Layer provides definitions for the representation of information that application entities either communicate or refer to in their communication. The presentation layer is responsible for the common formats and rules for representing data.

7. The Application Layer is a task to be performed, such as file transfer, message handling, etc.

Communication using the OSI seven layer model employed a mode that involved providing the routing and destination details with every unit of user data (the packet). It was theoretically possible to use this type of information exchange even for speech. Digitisation is performed first, and the stream of bits is partitioned into packets, each of which has the routing and network details as part of the network layer. Although this type of communication, message-based, was much more efficient, the technology of the 1970s and 1980s, precluded the packetisation of speech, as the processing and switching technology was just not fast enough to enable the individual routing of each packet of information. The compromise was to use the message basis of the lower three layers of the OSI seven layer model to enable all the signalling to be carried in a single, common channel.

Common channel signalling

This type of signalling combined the best features of the channel associated and packetised signalling. The communication was layered, enabling the features of the link layer (network entity communication, error detection/correction, etc.) to be employed to allow the network layer to use a message-based approach. Messages in the signalling channel then *referred* to a particular user connection and the channel that would be used to provide the communication. There was no necessity to provide a signalling channel in every physical link between two network nodes; users' communication channels could be referred to by the link and channel number.

Common channel signalling did not entirely fit the seven layer model. Its initial use was within networks to provide interswitch signalling, more correctly called *internodal signalling* in modern systems (and the interface is referred to as the Network Node Interface, NNI). The target was not just a message-based system of communicating between adjacent switches, but a complete signalling system that would maintain its own network of communication (a network within a network). Instead of layers, the different functions provided were grouped into levels, mainly because there was an additional level to cater for the network-wide signalling system.

2.3 SIGNALLING SYSTEM NUMBER 7 (SS7)

SS7 is a common channel signalling system supporting both call and non-call related signalling in existing networks. Underpinning the procedures and protocols for exchanging control information is a reliable and secure set of links that enable the signalling messages to be transferred. The links themselves, the procedures and protocols for message transfer, together with the mechanisms for routing those messages, can be collectively termed the *Message Transfer Part*. This comprises the first three levels of the SS7 protocol stack, MTP1, MTP2 and MTP3. A typical SS7 protocol stack is shown in Figure 2.2.

Message Transfer Part level 1 (MTP1) defines the physical, electrical and functional characteristics of the transmission path for signalling. Within narrowband digital exchange systems this is normally a 64 kb/s digital channel within a PCM system. Message Transfer Part level 2 (MTP2) is also specific only to narrowband, and defines the functions and procedures for the transfer of signalling messages over a signalling-data link between two directly connected signalling points. A thorough description of these two levels is contained in Richard Manterfield's excellent book on Common Channel Signalling (Manterfield 1991).

MTP3

The Message Transfer Part level 3 (MTP3) is responsible for the reliable transfer of signalling information from one signalling end point to another, taking into account Level 1 or Level 2 failures, and is virtually the same in both narrow- and broad-band environments. MTP3 can be divided into two basic functions, that of signalling network management and signalling message handling. The former function includes signalling traffic management, signalling link management and signalling route management.

The signalling message handling function is responsible for the actual routing of messages within the signalling network. Signalling messages initiated from Level 4 are passed to MTP3, and included in the message is the address of the signalling end point that the message is directed to. This address is termed the *Destination Point Code* (DPC), and forms part of the routing label. The rest of the label provides information on the sender's address, the *Originating Point Code* (OPC) and the *Signalling Link Selection* (SLS), which MTP3 uses to load-share signalling traffic when more than one link to the required destination exists. The DPC and OPC are each 14 octets (bits) long, and each point code uniquely identifies a node in the signalling network. Signalling messages will generally be destined for an adjacent node, being the next exchange in the routing taken by the call itself. MTP3 makes a signalling link selection and sends the message off to the node indicated in the DPC.

The route taken by connections through the network does not need to be followed by the signalling messages. Every node that the connections passes through must be reached by the signalling, but if there is insufficient

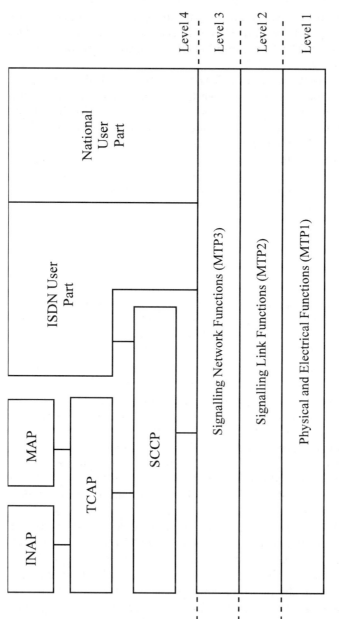

Figure 2.2 Typical SS7 Protocol Stack

2.3 SIGNALLING SYSTEM NUMBER 7 (SS7)

signalling traffic between two nodes to justify allocating a 64 kb/s channel for a signalling link, then it may be more efficient to concentrate the signalling traffic with messages for another node. This type of control is also known as 'quasi-associated signalling' (see Figure 2.3), and is encountered in providing an alternative signalling route for use in the event of failures or congestion. The signalling messages from exchange A and destined for exchange B are first sent to exchange C. It is the DPC that identifies them as destined for exchange B, and for these types of forwarded messages, the MTP3 at exchange C is performing the function of a *Signalling Transfer Point* (STP) (Knight 1995).

The Signalling Transfer Point (STP)

The STP demonstrates the flexibility and security that a signalling network provides. On receiving a message that satisfies the Level 2 procedures (the secure and intact transfer of the message across the transmission medium), the message must be acknowledged back to the peer Level 2 and passed to the MTP3. The first action taken by MTP3 is to determine if the DPC in the received message is identical to the Point Code address of itself. If it is not identical, and in the case of an intermediate node it is not, the routing table must be consulted to determine where next to send the message. The passage of the message is represented by the broken line in Figure 2.4.

In the event of a failure of a link, an alternative route should always be available, and to ensure security the alternative route must be provided over a physically separate path. Generally, therefore, only a single signalling channel is used in any single physical interface, further channels being provided over a separate interface. The MTP3 has to be provided as a central, concentrating signalling processor, since it must be able to direct

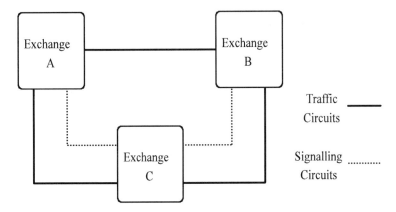

Figure 2.3 Quasi-Associated Signalling

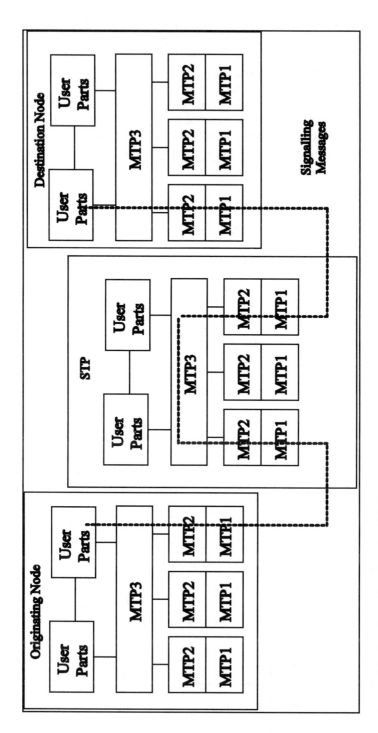

Figure 2.4 The Signalling Transfer Point

signalling messages onwards to other physical interfaces, as well as to the appropriate Level 4 user part, as in Figure 2.4. The mechanisms of MTP3 and the STP provide modern telecommunications systems with a signalling system that are able to exchange information even under conditions of failure.

Broadband internodal signalling systems defined by the ITU-T are based upon the same principles as the narrowband counterpart; there is a different link layer (more appropriate for the virtual channel environment of ATM) which replaces MTP2, and of course, the physical layer and interface is different, so MTP1 is replaced. The signalling ATM adaptation layer (SAAL) (Law 1994) and the ATM layer offer the same services to MTP3, enabling the same level to be used, and thereby easing the interworking between narrowband and broadband, networks and offering a migration path for network operators. Much work was performed on the application of SS7 to the broadband environment by the European collaborative project RACE MAGIC (research into advanced communications in Europe programme, multi-services and applications governing integrated control project; see Chapter 14) and the signalling network is described in their ninth deliverable (RACE 1993) (all 14 deliverables are freely available).

2.4 SIGNALLING ATM ADAPTATION LAYER

Signalling can be regarded as a user of an ATM, and must be matched to its cellular structure (de Prycker 1991). However, the network layer (either MTP at Level 3 or the Layer 3 signalling between the user and the network) is very much message-based, and needs a data link layer to reliably deliver messages as signalling data units to a peer entity at the other end of a physical link. Two distinct functions are provided by the SAAL:

- adaptation to the ATM cell structure, and
- data link layer functionality.

Each of the two distinct functions can be divided into two sub-functions, as shown in Figure 2.5.

The lower half of the SAAL is known as the *common part*, and consists of a standard AAL Type 5, as described in Chapter 1. This allows the upper half, the service specific part, which performs the data link functionality, to exchange signalling data units without performing any segmentation or reassembly, and without the necessity of error detection due to bit corruption. The two functions contained in the service specific part are:

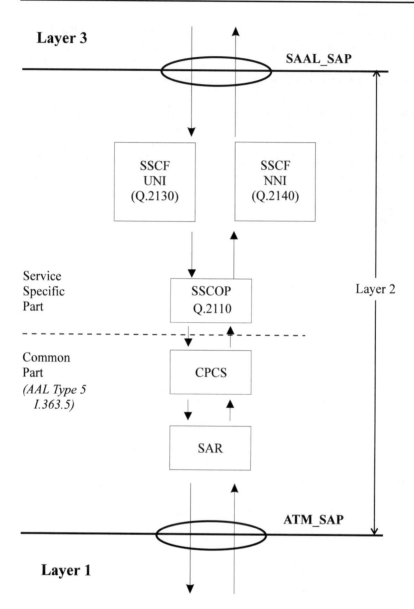

Figure 2.5 Structure of the SAAL

2.4 SIGNALLING ATM ADAPTATION LAYER

- Service Specific Connection Oriented Protocol (SSCOP)—which provides the generic data link protocol functions to allow for an assured data transfer between two peer entities
- Service Specific Co-ordination Function (SSCF)—which provides the functionality to enable a single SSCOP to provide the data link layer services for both the User Network Interface (UNI) and the Network Node Interface (NNI).

SSCOP

The SSCOP has internal interfaces to the SSCF and to the common part (AAL5). These interfaces are standardised using primitives, which can also be shown as signals in the Specification Description Language (SDL)—a method of formally describing dynamic interactions in an easy to follow graphical format. The signals used by the SSCOP to the SSCF are:

- *AA-ESTABLISH*—used to establish a point-to-point signalling association for assured information transfer between peer user entities.
- *AA-RELEASE*—terminates a point-to-point signalling association between peer user entities.
- *AA-DATA*—used for the assured point-to-point transfer of Signalling Data Units (SDU) between peer user entities.
- *AA-RESYNC*—used to resynchronise the SSCOP connection.
- *AA-RECOVER*—used for protocol error recovery.
- *AA-UNITDATA*—used for the non-assured, broadcast and point-to-point, transfer of SDUs between peer user entities.
- *AA-RETRIEVE*—to retrieve SDUs submitted by the user for transmission but not yet released by the transmitter.
- *AA-RETRIEVE COMPLETE*—indicates that there are no additional SDUs to be returned to the SSCOP user.
- *MAA-ERROR*—reporting of SSCOP protocol errors and certain events to layer management.
- *MAA-UNITDATA*—used for the non-assured, broadcast and point-to-point, transfer of SDUs between SSCOP and peer layer management entities.

SSCOP has just two primitives to the common part:

- UNITDATA.invoke—which is used by SSCOP to carry the SDUs for transmission to the peer entity, and
- UNITDATA.signal—which contains the SDUs from the peer entity.

Primitives provide the structured means for information to be passed from into and out of a function within a layer or sub-layer. The function itself contains states, actions, decisions, etc., which comprises the required capability. The SSCOP performs the following functions:

- *Sequence Integrity* preserves the order of SSCOP SDUs that were submitted for transfer by SSCOP.
- *Error Correction by Selective Retransmission* through use of a sequencing mechanism, the receiving SSCOP entity can detect missing SSCOP SDUs and perform error correction by selective retransmission.
- *Flow Control* allows a receiver to control the rate at which the peer transmitter sends information.
- *Error Reporting to Layer Management* provides a comprehensive indication of errors which have occurred.
- *Keep Alive* verifies that the two peer SSCOP entities are remaining in a link established state even when there has been no transmission for a lengthy period.
- *Local Data Retrieval* allows the local SSCOP user to retrieve in-sequence SDUs which have not yet been transmitted (or transmitted but not yet acknowledged).
- *Connection Control* performs the establishment, release, and re-synchronisation of an SSCOP connection and also the unassured transmission of variable length user-to-user information.
- *Transfer of User-Data* provides the conveyance of user data peer to peer as either assured or unassured data transfer.
- *Protocol Error Detection and Recovery* detects and recovers from errors in the operation of the protocol.
- *Status Reporting* allows the transmitter and receiver peer entities to exchange status information, assisting the error detection process and contributing to the protocol's reliability.

The SSCOP actions in exchanging signalling data units can be split into four phases.

Establishment phase

The establishment phase is invoked by one SSCOP sending a BGN (begin) SDU to its peer. Two outcomes are possible—the establishment, indicated by a BGAK (begin acknowledgement), or a rejection, indicated by a BGREJ (begin reject).

2.4 SIGNALLING ATM ADAPTATION LAYER

Data transfer phase

The data transfer phase enables the peer SSCOP entities to exchange information on behalf of their respective higher layer protocols (data received and sent through the SSCF). Data may be either assured or unassured. The unassured data is relatively simple, indicated by a UD (unnumbered data) SDU. If the error detecting logic discards a UD SDU, then this will not be indicated to either of the SSCOP entities. It is up to a user of the AAL to perform error recovery.

SSCOP provides the ability to exchange management information, using an unassured MD (management data) SDU.

The assured data transfer is used to transfer, across an SSCOP connection, information fields provided by the SSCOP user with an indication from the peer entity as to its reception. Assured data transfer, in any data link layer, relies upon sequence numbering in both entities. Entity 1 sends a unit of data and a sequence number of N. If it is received by the far end, it returns some form of response to indicate that it is now expecting data with a sequence number of $N+1$. If the transmitting end does not receive the acknowledgement, it assumes that the transmitted data was lost and therefore re-sends the unit of data with sequence number N. At the far end it knows, from its own counter, that this piece of data has already been received, and assumes (rightly) that the response was lost. The response is resent and, provided it is received, both ends are back in synchronisation again. If the maximum value of N is greater than 1, then up to $N-1$ data units can be sent before an acknowledgement, containing the next expected data unit number, needs to be received. The sequence numbers in SSCOP cycle through the range 0 to 16,777,215 ($2^{24}-1$) inclusive, although the number of data units that may be transmitted without an acknowledgement is constrained by such factors as the end-to-end transit delay, the POLL timer value for the peer entity, the SSCOP throughput (bits/s) and the data frame size. The SSCOP is slightly more complex than this functionality, because it incorporates the flow control by using a system of credits to prevent transmitters overloading receivers buffers.

SSCOP provides the data transfer using a SD (sequenced data) SDU, can request the current status of the peer entities counters' using the POLL (which is really a status request SDU) and receives acknowledgements as either STAT or USTAT (the former being a solicited status response, whilst the latter is an unsolicited status response). A USTAT will generally be sent when out of sequence data is received, while the POLL and STAT interchange enables an SSCOP entity to audit its peer, and also provides the 'keep alive' function in the event that no signalling data has been exchanged for some time.

Error recovery and resynchronisation phase

From the description of the assured data transfer, it would appear that the two entities cannot become 'out of step' with each other. However, any implementation exchanging information across a physical medium that has inherent unreliability (even if very small) can be inconsistent. This can be detected when an entity asks its peer to retransmit a piece of sequenced data, and the transmitter no longer holds the data because it believes that an acknowledgement has already been received. Under these circumstances, error recovery needs to occur. An entity detecting the error uses the ER (error recovery) SDU, and recovery is completed by the ERAK (error recovery acknowledge) SDU.

Under extreme conditions, an SSCOP entity may detect that both transmit and receive sides have become confused and bi-directional resynchronisation is performed by exchanging RS (resynchronisation) and RSACK (resynchronisation acknowledge) SDUs. Resynchronisation will take precedence over error recovery.

Release phase

Protocols, and especially signalling, have no concept of refusal of a release. The procedures for SSCOP release are therefore fairly straightforward—the exchange of END and ENDACK provides a self-explanatory release.

SSCF at the UNI

It is the SSCF that initiates the establishment and release of an SSCOP association (also known, rather confusingly, as a connection). The SSCF also transfers data either in the assured or the unassured mode. The interface between the SSCF and the network access signalling (Layer 3) is defined as a set of primitives, which are covered in Section 2.5.

SSCF at the NNI

The internodal signalling for broadband re-used the functionality of SS7. MTP3 in broadband has just three small differences:

- it identifies that the SS7 user part is broadband (B-ISUP)
- bit transmission order is defined in the lower levels
- parameters relating to changeover are different to accommodate the larger sequence numbers defined in SSCOP, and
- maximum MTP3 user data size (4096 instead of 272).

Only the third item above has any real impact on MTP3, and so it is the function of the SSCF to adapt the SSCOP interface to be suitable for use by MTP3. These functions are either dependent on peer to peer messages or are independent of any message exchange. The latter functionality performs

- mapping between primitives (see Figure 2.5)
- procedures concerning the local retrievement by MTP3 of unsent data, and
- congestion—which the SSCF detects when the transmission queue exceeds a preset value.

The SSCF supports MTP3 peer to peer messages by providing the following functionality:

- change remote processor status—supports the 'processor outage procedure' of narrowband MTP2, by SSCF remote communication to report to the local MTP3 any remote MTP3 failure and recovery,
- change of link status—which enables MTP3 to 'move' all the local state variables from one link to another after a period of analysing a links suitability for signalling traffic (the narrowband proving procedures, both normal and emergency).

The service access point

The SAAL demonstrates some important features of signalling. At the top and bottom of Figure 2.5 is a pattern denoting the SAP, or Service Access Point. The SAP provides a defined point in a protocol architecture over which will be a standardised internal interface. Standardisation, which is covered in Chapter 3, is normally concerned with the exchange of information across a physical interface between equipment that may be provided by different suppliers. The SAP is clearly an internal interface between the layers in the OSI stack (or levels in SS7). To partition the design of signalling protocols, it is necessary to define how one layer or level interacts with the next. The ATM_SAP is specific only to ATM—it provides the same functionality to all of the AAL types. The SAAL_SAP has two variants—the UNI SAAL and the NNI SAAL.

2.5 SIGNALLING AND PROTOCOL PRIMITIVES

Signalling and protocols provide a means to deliver information from one protocol service user to another. The means of information being passed

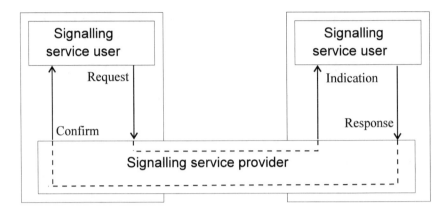

Figure 2.6 Primitives between Signalling Service Provider and Signalling Service User

between protocol service user and protocol service is through primitives, as illustrated in Figure 2.6.

The signalling service provider performs the complete transfer of the signalling service users information over a physical interface or interfaces that may comprise a transmission medium or concatenation of transmission mediums. The information transfer includes any assurance of delivery by signalling service provider, perhaps involving retransmission, but this is invisible to the signalling service user. The user sends information in a *request*, which the signalling service provider then delivers to the remote user as an *indication*. Actions at the remote end may result in a *response* being elicited, which will be delivered back to the local user as a *confirm*. In some signalling and protocol standards, this may be represented as the information type and then the request, indication, response or confirmation as an abbreviated suffix. An example would be a call establishment information primitive, that is a complete confirmed service. The information is carried by the signalling service provider, in SDUs, and the logic in establishing a signalling association for transfer of information is also the responsibility of the signalling service provider. The signalling service user need not be aware of this, under normal, error free acceptance by the remote end. A CALL-ESTABLISH.req is invoked by the local user and as a result a CALL-ESTABLISH.ind is delivered to the remote user. If the call is accepted then the remote user invokes the CALL-ESTABLISH.resp which is delivered as a CALL-ESTABLISH.conf. It should be noted that a user, in this case, is the user of a service, and not necessarily a user application or real (human) user. In a real protocol, of course, either end may begin the call establishment and so all four primitives are available to both service users.

REFERENCES

de Prycker M, *Asynchronous Transfer Mode*, Ellis Horwood, 1991.
Knight R R, The Evolution of the SS7 STP in a Broadband and IN environment, *IEE Tutorial Colloquium on Signalling for Broadband*, The Institute of Electrical Engineers, December 1995.
Law B, Signalling in the ATM Network, *BT Technology J.* Vol. 12 No. 3, July 1994.
Lesher S, Almon Strowger: Dialing for Dollars, http://www.siemen-scom.com/customer/9801/11.html, 1999.
Manterfield R, *Common Channel Signalling*, IEE, 1991.
RACE Project R2044, Signalling Network Architecture, 'MAGIC' Project Deliverable Number 9, September 1993.

FURTHER READING

Manterfield R, *Common Channel Signalling*, IEE, 1991.

STANDARDS

Q.700 *Introduction to CCITT Signalling System No. 7*, ITU-T (03/93).
Q.701 *Functional description of the message transfer part of Signalling System No. 7*, ITU-T (03/93).
Q.702 *Signalling Data Link*, ITU-T (11/88).
Q.703 *Signalling Link*, ITU-T (07/96).
Q.704 *Signalling network functions and messages*, ITU-T (07/96).
Q.705 *Signalling network structure*, ITU-T (03/93).
Q.709 *Hypothetical signalling reference connection*, ITU-T (03/93).
Q.2100 *B-ISDN signalling ATM adaptation layer (SAAL) overview*, ITU-T (07/94).
Q.2110 *B-ISDN—ATM adaptation layer—Service specific connection oriented protocol (SSCOP)*, ITU-T (07/94).
Q.2130 *B-ISDN—signalling ATM adaptation layer—Service specific coordination function for support of signalling at the user-network interface (SSCF at UNI)*, ITU-T (07/94).
Q.2140 *B-ISDN—signalling ATM adaptation layer—Service specific coordination function for support of signalling at the network node interface (SSCF at NNI)*, ITU-T (02/95).
Q.2144 *B-ISDN—signalling ATM adaptation layer (SAAL)—Layer management function for the SAAL at the network node interface (NNI)*, ITU-T (10/95).

Q.2210 *Message transfer part level 3—functions and messages using the services of ITU-T recommendation Q.2140*, ITU-T (07/96).

QUESTIONS

1. Describe an in-band signalling mechanism in common use today.
2. How does a signalling system differ from a signalling protocol?
3. Why is it necessary to protect a signalling network?
4. What are the disadvantages of the Signalling Transfer Point (STP)?

3
Signalling Standards

Bryan Law and Dick Knight

In any technology as complex as telecommunications, it is essential that there is a mechanism which enables the design, production and operation to be consistent. This is especially important where there is a need for at least two realisations to be compatible to achieve their main purpose—an effective communication. This is achieved by the use of standards, which comprise documents containing agreements reached by bodies responsible for that particular aspect of telecommunications. Membership of the body will be drawn from operators, suppliers, customers and consultants with an interest in the area under study. Increasingly in the modern telecommunications world, the national or regional regulator may also be involved in the standardisation process (Clarke 1994). The documents produced will be the result of considerable study, discussion and analysis enabling endorsement at company, national, regional and world level, as appropriate.

While the importance of standards within the whole area of telecommunications can be gauged, it is fundamental for signalling. It is also crucial that any signalling standard produced must be clear, comprehensive and unambiguous. In addition, signalling serves the requirements of the telecommunications service being delivered, not as an end in itself. This means that the overall service required, and delivered, must not be modified by the signalling mechanism that supports it. These stringent requirements mean that signalling studies must be conducted with considerable care and awareness, and therefore results in a process that takes some time. However, the pace of modern telecommunications developments and market needs mitigates against a lengthy process. Therefore, those working in the area of signalling standards have a number of conflicting pressures that they have to satisfy so that the right standard can be produced at the right time.

B-ISDN is considerably more complex and has greater scope than either Narrowband ISDN (N-ISDN) or the Public Switched Telephone Network

(PSTN). Hence, from relatively simple signalling mechanisms produced to support mainly telephony-based services in the PSTN, there has been a gradual increase in the required capabilities. In N-ISDN, based on the 64 kb/s switched fabric, the scope of services was increased and, while the telephony based circuit switched ancestry could still be clearly seen, the impact of both data and packet transfer requirements became increasingly significant. The evolution of telecommunications to the cell-based ATM environment has provided the ability to support an even greater range of services and network capabilities. This means that there are demands on the signalling mechanism to be powerful, flexible and resilient in order to use a B-ISDN to its maximum potential. To ensure that this complex and wide ranging capability achieves the greatest efficiency, incompatibility and ambiguity must be kept to the minimum. Signalling enables services within the network, gaining access to the network and between networks. Hence, the need for standardisation in B-ISDN signalling is essential.

3.1 VALUE OF STANDARDS

The value of standards to those involved in telecommunications can be identified under four main headings:

- financial advantage
- widespread acceptance
- common understanding, and
- stable environment.

Financial advantage is achieved by a standard approach to all aspects of telecommunication enabling economies of scale. From the supply side of the industry, the need to invest in customised development for each individual network operator for a probable limited market is avoided. Instead, with a common basis for specification the development can be far more generic and still satisfy a wide range of customers. This reduces the development costs and allows savings to be made in production. A long production run of a relatively common product has much greater benefit than shorter production runs of specialist products. From the operators' perspective, a large number of suppliers offering the same product increases choice, and therefore the ability to negotiate a competitive price, is increased. Further advantage is accrued in operation when the cost of running and maintaining relatively common network fabric will be reduced compared to cost if customised builds are present in the network. In addition, procurement costs are also reduced if a common specification forms the basis for the preparation of documentation.

Widespread acceptance of product specification, operation and management procedures also has considerable benefit. For the supplier and procuring operator contractual negotiations will be greatly reduced if there is only a relatively small area of the product that has to be customised for any specialised operator. For operation both within and between operators networks, acceptance of the agreed standards will make for easier operation and co-operation when providing the customer with an end to end service. This is particularly important in the regulated multi-vendor environment, where it is essential that a commonly agreed basis for interconnect is available (Wright 1996). In a similar way to aiding co-operation in operational matters, it is equally essential that widespread agreement is also present in the area of network management.

While benefit from cost reduction and agreement is essential, there would be little value if the standard that was produced was ambiguous. A clearly understood standard will aid the supplier in design and production. The operator will find it easier to negotiate the procurement of the product and operate and manage the network when the base standards which are used in these areas are clearly understood. Inter-operation between operators will be easier when there is a commonly understood basis for interconnect. There will always be the need to strike the right balance for the standard between flexibility and service optimality and between being comprehensive while allowing for service differentiation.

The presence of a widely accepted and understood standard will also lead to stability in the telecommunication environment. This does not mean that this is a sterile and unmoving situation rather that any changes/moves that are required will also have to be as a result of common agreement. Hence, any change or new standard introduced will have a well understood and accepted basis from which it is developed rather than one off or quick fix solutions. This will aid the supplier in developing his product, and the operator in the evolution of the network from understood and accepted starting points.

Signalling in the B-ISDN demonstrates how the attributes of financial advantage, widespread acceptance, common understanding and the stability that a standardised approach brings are essential and bring value to supplier and operator. Of course, the customer also benefits from B-ISDN signalling standards through a more efficient and cost effective service.

3.2 INTERNATIONAL STANDARDS AND INDUSTRY BODIES

The international bodies involved in B-ISDN signalling standards range from worldwide through regional down to national level. There may also be an additional stage of standards activity within individual organisations and companies, but these are not considered in this book.

For telecommunications signalling standards there are two bodies whose output is recognised worldwide, the International Telecommunications Union—Telecommunications Standardisation Sector (ITU-T) and the International Standards Organisation (ISO). Both are eminent bodies, but the ITU has by far the greater role in studying and producing telecommunications network standards. In addition, while not actually a true standards body, the work and output of the ATM Forum also has wide recognition in the ATM and broadband world, and their work has had a great impact on B-ISDN signalling studies. No work on modern communications would be complete without mentioning the Internet Engineering Task Force (IETF), and the fifth major player in the world's standards is the Digital Audio Visual Council (DAVIC).

Standardising the world's telecommunications

The United Nations are ultimately responsible for the standardisation of the world's telecommunications. Their approach to this type of task is to set up a small committee to oversee the working method of a recognised body that actually does the work. The World Telecommunications Standardisation Committee (WTSC) is roughly the telecommunications counterpart of the World Health Organisation, and devolves the work of detailed specification to the International Telecommunications Union (ITU). Prior to 1993, the standardisation body responsible for telecommunications worldwide was the International Telephone and Telegraph Consultative Committee (CCITT). The CCITT working method was a fixed four year study period producing standards (titled recommendations) within Study Groups (SG). Even if a study group completed and agreed the technical content of a recommendation before the study period elapsed, it could not be endorsed as a worldwide standard until approved by the full formal meeting of the CCITT (plenary assembly meeting) at the end of the study period. Recommendations were published as sets, and the covers were in a different colour for each study period. It is still possible to get people to talk about the differences between Q.931 red book and blue book. CCITT study groups were numbered using roman numerals.

3.2 INTERNATIONAL STANDARDS AND INDUSTRY BODIES 47

International Telecommunications Union working method

A standards making process geared to a single output every four years slowed down the increasingly expanding and rapidly changing telecommunications industry, and so the CCITT and its radio counterpart, CCIR, were replaced by new ITU bodies; the ITU Telecommunication Standardisation Sector (ITU-T) and the ITU Radio Standardisation Sector (ITU-R), and a third portion, the ITU-D, to address the needs of developing countries. These bodies would again be supported by study groups as before, but now the study groups would have far greater autonomy. Rather than waiting for the full four year cycle to elapse before presenting candidate recommendations to a full ITU meeting, study groups would be able themselves to approve those for which they were responsible. Table 3.1 lists the ITU-T study groups and the broad area of responsibility. There is no study group 1 or 14 at present.

The process has now been refined such that when the working party that has undertaken the detailed study and is satisfied that the draft recommendation is sufficiently sound, mature and stable, it may put the draft for determination. Determined drafts then undertake a period of scrutiny, allowing time for ITU-T members to consider the candidate recommendation. They have until the next study group meeting to comment against it (two months) then, before the formal study group meeting, the ITU-T secretariat will announce if members are happy for the approvals process to go ahead. At the full study group meeting, if there are no serious objections to the draft text, it is formally approved by the study group deciding to adopt the new recommendation. This two step method of approval is used to refer to recommendations as 'determined' and 'decided'. The one remaining process is that of publication by the ITU secretariat, including translation into both French and Spanish. The recommendations themselves are divided into 'series', with each series being identified by a single letter of the alphabet, as shown in Table 3.2.

This book is primarily concerned with broadband signalling standards, which are contained in the Q series of recommendations, numbers in the range 2000 to 2999. The I series of recommendations provide the major requirements for the Q series.

ITU-T Study Group 11

Within the ITU, the study group responsible for signalling recommendations is Study Group 11 (SG11). Study Group 11 is responsible for signalling protocols for ISDN, mobility, network intelligence and signalling transport

Table 3.1 ITU-T Study Groups

Study Group	Title	Responsibilities
2	Network and service operation	General aspects of service definition related to telecommunication services: PSTN based, ISDN(s), mobile and UPT services; including principles of their interworking and relevant user Quality of Service (QoS). Network operations including routing, numbering, network management and service quality of networks; human factors; service and operational aspects of fraud prevention. Lead Study Group on Service definition, Numbering, Routing and Global Mobility.
3	Tariff and accounting principles including related telecommunications economic and policy issues	Studies relating to tariff and accounting principles for international telecommunication services and study of related telecommunication economic and policy issues, as well as policy issues related to carriage and content.
4	Telecommunication management network (TMN) and network maintenance	Maintenance of networks, including their constituent parts, identifying needed maintenance mechanisms and for applications of specific maintenance mechanisms provided by other Study Groups. Lead Study Group on TMN.
5	Protection against electromagnetic environment effects	Electromagnetic compatibility (EMC) of telecommunication systems, including precautions to avoid hazard to human beings.
6	Outside plant	Construction, installation, jointing, terminating, protection from corrosion and others forms of damage from environment impact, except electromagnetic processes, of all types of cable for public telecommunications and associated structures.
7	Data networks and open system communications	Data communication networks, and for studies relating to the development of open system communications and to the application of open system communications including networking, message handling, directory, security and open distributed processing. Study Group 7 is the Lead Study Group on Open Distributed Processing (ODP), Frame Relay and for Communication System Security.
8	Characteristics of telematic systems	Telematic terminal characteristics and related service aspects. Lead Study Group on Facsimile.

3.2 INTERNATIONAL STANDARDS AND INDUSTRY BODIES

Table 3.1 (cont.)

Study Group	Title	Responsibilities
9	Television and sound transmission	Studies of the specifications to be satisfied by telecommunication systems used for contribution, primary distribution and secondary distribution of video, audio and the associated data signals, related to television, sound-programme and associated services, including interactive ones.
10	Languages and general software aspects for telecommunication systems	Technical languages, the methods for their usage and other issues related to the software aspects of telecommunication systems.
11	Signalling requirements and protocols	Signalling requirements and protocols for telephone, N-ISDN, B-ISDN, UPT, mobile and multimedia communications. Lead Study Group on Intelligent Network and FPLMTS.
12	End-to-end transmission performance of networks and terminals	End-to-end transmission performance of networks and terminals in relation with the perceived quality and the acceptance of text, speech and image signals by the users and for the related transmission implications.
13	General network aspects	General network aspects and the initial studies of the impact of new system concepts and innovative technologies on telecommunication networks with far-reaching consequences, including broadband ISDN and global information infrastructure studies, taking into account the functional responsibilities of other Study Groups. Lead Study Group on General network aspects, Global Information Infrastructures (GII) and Broadband ISDN.
15	Transport networks, systems and equipment	Transport networks, switching and transmission systems/equipment including the relevant signal processing aspects. Lead Study Group on Access Network Transport.
16	Multimedia services and systems	Multimedia service definition and multimedia systems, including the associated terminals, modems, protocols and signal processing. Lead Study Group on Multimedia services and systems.

Table 3.2 Organisation of ITU-T Recommendations

A	Organisation of the work of the ITU-T
B	Means of expression (definitions, symbols, classification)
C	General telecommunication statistics
D	General tariff principles
E	Overall network operation, telephone service, service operation and human factors
F	Telecommunication services other than telephone
G	Transmission systems and media, digital systems and networks
H	Line transmission of non-telephone signals
I	Integrated Services Digital Networks (ISDN)
J	Transmission of sound programme and television signals
K	Protection against interference
L	Construction, installation and protection of cable and other elements of outside plant
M	Maintenance: transmission systems, telephone circuits, telegraphy, facsimile and leased circuits
N	Maintenance: international sound programme and television transmission circuits
O	Specifications of measuring equipment
P	Telephone transmission quality, telephone installations, local line networks
Q	Switching and Signalling
R	Telegraph transmission
S	Telegraph services terminal equipment
T	Terminal characteristics and higher layer protocols for telematic services
U	Telegraph switching
V	Data communication over the telephone network
X	Data networks and open system communication
Y	Global information infrastructure
Z	Programming languages

mechanisms, as well as those for broadband. It is divided into constituent working parties of technical experts for a particular field or area of work, and Working Party 1 (WP1) has sole responsibility for all aspects of broadband network signalling protocols. Study Group 11 WP1 produce recommendations covering signalling requirements, user access and internodal network signalling as a basis for world-wide implementation by operators and suppliers. The working party is itself split into smaller sub-working parties, each headed by a rapporteur. Sub-working parties receive their terms of reference for study by being set 'questions', such as 'What are the signalling requirements for broadband services?'

Study Group 11 relies on experts from other study groups to initiate and to complete B-ISDN signalling studies. Study Group 13 is the main provider of requirements for signalling, providing details of the network services, such as switching capabilities, that need to be signalled. The other main

3.2 INTERNATIONAL STANDARDS AND INDUSTRY BODIES 51

study groups that Study Group 11 interacts with are Study Groups 2, 4, 7, 15 and 16.

Broadband standards—releases, capability sets, steps and capabilities

Since the activities in Study Group 11 are very much dependent on the requirements from ITU-T Study Group 13, no signalling work could be really progressed until Study Group 13 work was sufficiently mature. Study Group 13 realised that their studies would be extensive in scope, and would take considerable time to complete. However certain aspects could be quickly determined and documented as recommendations. The growing market pressure for world standards caused Study Group 13 to produce these recommendations in releases. Each release was planned to contain a complimentary menu of ATM network capabilities which operators could utilise to deliver broadband services to their customers. Recommendation I.310 identified the network capabilities which were required to deliver an overall broadband service and partitioned them into three releases. Study Group 11 examined the Study Group 13 recommendation, and in determining the signalling requirements to support each release, decided that it would also use a release approach.

This was fairly satisfactory when applied to the simpler requirements of basic connectivity for Release 1, but it was soon realised that Releases 2 and 3 were much too complex for this approach. Growing pressure for signalling standards caused further partitioning to enable more rapid production of recommendations. The partitioning was again based on complimentary sets of signalling capabilities, and so Capability Sets (CS) were adopted by Study Group 11. The order in which each capability set was studied was determined by the priority set by the world telecommunication industry for particular broadband signalling capabilities. Capability sets produced some confusion at first, especially as the Intelligent Networks (IN) studies had also been broken down into capability sets. The IN capability sets and the broadband signalling capability sets, whilst both part of Study Group 11, are not aligned. To clarify the situation, Working Party 1 adopted the practice of referring to their capabilities as 'signalling capability sets', or SCS, and IN CS as referring to IN capability sets. In theory, IN CS3 and broadband SCS3 should have been aligned, to provide the point at which broadband would inherit an intelligent network capability, but over-optimism in the elapsed time to produce the capabilities (in both IN and broadband studies) has prevented this achievement. IN CS4 and SCS3 will now provide an intelligent broadband network, and is covered in detail in Chapter 16.

The use of the capability set approach proved very successful in partitioning and pacing studies to make the best use of the signalling expertise available. However, even when broken down into capability sets, given the pace of broadband service requirements, the time scale for delivering signalling standards was still too long. To be fair to the ITU-T and Study Groups 11 and 13, they had been used to larger investment and consequent slower pace of change in the development of large scale public telecommunications networks. Broadband services represented the sharper end of telecommunications, data and computer communications services, and some of the industry sectors survived in a market of perpetual change.

Study Group 11 had already introduced, internally, the concept of 'steps' within a capability set to reduce the size of each study bite. This helped ease the delay in producing the required standards, by grouping together families of signalling recommendations, but even so, it was not considered fast enough. It was eventually decided that as the market required a signalling capability, this would be studied as a specific item. While this seemed to meet the time scale constraints, there are some who remain concerned that overall compatibility with supporting standards has been lost.

International Standards Organisation (ISO)

ISO have only recently embarked upon study leading to the production of B-ISDN signalling standards While not involved in public telecommunications, ISO have a long history of producing standards for data communication based on inputs from the computing and private network supply industries. Recent studies of B-ISDN signalling by both these industries has resulted in ISO studies being initiated, and the first ISO standards for B-ISDN signalling being produced. ISO have a co-operation agreement with ITU-T that enables the publication of jointly produced standards. An example of this co-operation is the Information Technology Open Systems Interconnection Application Layer Structure, ITU-T recommendation X.207, which contains identical text as ISO 9545.

ISO working method

Study has only recently commenced in ISO, initiated by input from ECMA (described later in this chapter) of the first version of BQSIG. ISO are endorsing the ECMA input, and this is published as an ISO standard.

In future, the output of the collaboration between ETSI and ECMA on B-ISDN access signalling will also be put into ISO with the aim of having this also endorsed as an ISO standard activity.

ATM Forum (ATM-F)

The ATM-F, while not strictly a standards body, is recognised as one of the main players in the production of broadband signalling standards. Set up as an interest group representing mainly terminal equipment manufacturers and private network suppliers, it has opened into a wider participation and is attended by many organisations also involved in ITU-T. The original purpose of the ATM-F was to sharpen the standards processing mechanism to represent both the requirements and the timescales required in the private network environment. The ITU-T output formed the basis for modifications to provide industry standards for private network operation. These standards, as initially produced by the ATM-F, provided a complete network interface specification (e.g. User Network Interface 3.1), comprising all aspects (physical, transmission and management, as well as signalling).

ATM-F activity extended into defining the interface for communication between private network switches—their Private Network-Network Interface (P-NNI). Taking their UNI 3.1 as the starting point, the ATM Forum defined a new signalling protocol for interswitch communication, and then went on to show how this could be used as a signalling protocol for communication between networks. Despite the term *private* in the title, PNNI is also used as a public network signalling protocol, and this led to the need to interwork with the public network signalling protocol, B-ISUP produced by ITU-T. The ATM-F started their studies of an inter-carrier interface B-ICI based on the ITU-T SS No. 7 B-ISUP and UNI network interworking protocols. More recently, the ATM Forum have moved towards the production of (yet) another signalling protocol, which they have named the ATM Inter-Network Interface (AINI). This is intended to make it easier to interwork PNNI with B-ISUP, and is more closely based upon PNNI, without the ability to export routing information between the networks.

Internet Engineering Task Force (IETF)

IETF is a relatively new experts group directed to identifying problems and providing solutions for this increasingly popular and expanding means of communication. Both public operators and suppliers recognise the importance of the internet, and ITU-T not only has a specific item as part of its B-ISDN signalling studies to address this topic, but has established a mechanism for exchanging information.

Digital Audio Visual Council (DAVIC)

DAVIC is an experts group established to develop a common understanding of the requirements to support interactive multimedia services. The initial starting point was to interpret existing standards and show how these could be combined to form advanced services. Their output also ensures that the standards bodies take due account of the needs of multimedia services as they become increasingly important in the modern telecommunications environment. DAVIC has studied and produced application protocols which need to be taken into account when producing B-ISDN signalling standards. In addition, DAVIC provides another source of service requirements to develop B-ISDN signalling requirements.

3.3 REGIONAL STANDARDS BODIES

North America, Europe and Japan tend to be the main players in ITU-T and it is these regional organisations that are the main contributors and users of B-ISDN standardisation studies.

American National Standards Institute (ANSI)

Within the American National Standards Institute, telecommunications are the responsibility of committee T1. ANSI T1 co-ordinate studies in North America that lead to the production of standards for use within the North American regulatory environment, and determine the North American position for input to ITU-T standards activities. The detailed study on signalling and switching issues is conducted by the subcommittee S1. The Alliance for Telecommunications Industry Solutions (ATIS) also generate activity in the US and Caribbean in the area of signalling, contributing to ITU-T via the US delegation. Both ANSI T1 (T1 S1) and ATIS undertake considerable study in the area of B-ISDN signalling, which leads to north American standards being produced.

European Telecommunication Standards Institute (ETSI)

ETSI is the European body responsible for developing telecommunications standards for use in the European Union's harmonised market. Just as the

3.3 REGIONAL STANDARDS BODIES

ITU-T divides its work amongst study groups, ETSI undertakes specific areas of telecommunications studies through specialist Technical Committees (TC) and Working Groups (WG). For all signalling, including requirements and also services, the technical committee is Services Protocols and Networks (SPAN), an amalgamation of the old Signalling and Protocols and Switching (SPS) technical committee with the Network Architecture (NA) technical committee. Instead of then sub-dividing according to general capabilities, ETSI works rather upside down compared to ITU-T, with working groups arranged according to interfaces—SPS/SPAN1 (internodal signalling), SPS/SPAN3 (signalling requirements), SPS/SPAN5 (access signalling), etc. Working groups then have sub-working groups to deal with particular areas, such as the B-ISDN standards. Study in these working parties leads, as with ANSI T1, not only to a regional input into ITU-T, but also regional standards. ETSI designate these as European Norms (EN), although they were previously known as European Telecommunication Standards (ETS).

At the same time that the ETS became the EN, ETSI changed the naming of the groups. Technical committees used to be sub-divided into sub-technical committees, but, so it was argued, they should really have been technical sub-committees. This was corrected by changing the name to working groups, but no-one appears to have remembered why, when the working groups were themselves sub-divided. Perhaps they should have been named working sub-groups rather than sub-working groups!

The European standards are often based on the relevant ITU-T Recommendations, and the ETSI target is to have the EN as close as possible to the ITU-T Recommendation, while still supporting European requirements. This is generally achieved in ETSI by producing documents that detail any differences with the ITU-T recommendation (known as a delta document). This has not always been true in ETSI, and they have occasionally produced standards that have become accepted worldwide. The best known example of this is from the mobile world, where ETSI's global system for mobile communications (GSM) standards for digital mobile communication have been implemented in all continents (but not yet all countries).

One aspect that an EN series also covers that is absent in the companion ITU-T Recommendations is conformance and test specification. This important element is comprehensively covered by ETSI studies, and there is a possibility that in the future this will be imported by ITU-T.

While the current method of working in ETSI will continue for some time, there is now a move towards a more flexible, resilient and quicker mechanism through the use of smaller project teams to support technical committees. This is in response to the needs of the telecommunications market which has imposed far more stringent time scales on standards production than in the past. One ETSI project team particularly worth noting is the EASI project, Enhanced ATM System Interoperability, attempting to 'ease' the technicalities for interworking services, management and signalling between networks in different regulatory environments. At the UNI, ETSI have designated the signalling protocol to be Digital Subscriber Signalling number two (DSS2), number one being the ISDN counterpart.

ETSI published documents are now available for download from their web site for free.

ETSI working method

ETSI have been a source of considerable study, and their results input to ITU-T Study Group 11 in the area of both signalling requirements, architecture and signalling protocols. The aim of ETSI is to align the ETSI standard (EN) as closely as possible with the equivalent ITU-T Recommendation. Whilst a technical committee, such as SPAN, is responsible for the approval of documents, the technical work takes place in smaller groups, termed working sub-groups, and an approval step is performed by the main working group (e.g. SPAN5) before the technical committee approval is sought. There then follows a period of public enquiry, whereby the formal national standards organisations (government designated bodies) of member countries have an opportunity to scrutinise the draft European standard. Any comments are forwarded back to the committee, the working group and often the original working sub-group for resolution before publication.

Computer Manufacturers Association (ECMA)

Once a European only organisation, the European Computer Manufactures Association realised the importance of global standards and dropped the European designation from their title, although they kept the original abbreviation of ECMA. ECMA responsibility lies in the area of private network standards. Using the basic output from ITU-T Study Group 11 on B-ISDN signalling, ECMA have produced a signalling mechanism especially tailored for the requirements of private broadband networks, known as BQSIG. As its narrowband counterpart, QSIG is complimentary to ETSI DSS 1, so there is a similar relationship between BQSIG and ETSI DSS 2.

However, both ETSI and ECMA realise that in the standardisation of B-ISDN signalling, the best way forward is to harmonise their future studies. Therefore, there is a co-operation agreement between ECMA and ETSI for the production of signalling standards that are suitable both for the private and public network. A special joint working group formed from ECMA and ETSI B-ISDN signalling experts was established to undertake this study.

ECMA have also been providing inputs to ISO (to drive those studies) and to the ATM Forum.

ECMA working method

ECMA have recently completed their initial B-ISDN signalling studies, which have led to the production and approval of the initial version of the BQSIG private network signalling standard. As with its N-ISDN counterpart QSIG, BQSIG is based upon the ETSI version of the equivalent public network access signalling system. BQSIG is therefore an enhanced version of ETSI DSS 2 Release 1 and CS2 step 1 ETSs/ENs, with suitable additions and modifications to meet private network signalling criteria.

As stated under the section on ISO, the ECMA standard is now being adopted and endorsed by ISO, and for future standards there will be very close collaboration with ETSI.

Japan ITU-T co-ordination body

This body provides a forum for obtaining a common approach to standards between Japan and ITU-T. By influencing ITU-T studies in a manner complimentary to the position adopted by the main Japanese players, it is reasonable to assume that resulting ITU-T SG 11 B-ISDN signalling recommendations will form the basis for the Japanese signalling standards.

National and industry bodies

Modern telecommunications networks are no longer the sole concern of governmental departments or public industries. Our communications needs are now served by competing network providers, regulated mainly on a national basis. National standards ensure the widest choice to the customer with service as seamless as possible. It is becoming harder and harder to determine just whose network a call does or does not pass through, and yet the service remains unaffected.

The liberation and growth of the terminal market provides more and more applications requiring network and signalling support. This has led to the growth of both national standards bodies and industry bodies representing a particular application interest.

3.4 APPROACH TO SIGNALLING STANDARDISATION

While the bodies directly concerned with signalling standardisation are moving towards a more collaborative approach to their work, it is important that the 'users' of signalling are involved in this process. The aim is to have wherever possible generic signalling capabilities rather than specific signalling protocol for each and every 'user' application. Therefore, within ITU-T and ETSI discussions have commenced with the appropriate groups studying IN and mobility. This is essential if future studies in B-ISDN signalling are to produce as a generic as possible a mechanism. Such a generic mechanism is essential to support the wide range of services and capabilities that it is envisaged will require to be offered by operators and produced by the supply side of the industry. In addition, further input is required from bodies such as DAVIC and the IETF, who represent the end user applications that will be supported in a future B-ISDN.

ETSI—ITU-T

This is perhaps the oldest form of collaboration (albeit implicit) in which ETSI having reached an agreed position on signalling for Europe, if possible, inputs to ITU-T in order to influence the output Recommendations to reflect the European requirements. If achieved, all that is required for a European standard is a simple endorsement of the ITU-T Recommendation. However, if a European complexion still has to be documented, a delta document is produced to the base ITU-T Recommendation. The aim is to make this delta document as 'thin' as possible. From the point of view of signalling, ETSI adds significantly to the standardisation process by insisting upon the production of Protocol Implementation Conformance Statements (PICS). The PICS provide a simple means for a supplier to indicate in a tick sheet the capabilities that have been implemented in the product. ETSI use the PICS to produce testing documentation and abstract test suites, further assisting the procurement process.

ECMA-ETSI

There is now full collaborative activity by the two groups responsible for B-ISDN standards in ECMA and ETSI. From CS 2 step 2 onwards, joint studies and meetings are being held resulting in common inputs to both ITU-T and ISO. In addition, a common set of standards will be produced with minimal departures to satisfy the particular public or private network signalling requirements. There is also a joint input into ATM-Forum studies in an attempt to bring all branches of B-ISDN signalling activity into as close alignment as possible.

ISO—ITU-T—ATM-F

Full co-operation between ETSI and ECMA in the area of B-ISDN signalling studies and standards work with both bodies feeding the output of studies into the appropriate worldwide body, ITU-T and ISO, respectively, there is a move to achieve convergence in signalling standards. In addition, there has already been an initial meeting of ITU-T and ISO and a common understanding that it is desirable to ensure this world standards convergence was reached. It is then hoped that commercially driven bodies such as the ATM-Forum will be encouraged to move their work closer to ITU-T and ISO.

3.5 CONCLUSION

Signalling standards in the telecommunications world impact upon customers service perception, companies, industry sectors, nationally regulated and international networks, so it is essential that there is a widely recognised and well understood method of producing standards. Many organisations use consensus procedures wherever possible to produce their standards. Such procedures ensure the resulting standard is stable and acceptable to the widest possible community. By its very nature such a mechanism will be structured, formal and tending to the bureaucratic. It is also very time consuming. While there is an ongoing activity to minimise the inflexibility such characteristics introduce to the standards making process, this is the price that has to be accepted for standards that have to have the degree of maturity and stability to be accepted worldwide and as a basis both for contractual and regulatory commitments.

This chapter has shown why standards are considered to be of importance in the B-ISDN signalling environment, and how a number of bodies worldwide are involved in the study. It has illustrated how the evolving needs of world telecommunications has influenced a change in studies, and led to a proliferation of standards to meet the industry's requirements. However, standardisation is not a static process and, given its importance to telecommunications, it needs to continue to develop. This is reflected in the increase in the collaboration by the bodies involved in the study of B-ISDN, signalling. Such collaboration will hopefully increase to include all the main players involved in B-ISDN and so lead to a set of standards that has a truly worldwide application assisting both the operators and suppliers, and ultimately benefiting the telecommunication customer.

REFERENCES

Clarke P G, ISDN Standards: An Overview, *BT Eng. J.* Vol. 12, No. 4, 1994, pp. 47–55.
Wright T, Why Standards and Architectures: Structured Information Programme 14.1, *BT Eng. J.* Vol. 14, No. 4, 1996.
Popple G W and Glen P J, Specification of the broadband user/network interface, *BT Technol. J.* Vol. 11, No. 1, 1993, pp. 86–92.

FURTHER READING AND INFORMATION

Web pages

ANSI   http://www.ansi.org/
ATIS   http://www.atis.org/
DAVIC   http://www.davic.org/
ECMA   http://www.ecma.ch/
ETSI   http://www.etsi.org/ (ETSI standards are available free, follow the links or try going straight to http://webapp.etsi.org/publicationssearch/)
IETF   http://www.ietf.org/
ISO   http://www.iso.ch/
ITU   http://www.itu.int/

STANDARDS

Q.767 *Application of the ISDN User Part of CCITT Signalling System No. 7 for international ISDN connections*, ITU-T (02/91)
I.310 *ISDN—Network Functional Principles*, ITU-T (03/93).
ATM User-Network Interface Specification 3.1, ATM forum af-uni-0010.002.

QUESTIONS

1. Why are standards important?
2. What added value does ETSI provide to the standardisation process?
3. Which sector of the telecommunications industry started the ATM Forum, and why?

4
Requirements for Signalling—Broadband Services

Roger Morley and Dick Knight

4.1 THE B-ISDN

The capabilities that ATM switching offers are not sufficient in themselves to determine the signalling and protocol requirements for complete network solutions. The services that can be offered by network operators must also be defined to enable signalling protocols to carry information other than simple connection types, and thus to provide a co-ordinated network that offers services other than simple connectivity—a complete communications solution—the Broadband ISDN (B-ISDN).

B-ISDN is a single network which is capable of supporting a wide range of audio, video and data applications. However, the B-ISDN can only justify its introduction as the preferred network for supporting such a diverse range of applications if it offers advantages in terms of cost, convenience and flexibility, over the re-use of existing networks, to both user and network operator alike. The B-ISDN must clearly, therefore, offer more than just a straightforward evolution of the point-point 64 kb/s circuit and packet transfer modes provided by its narrowband predecessor.

This chapter explains the method used in ITU-T to derive signalling requirements, the I.130 method, and considers initial potential user applications proposed in the standards bodies, especially ITU-T, that have been characterised to derive services that are the starting point for identifying the key network capabilities that B-ISDN signalling systems must support. The network services and connection types are used to illustrate a potential B-ISDN service.

4.2 ITU-TREC. I.130 METHOD FOR SIGNALLING PROTOCOL SPECIFICATION

The job of specifying a method for characterising services was undertaken by ITU-T, and their resultant three stage method was published in 1988 in recommendation I.130. The I.130 method is a top-down approach explicitly incorporating service-related requirements. Although originally aimed at Narrowband ISDN services, it has been shown (RACE 1993) to be extendible for capturing service characteristics and behaviour of many multimedia services identified for the B-ISDN. As well as providing a common framework for service definition, the starting point, I.130 demonstrates that the definition of network resources is a major stage in identifying the requirements of the signalling protocols for the services described.

The three stages of the method are the service aspects (stage 1), functional network aspects (stage 2) and network implementation aspects (stage 3). The application of the method for stage 1 results in a description of the service, stage 2 results in one or more implementation independent scenarios, and stage 3 results in a set of protocol and switching standards implementing the service for each scenario.

Service description

The first of the three stages involves service definition from the user's viewpoint, and it has been resolved into three steps:

- *Prose description.* Extends the textual description, using a template, to capture the involvement of the service users with each other, and the types of information that flows between them. The description must include the networking aspects and interactions with other services in a generic way that does not constrain either terminal or network design. The aim of the prose description is to provide an understanding of the service without describing an implementation. The complete requirements for the prose description are described in ITU-T recommendation I.210.

- *Static description.* Identifies as far as possible a complete set of common connection and service user attributes. The characteristics of the service captured within the prose description are then used to derive values for the common set of connection and service attributes to create a static instance of the service to be described. This step specifies the service entities, attributes and attribute values to be carried by the signalling system. This attribute technique is fully described in ITU-T recommendation I.140.

- *Dynamic description.* Captures the information that must flow between users and the network to create, modify and delete an instance of the service. The description must identify all possible actions relevant to the service, from the user's point of view. Interactions within the network are ignored by treating the network as a single entity dealing with the user's actions. I.130 recommends that interactions are presented in the form of an overall Specification and Description Language (SDL) diagram. This type of diagram is used extensively in signalling and is a type of flow chart, based on a finite state machine, that identifies all the possible actions for every possible stimulus. The use of overall SDL diagrams in a service description is given in annex D of ITU-T recommendation I.210.

Network capabilities

Stage 2 deals with the first level of abstraction of the network. It identifies functional capabilities and derives information flows to support a stage 1 service description. Stage 2 helps to formulate how a network can support the defined static and dynamic behaviour, in an abstract manner. Stage 2 is decomposed into five steps:

- *Derivation of a functional model.* Step 2.1 groups the functions that are required to provide a service into functional entities. Recommendation I.310, ISDN functional principles, describes the concept of functional entities. The functional entities and their relationships are aggregated into a functional model, derived for each basic and supplementary service. Since supplementary services have a relationship with basic services and, especially in the case of B-ISDN, the basic services can interact with each other, it is better to derive a composite functional model. To take account of all service requirements, a complex task, a unified functional model must be derived.

- *Information flows.* Once the functional model or models have been derived, the distribution of the service information to the functional entities must be performed in step 2.2. These interactions are termed information flows. It is important to understand that there is a great difference in an information flow, and a protocol message. A protocol message may transfer some or all of the information defined in a step 2.2 information flow, but the information flows take no regard of how the information is actually transported and acknowledged, nor is there any definition of how to deal with unexpected cases.

- *SDL diagrams for functional entities.* Step 2.3 requires the functions performed within the functional entities to be identified. This is represented by an SDL diagram showing the users inputs and outputs (from step 1.3) and resulting information flows from step 2.2

- *Functional entity actions.* Whilst step 2.3 describes the distribution of information, there must be some actions taken within functional entities to realise the service. These actions are described in step 2.4, and are entirely within a functional entity. The actions assist the designers to determine the information needed at each entity, its function or purpose and assist in understanding the switching requirements. Step 2.4 is represented by a list or sequence of actions by a functional entity in a prose, or textual form.

- *Allocation of functional entities to physical locations.* The final step of stage 2 is to provide the specific types of physical locations for the functional entities and the relationships that support the information flows. A specific type of physical location could be a telephone exchange or a PABX, and each incorporates one or more functional entities. The relationship between physical locations must be realised using one or more protocols. Step 2.5 requires that each allocation and definition of relationships be treated separately in a scenario.

Although the stage 2 steps are not always performed, there has been some work undertaken for B-ISDN, which is covered in later chapters. Examples of information flows and functional modelling can be found in Chapters 14 to 16. The complete details of all the steps of stage 2 are contained in recommendations Q.65 and I.310.

Signalling and switching

Stage 3 of the I.130 method results in the recommendations concerning signalling and switching using the stage 2 output as the basis for the final two steps:

- *Protocols and formats.* Messages, message elements and procedures are designed into the relevant signalling systems to support the information flows between the nodes as identified in the second stage.

- *Switching and service nodes.* The functional entity actions identify the requirements for the switching functions and the service nodes. These are defined in the Q.500 series recommendations, and are outside the scope of this book.

4.3 NETWORK ARCHITECTURE

A communications network comprises different elements and different types of interfaces. As a generality, most interfaces in public networks are classified as either a User Network Interface (UNI) or a Network Node Interface (NNI) for the purposes of signalling. However, a number of different standards bodies have classified interfaces between network elements using letters of the alphabet. These have also been applied to the broadband environment and are distinguished from their narrowband counterparts by using a 'B' as a subscript. The User Network Interface is described in rec. I.413. The interface can be broken down into reference points, ITU-T define the S_B and the T_B interface, and ECMA refer to the Q_B interface, as shown in Figure 4.1. ETSI provide standards for the V_B interface, and there is an electrical 'wires only' interface, the U_B interface as shown in Figure 4.2. All of these interfaces can be considered to be part of the UNI, as they separate access network elements.

The S_B, T_B, and Q_B reference points are all functional references over which different types of signalling are applied. This ignores the fact that there is a real, physical interface between the public network and the network termination equipment, and also that there may be an access network, enabling the broadband services to be offered over the full range of physical media that may be present between the public or private network and the customer. There are no signalling issues for the U_B interface and the V_B interface is explained in Chapter 5 (V_B chapter).

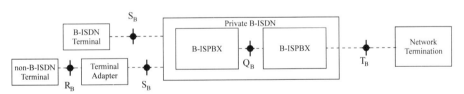

Figure 4.1   Signalling Interface Reference Points in a Broadband Network

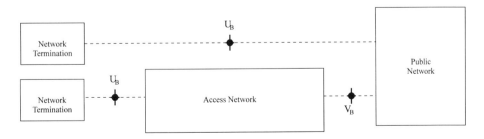

Figure 4.2   Physical Reference Points in a Broadband Network

4.4 POTENTIAL B-ISDN USER APPLICATIONS

Advances in data and image processing capabilities have resulted in a plethora of advanced applications potentially requiring high bandwidth and flexible network connectivity promised by the B-ISDN. The number of such applications is practically endless, and is limited only by our imagination and technological constraints in terms of affordable transmission bandwidth, etc. So far, the majority of applications identified for the B-ISDN have been audio-video based, but these are not viewed as the only service drivers shaping the definition of B-ISDN network capabilities. Several organisations, such as the International Telecommunication Union—Telecommunication Sector (ITU-T), the European Telecommunications Standards Institute (ETSI) and the Digital Audio Visual Council (DAVIC), have been examining what range of applications could conceivably use the B-ISDN, and a representative sample is shown in Figure 4.3.

Although ITU-T defines broadband, in Recommendation I.113, as offering connectivity at greater than 2 Mb/s, the B-ISDN is by no means limited to only supporting applications requiring connectivity at 2 Mb/s and above. Building security and traffic monitoring applications can work with connection bandwidths of 64 kb/s or lower, depending upon how frequent information updates are required.

4.5 NETWORK SIGNALLING SERVICES

Although the B-ISDN must be capable of supporting a wide range of applications, it cannot offer each application an individual set of connectivity services. Such an approach would result in a separate set of B-ISDN network capabilities for each application, and a separate signalling protocol for those applications which wanted 'on demand' connectivity. In order, therefore, to keep the range of network capabilities that are needed down to a managed set, a smaller sub-set of B-ISDN services must be defined, from which the network capabilities can be derived, which can support a group of applications sharing some common characteristics. ITU-T recommendation I.211 classifies services into service categories and service classes. The service categories described within Rec. I.211 have also been included in Figure 4.3, together with the service classes, within each category, relevant to the applications mentioned. The specification of the I.211 service classes, and the information types they contain (e.g. audio, video etc.), was the first step taken by ITU-T towards defining what limited set of B-ISDN services should be supported. If, for example, as shown in see Figure 4.3, the B-ISDN could support a video-on-demand service, then it can support tele-shopping, Karaoke-on-demand and other

4.5 NETWORK SIGNALLING SERVICES

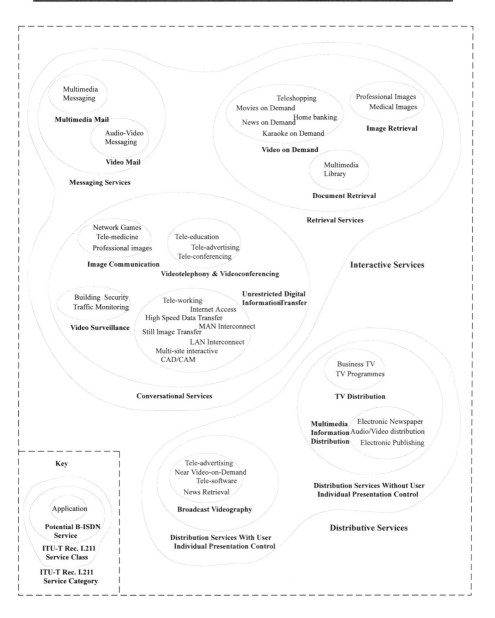

Figure 4.3 Potential B-ISDN Applications, Service Classes and Service Categories

multimedia interactive applications. Although I.211 contains a comprehensive set of B-ISDN services, it is by no means complete and must evolve to meet the needs of emerging applications.

Bearer and teleservices

From a signalling perspective, the limited set of services to be supported by the B-ISDN may be internationally standardised and offered by an administration as bearer services or teleservices, illustrated in Figure 4.4. A bearer service is assumed to just provide connectivity and takes no account of, or offers support for, an application. The ATM Adaptation Layer (AAL) could be empty and cell payloads could, for example, be passed directly to the application. However, it is more common to include the specifics of the AAL within the a bearer service, especially if the B-ISDN is required to support interworking between networks. A Telecommunications Service, or Teleservice, on the other hand requires all application characteristics, expressed in ISOs OSI seven layer protocol model, to be specified. The list of standardised bearer and teleservices identified so far for the B-ISDN, or could be supported by the B-ISDN, are listed below.

The bearer services currently defined for the B-ISDN are:

- Broadband Connection Oriented Bearer Service
- Broadband Connectionless Data Service
- Virtual Path Service for Reserved and Permanent Communications.

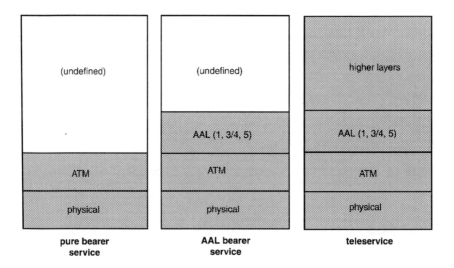

Figure 4.4 Bearer and Teleservices in the B-ISDN

4.5 NETWORK SIGNALLING SERVICES

B-ISDN standardised teleservices are:

- Tele conferencing
- AudioGraphic conference
- Video Telephony
- Video conferencing
- Broadband Videotex
- Multimedia Conferencing
- Audio Visual Interactive Services
- Tele-writing
- Broadband Teletex
- High Speed Telefax.

The involvement of the network in a teleservice depends upon the degree to which is has been standardised. Four levels of teleservice standardisation have been proposed (RACE 1003) for the B-ISDN and are described below.

Standardised teleservices

All of the attributes of the standardised telecommunication service assume standardised, default values which cannot be changed. The network is assumed to allocate all of the resources for the service and ensure that semantic rules given to the service are obeyed. In addition, the network will also ensure terminal-to-terminal compatibility relating to Higher Layer Functions (HLFs) and Lower Layer Functions (LLFs) of the teleservice. The Lower Layer functions cover the Physical Layer, ATM Layer and AAL aspects of the service. The Higher Layer Functions cover all aspects of the service above the AAL functionality. If the standardised service is a Bearer Service, the network will only ensure LLF compatibility. ITU-T has undertaken standardisation activity on several multimedia target services for the B-ISDN, listed above.

Although strictly confined to the functionality defined within the standards-based service description, standardised services can prevent the proliferation of a large number of standards for multimedia information types. However, if a change is required to a standardised service, this can only done via the appropriate standards body which can take a long time relative to the speed at which applications are being developed.

Semi-standardised teleservices

Semi-standardised teleservices permit a greater degree of flexibility over fully standardised services. Semi-standardised services allow the user to select or discard some parts of the service according to a general, standardised service description. Within a multi media context, two types of flexibility are possible: selection of a specific value of an attribute from a set of allowed values and/or selection of which parts of a multimedia service should be used in an actual call. An example of the first kind of flexibility could be the selection of the bandwidth value for a data transfer operation. The second kind of flexibility could allow a user to select the audio but not the video component of a videotelephony service.

The user is able to select the required service configuration at call set-up and/or during the lifetime of the call and the network, in response, allocates all of the necessary resources. The network will also check for terminal–terminal compatibility relating to both the higher and lower layers of a teleservice, or just the lower layer functions only of a bearer service.

The presence of semi-standardised services starts to give the user some choice over what aspects of a teleservice or bearer service are used. However, they are still reliant on standards bodies to introduce new service options or modify existing ones.

User-defined teleservices

This service category allows users to pre-define the characteristics of a service which is required *a priori* with the network provider. Once pre-defined, users can then establish actual calls using their own user defined service. The network will allocate all of the required network resources for the call but is not responsible, in principle, for ensuring complete terminal-to-terminal compatibility. However, the network may be responsible for ensuring terminal compatibility relating to individual standardised components of the service.

Although user-defined teleservices can only be used by users who have previously agreed the service definition with the network, the service required for an application can be built from a set of standardised and user-defined components. The network operator only has control over parts of the service based upon standardised components, with respect to compatibility checking, etc.

Open teleservice

The open teleservice permits a user to have a service consisting of standardised and non-standardised components. The selection of the

4.5 NETWORK SIGNALLING SERVICES

required components is specified only at call set-up time and can be modified during the lifetime of a call.

The network will allocate all of the required resources for the call, but the user is responsible for terminal-to-terminal compatibility. The network may, however, assist in the compatibility checking related to standardised components used for the call.

The open service allows the user to obtain the desired service from the network as and when required. The user is also able to freely modify the characteristics of an active call as desired, and only makes use of the network resources and service elements that are really required.

B-ISDN supplementary services

An additional 'supplementary' set of B-ISDN service functionality has also been defined and standardised by ITU-T to complement the specified range of bearer services and teleservices. Supplementary services cannot exist on their own and hence are used in conjunction with a bearer service or teleservice either on a per call basis or under a subscription option. The existence of supplementary services offer to the application additional information or services, which can be used to create value added features such as notification to the user of who is calling. The supplementary service set was based upon those previously defined for the narrowband ISDN.

The B-ISDN supplementary services are:

- Calling Line Identification Presentation
- Calling Line Identification Restriction
- Connected Line Identification Presentation
- Connected Line Identification Restriction
- Closed User Group
- Priority Call
- Sub-addressing
- Direct Dialling In
- Multiple Subscriber Numbering
- User–User Signalling.

4.6 IDENTIFICATION OF NETWORK CAPABILITIES

An analysis of the potential B-ISDN bearer and teleservices by ITU-T for signalling requirements has so far identified five connection types, on a switched, semi-permanent and permanent basis, that the B-ISDN should support from a signalling point of view. The five basic connection types are described in Chapter 1. A sixth type, providing the emulation of Type 5 by a server having multiple point-to-point connections to a number of end points, is really a service that can be offered either within a network or by a service provider.

It was realised, however, that further network capabilities will be incorporated into B-ISDN, in evolutionary steps, to meet new user requirements and accommodate advances in network developments and progress in technology. Ideally, connections must support both circuit mode and packet mode services of a mono- and/or multimedia type, and of a connectionless or connection-oriented nature, and in a bidirectional or unidirectional configuration.

Mixtures of connection types

The CS2 requirements would be regarded as being fully met when a complex call may be established consisting of any mixture of Type 1 and Type 2 connections, as represented by the configuration shown in Figure 4.5. This should include a party which may choose to participate in the call, but which is not attached to any connections—and therefore does not directly communicate with any of the other parties (i.e. has no user plane

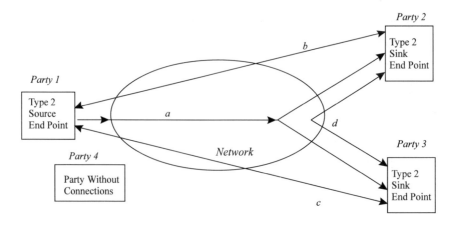

Figure 4.5 A Complex CS2 Call from the Signalling Requirements

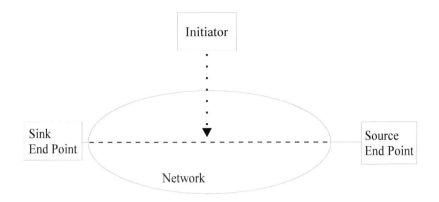

Figure 4.6 Third Party Set-up

communications). All parties and connections within the call may be replicated (i.e. there may be more than one connection of type 2 (labelled 'a' in Figure 4.5), in which case there may be more parties who are Type 2 source endpoints).

Third party set up

The B-ISDN CS2 requirements also identify the need to support an endpoints ability to set up a connection between two other endpoints, without the connection owner being involved in the sending or receiving of any information over that connection. This type of facility is known as a third party set up, and is shown in Figure 4.6.

4.7 THE TRAVEL AGENT SERVICE

This service scenario has been included as a theoretical example of an open teleservice which the B-ISDN needs the future capability to support. It illustrates a number of problems for B-ISDN signalling which is required to support and integrate, on demand, several different teleservices. Taking part in the service are four end users: two service providers, a retailer and a customer. It is assumed that all have some form of an intelligent broadband terminal.

The service starts with the customer contacting the travel agent with a view to booking a holiday. The initial communication is telephony (speech only), but since both parties have broadband terminals, this initial communication is upgraded to videotelephony. The travel agent determines the customer's requirements, such as price restrictions, type of holiday, general region, and then initiates a search of available holidays with the travel companies 'agents only' database. The results of the search are then provided to the customer as data, and the customer selects those that look promising. The travel agent then offers to provide video clips to the customer of the selected holidays—both resorts and hotels could conceivably be available. However, the customer's network connection could be xDSL, based and consequently may not have enough bandwidth to simultaneously support a video connection, for viewing the holiday destinations, and a videotelephony connection for communicating with the travel agent. Therefore, the videotelephony link to the travel agent must be dropped, but the travel agent still needs to remain in the call, however, to be reconnected at the end of the video clips.

The customer, in this theoretical scenario, is wholly convinced that one of the holidays will be suitable and decides to book. The travel agent must make the booking with the travel company—an automatic transaction on a data link—and then ask the customer for payment. Fortunately, the theoretical customer has an acceptable credit card, but since this is a late booking, must pay the entire amount. A double link to the credit card company is therefore required, enabling the travel agent to provide the retail outlet number and receive an authorisation code. The customer, meanwhile, must provide the important PIN or similar form of electronic signature to confirm the payment.

Communication configurations

From the network point of view, the call begins with a single telephony link between two parties (Figure 4.7), and is then upgraded to a videotelephony link. The signalling requirements are therefore to provide a single Type 1 connection between two parties, and subsequently allow an on demand modification of the connection's traffic characteristics to support the extra bandwidth requirements of videotelephony.

The introduction of data links to the travel company enables the travel companys database to be searched and the results provided to the customer.

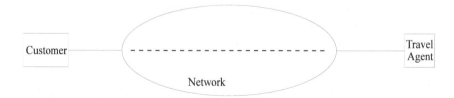

Figure 4.7 Initial Configuration

4.7 THE TRAVEL AGENT SERVICE

This requires the addition of a third party and two connections (both data types). The signalling requirements must consequently allow an upgrade from a two party call to a multiparty call, in which each party is connected to each other using a Type 1 connection. The call must allow each connection to have different ATM Layer transfer capabilities (such as Continuous Bit Rate/Variable Bit Rate) and bandwidths (Figure 4.8).

Having found some suitable choices, the travel agent offers the customer video clips to enable a final selection. Assuming that the customer's access point has limited bandwidth, the size of the connection between the customer and the travel agent has to be reduced, with a consequential downgrading of the teleservice from videotelephony back to telephony. At the same time, the video server is introduced into the call, and the customer's available bandwidth is reassigned from the travel agent's connection to the new connection with the video server (Figure 4.9).

The signalling requirements of this portion of the call are quite complex. As well as the multiparty aspects of adding a new endpoint, the transfer of bandwidth from one connection to another is a radical new requirement. If the bandwidth is not retained by the customer, it may be seized by another application in the same household, or worse, on the same access network.

Once the final holiday has been selected and the video clips have

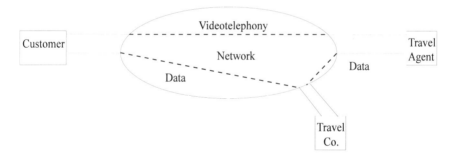

Figure 4.8 Configuration for Searching the Travel Agent's Database

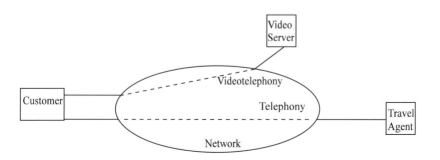

Figure 4.9 Configuration for Viewing the Travel Agent's Holiday Destinations

finished, the call needs to revert back to the original configuration shown in Figure 4.7. The holiday is then booked with the travel firm, and payment is expedited by another data connection to the credit card company, enabling the travel agent to enter the details and the customer confirm the transaction using a PIN or password (Figure 4.10).

The signalling requirements for making payment are the same as those described above for searching the travel agents database of holiday destinations, as shown in Figure 4.8.

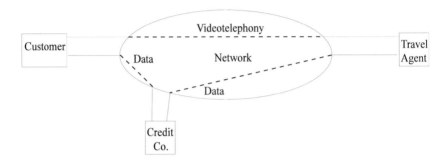

Figure 4.10 Configuration for payment

4.8 CONCLUSION

Consideration of the types of application that the B-ISDN is to support is an essential first step towards the definition of the bearer, teleservice and supplementary services that the network needs to support. From these bearer services, teleservices and supplementary services, the network capabilities needed can be derived which ultimately must be supported by the signalling systems. So far, B-ISDN signalling systems that have been produced can support bearer services, teleservices and supplementary services needing point-point or unidirectional point-to-multipoint network connections. Clearly, these capabilities are insufficient to accommodate the applications cited in the chapter, and hence further work toward the target solution for B-ISDN signalling is taking place, based upon the B-ISDN Signalling CS2 work.

There is not perceived to be a single killer application or service that the B-ISDN should be tailored to support. The B-ISDN can only be successful if it is developed as a general purpose platform, capable of supporting a variety of applications, such as the travel agent service. Of key importance will be the ability to seamlessly interwork with or support, in a complementary manner, similar applications that can run over the internet.

REFERENCES

RACE MAGIC Deliverable 3, *Service Description Framework and B-ISDN Service Descriptions*, R2044/BTL/DP/DS/P/003/b2, June 1993.

STANDARDS

I.121 *Broadband aspects of ISDN*, ITU-T (04/91).
I.211 *B-ISDN service aspects*, ITU-T (03/93).
I.130 *Method for the characterization of telecommunication services supported by an ISDN and network capabilities of an ISDN*, ITU-T (1988).
I.140 *Attribute technique for the characterization of telecommunication services supported by ISDN and network capabilities of an ISDN*, ITU-T (03/93).
I.210 *Principles of telecommunication services supported by an ISDN and the means to describe them*, ITU-T (03/93).
I.325 *Reference configurations for ISDN connection types*, ITU-T (03/93).
I.375 *Network capabilities to support multimedia services: general aspects*, ITU-T (06/98).
Q.65 *The unified functional methodology for the characterization of services and network capabilities*, ITU-T (06/97).
Supplement 7 to Q Series Recommendations, *Technical report TRQ.2001— General aspects for the development of unified signalling requirements*, ITU-T (03/99).
X.200 *Information technology—Open Systems Interconnection—Basic reference model: The basic model*, ITU-T (07/94).
DAVIC 1.0 Specification Part 1, *Description of DAVIC Functionalities*, 1995–1996.
Z.100 *CCITT specification and description language (SDL)*, ITU-T (03/93).

QUESTIONS

1. Briefly describe the I.130 method for B-ISDN service categorisation, and why such a model was thought to be necessary to aid signalling protocol specification.
2. Why doesn't the B-ISDN have a single signalling protocol for each service it supports?
3. Explain the difference between a bearer service and a teleservice.
4. What are supplementary services? What are the difficulties of having more than one supplementary service active on the same call at any one time?
5. Compare the merits and drawbacks (from a network operators viewpoint) of offering standardised teleservices.

5
Access Signalling

Bryan Law and Dick Knight

5.1 INTRODUCTION

The access signalling mechanism that has been adopted worldwide for the support of B-ISDN in an ATM cell-based environment is based upon the ITU-T access to the B-ISDN, titled Digital Signalling System No. 2 (DSS 2). Whilst this chapter concentrates on the ITU layer 3 signalling protocol defined in recommendation Q.2931, it should be noted that there are few differences between this signalling protocol and the competing standard (technically an implementation agreement) produced by the ATM Forum, generally referred to as UNI 4.0.

5.2 RELEASE 1 ACCESS SIGNALLING

It was widely felt that the introduction of B-ISDN on ATM called for a far reaching approach to be taken when studying signalling. A mechanism based on the OSI Application Layer Service (ALS) was seen as being well suited to an B-ISDN network and the services it would support. However, there was considerable commercial pressure in some parts of the world to provide a basic broadband service by 1993/1994. Hence, it was believed that any attempt to produce an advanced approach to signalling would not meet these service requirement dates. Instead, it was proposed to take a two stage approach in the specification of B-ISDN signalling:

1. An initial phase to meet the timescales for the Release 1 B-ISDN requirements.
2. Further studies of a signalling technique for advanced B-ISDN services to follow Release 1, leading to the target solution for B-ISDN signalling.

For Release 1 it was agreed that the existing N-ISDN signalling systems should be used as a basis for the support of initial broadband services, which had relatively simple signalling requirements. Therefore for access signalling the Release 1 B-ISDN signalling was based on Digital Subscriber Signalling System No. 1 (DSS 1). This resulted in a 'monolithic' protocol (i.e. call and bearer control bound together), but since the main requirement of Release 1 signalling was to support simple point-to-point bearer services, this was not a problem. The protocol stack for Release 1 B-ISDN access signalling is depicted in Figure 5.1.

Metasignalling

To provide a connection for the user, there must first be a signalling interaction in order to establish the attributes of the connection that will be required for this service. At first sight, there seems to be an oddity—how do you signal for a connection, when you don't yet have one? We have already seen that the layer 2 signalling is point to point, and, as it relies on sequence numbers, cannot establish a signalling association between multiple terminals and the network. There are two solutions to this problem—pre-allocated (standardised) virtual channels, and a mechanism for dynamic signalling channel allocation. The former solution is provided by a common agreement that virtual channel 5 in every virtual path will be reserved for signalling. The latter solution also requires a reserved VCI, one on which management requests for a signalling channel can be serviced. The size of these messages is considerably smaller than a signalling message, and there is no real layer 2 (unique reference numbers (randomised) are used to associate replies with requests), and so the bandwidth allocated

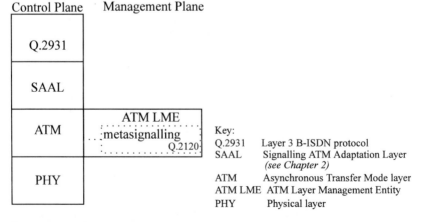

Figure 5.1 Protocol Reference Model for Release 1 Access Signalling

to the Layer Management VCI is just 167 cells per second. VCI = 1 in every VP is reserved for this purpose, although the bandwidth will only be reserved if the network supports the metasignalling procedures. To provide support for other management and signalling, and also to provide scope for future enhancements, virtual channels 0 through 31 are reserved.

The ATM layer and the SAAL have been covered in Chapters 2 and 3, respectively. Although, as can be seen from Figure 5.1, metasignalling is not part of the SAAL, and therefore outside layer 2 signalling, it has sufficient potential impact on B-ISDN access signalling to require some explanation in this book.

Unlike N-ISDN, where there is a permanent signalling 'D-channel' available, in the B-ISDN it is possible for certain configurations of the UNI to establish a signalling channel only when required. This saving of resource could be crucial on the B-ISDN UNI, where some 2000 plus terminals requiring signalling would seriously impinge on bandwidth available. Therefore, the broadband terminal (B-TE) uses metasignalling, to request the allocation of signalling resources. Metasignalling has a unique Virtual Channel Identifier (MSVCI), and uses a series of messages (maximum length 42 octets) and procedures to assign Signalling Virtual Channel Identifiers (SVCI) and allocate bandwidth, monitor signalling activity and remove the SVCI and resource when no longer required. There are three phases in the procedure: assignment, checking and removal

Metasignalling is not necessary for point-to-point UNI configurations, where a fixed low bandwidth point-to-point Signalling Virtual Channel (SVC) is available for signalling in every virtual path. It is this capability that has caused much debate on the necessity for metasignalling, even though it was standardised.

5.3 LAYER 3 SIGNALLING

Release 1 B-ISDN services provide simultaneous control of call and bearer connection, and the protocol has been closely based on the N-ISDN layer 3 access protocol (DSS1) as specified in ITU-T Recommendation Q.931. While there is great similarity in general, there are significant differences, and a different protocol discriminator is used for the DSS2 protocol. All layer 3 messages consist of the following parts:

- protocol discriminator
- call reference
- message type (including message compatibility instruction indicator)
- message length, and
- variable length information elements, as required.

The protocol discriminator is the first octet in every message, it is one octet long and for the protocol defined in Q.2931, its value is 9 (00001001). The call reference forms the next four octets and is split into three parts: the length, a flag and the call reference value. The call reference flag provides a simple contention resolution mechanism. If two peer layer 3 signalling entities both initiate a call establishment (described in Section 5.3.1) and, coincidentally, choose the same call reference value, then both calls may continue. This is because the flag is used to indicate which end picked the value, the call originator (for this association) or the other side of the association. The call reference value therefore only has significance between two peer signalling entities.

Layer 3 messages

As with N-ISDN access signalling, there are three phases related to the control of the B-ISDN call and connection. These are Call establishment, Call clearing and In-call control. Call establishment is supported by the following messages:

- SETUP
- CALL PROCEEDING
- ALERTING
- CONNECT
- CONNECT ACKNOWLEDGE

The necessary procedures, to enable these messages to establish a call, are based on those for N-ISDN, and the message sequence is shown in Figure 5.2. The call proceeding message is used by the network to prevent a timeout occurring in the terminal. This message need not be used at the destination side as the alerting message can be used to prevent the network timing out. The alerting message is optional as it is not needed for those applications (e.g. data) that do not need to interact with the human user to accept a call. In these cases, the call proceeding would only be used if there is any delay (e.g. within a customer network) before the connect is returned.

Call clearing is supported by the following messages:

- RELEASE
- RELEASE COMPLETE

The procedure is unlike N-ISDN, being based on a two message procedure and not the DISCONNECT message. The procedures associated with a

5.3 LAYER 3 SIGNALLING

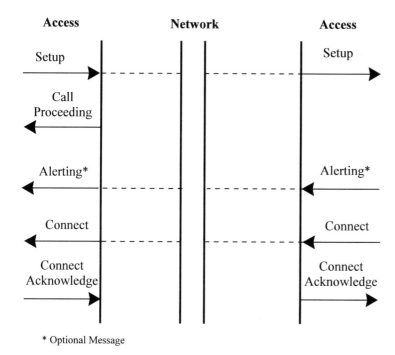

* Optional Message

Figure 5.2 Successful Call Establishment Access Messages

release ensure that any discontinuity between two peer layer 3 signalling entities does not prevent the release of associated resources. The release indicates that the signalling entity that sends this message has already released the communications resources (bandwidth, VPI/VCI, etc.) associated with the call, and only the signalling association identified by the call reference remains. The receipt of the release complete message cancels a timer and releases the call reference value. If the release complete message were not received for some reason, then when the timeout matures, the entity that sent the release message sends the release complete message and releases the call reference value anyway. The receipt of a release complete in the idle state is not therefore an error (Figure 5.3).

'In-call' control is supported by the following messages:

- STATUS

- STATUS ENQUIRY

The status is exchanged across the access link between peer layer 3 signalling entities, and are not designed for end to end exchange. The status procedures are similar to those produced for N-ISDN, but with significant enhancement. These enhancements provide an audit function

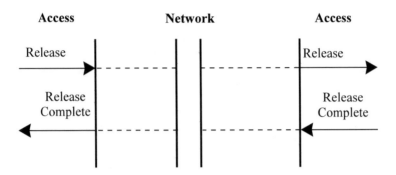

Figure 5.3 Successful Release Sequence

that is invoked whenever a layer 2 detects a discontinuity. In such a situation Layer 2 will in addition to informing layer 2 management, also inform layer 3. Layer 3 will then invoke the status enquiry and the response, carried in the status message, ensures that both peer entities are aware of the current call status.

Information elements

The parameters or Information Elements (IEs) again draw heavily on those defined for N-ISDN. The exceptions being; Call Reference, Connection Identifier (VPI/VCI), Broadband bearer capability, Broadband higher layer information, Broadband low layer information, Broadband repeat indicator and Broadband sending complete, which replace the N-ISDN equivalents. In addition there is the information required to support ATM characteristics these information elements being; ATM Adaptation Layer parameters and ATM Traffic Descriptors. These ATM traffic parameters have been evolved as the specification work progressed in this area and traffic characteristics defined by SG 13 in Recommendation I.371 are supported in signalling. There is also a B-ISDN layer 3 compatibility mechanism based on instruction indicators which is not found in N-ISDN. This mechanism requires a sending side to inform the receiving side on the severity of the action to take in the event that an information element, parameter, or message is not understood. The actions to be taken can result in the clearing of the call, at one extreme, through to ignoring the offending item.

Support of N-ISDN circuit mode services

To support 64 kb/s based circuit mode services, additional procedures, messages and information elements are required to those for B-ISDN basic call control. There are again the three phases of signalling to control the call; establishment, clearing and in-call. The additional messages, whose function and description are as for N-ISDN are:

- PROGRESS
- SETUP ACKNOWLEDGE
- INFORMATION

The additional information elements are the Narrowband Bearer Capability, Narrowband Low Layer Compatibility and Narrowband Higher Layer Compatibility. These information elements are used with the appropriate B-ISDN messages and procedures to achieve the same results in terms of service as the N-ISDN equivalents.

Support of B-ISDN release 1 supplementary services

B-ISDN release 1 will support supplementary services which are listed in CCITT (ITU-T) Recommendation Q.767 with the exception that Terminal Portability, which is not considered to provide any useful service for the user at the B-ISDN UNI, and is therefore not supported. To provide signalling support the B-ISDN access protocol includes an additional message NOTIFY which again is N-ISDN based and is used to convey information regarding the call/connection relating to control of the supplementary service.

5.4 POINT-TO-MULTIPOINT SIGNALLING

The support of point-to-multipoint connections had been contained in the original menu of services identified by SG 13 for Release 1 signalling. However, the time constraints already mentioned led to the decision being taken to delay the completion of the study of this protocol until after point-to-point access signalling had been finalised. In the event this was not a problem, since it was decided that the mechanism used for the establishment, in call control and release of a point-to-multipoint connection would be based on that used for control of point-to-point connections, Release 1 signalling as in Q.2931. The possibility of using a mechanism based on 'atomic actions', enabling simultaneous establishment of the root connection and all leaf connections in one action was not pursued.

Scope and overview

The point-to-multipoint protocol enables a point-to-multipoint connection to be established, controlled during the in-call phase, and released from both single terminal or private network. At present, the connections established using this protocol are restricted to unidirectional transmission (i.e. root to branch).

The method of establishing the point-to-multipoint connection is for the Calling (root) party to request the establishment of a connection to one of the Called (leaf) party, as in Figure 5.4. This makes use of the set-up procedure as for Release 1, but indicates in the Broadband Bearer capability IE requested that the connection is a point-to-multipoint connection.

Once this initial connection, is active or the alerting stage has been reached, it is possible for the root to add the required leaf connections to the existing connection by sending add party requests, as in Figure 5.5. During the progress of the call more leafs can be added by the use of the add party request and it is possible to use a multiple add party request (i.e. the root can send further add party requests without waiting for a response).

A leaf can be dropped at any stage during the call either as result of the root (see Figure 5.6) or leaf party requesting the deletion.

Should a deletion result in the last leaf connection being deleted then the call is released, as in Figure 5.7.

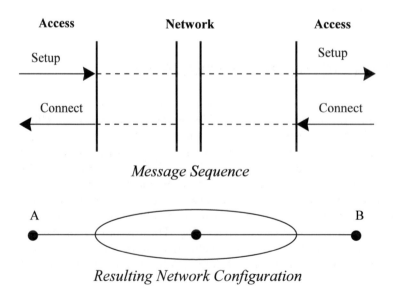

Figure 5.4 Establishing a Point-to-Multipoint Call—First Leaf Party Set Up

5.4 POINT-TO-MULTIPOINT SIGNALLING

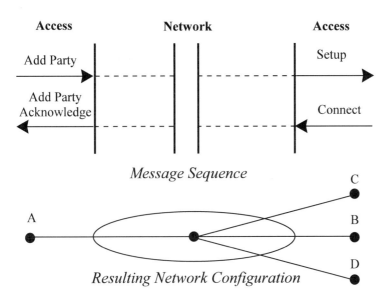

Figure 5.5 Establishing a Point-to-Multipoint Call—Subsequent Leaf Party Set Up

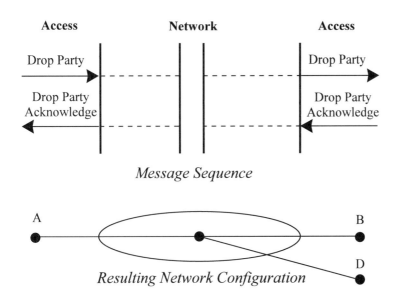

Figure 5.6 Dropping a Party from a Point-to-Multipoint Call

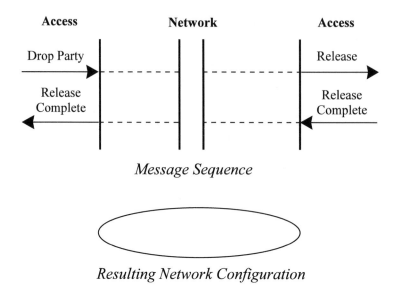

Figure 5.7 Dropping the Last Leaf Connection of a Point-to-Multipoint Call

Messages

The procedure makes extensive use of the point-to-point mechanism, therefore many of the messages specified in DSS 2 for Release 1 in Q.2931 are reused with enhancement to cater for the point-to-multipoint environment. In addition, there are new messages and Information Elements (IEs) specified to support the addition and deletion of the leaf connections during the life of the call.

DSS 2 (Q.2931) messages reused

The following DSS 2 Release 1 messages are reused with enhancement:

- ALERTING   Endpoint reference IE added
- CALL PROCEEDING   Endpoint reference IE added
- CONNECT   Endpoint reference IE added
- SETUP   Endpoint reference IE added
- STATUS   Endpoint reference IE and Endpoint state IE added

5.4 POINT-TO-MULTIPOINT SIGNALLING

- STATUS ENQUIRY   Endpoint reference IE and Endpoint state IE added

- NOTIFY   Endpoint reference IE added.

New DSS 2 (Q.2971) messages

The following new messages have been specified to support the point-to-multipoint procedure:

- ADD PARTY   sent from root to add a party to an existing connection

- ADD PARTY ACKNOWLEDGE   sent to root from network on successful addition of Party

- PARTY ALERTING   sent to root to indicate alerting initiated

- ADD PARTY REJECT   sent to root from network on unsuccessful request for addition of Party

- DROP PARTY   sent from root to network or network to root to drop a Party from an existing point-to-multipoint connection

- DROP PARTY ACKNOWLEDGE   sent from root to network or network to root to acknowledge a Party has been dropped from an existing point-to-multipoint connection.

New information elements to support point-to-multipoint

Two new information elements have been specified to support point-to-multipoint signalling, in addition to those in Q.2931:

- Endpoint Reference   identifies the individual endpoint in a point-to-multipoint connection and by means of a flag (set to 1) indicates if it is the first end party, and therefore able to negotiate. [Flag set 0 indicates subsequent party without negotiation rights]

- Endpoint State — indicates the state of the endpoint in the point-to-multipoint connection (i.e. Add Party initiated, Active, Drop Party initiated, etc.).

Procedures supported in point-to-multipoint

The following procedures are supported in DSS 2 point-to-multipoint:

- Adding a Party at the originating interface — set-up of first party, adding a party, action on add party received, party alerting, add party failure, add party connected, add party rejection, transit network selection

- Add party establishment at at the destination interface — release 1 procedures with some modification, leaf does not support multipoint procedures

- Party Dropping — exception conditions, party dropping initiated by user (root and leaf), party dropping initiated by network (root and leaf), drop collision, dropping of all parties

- Restart — enhance release 1 procedures, all parties associated with virtual channel and call shall be dropped

- Handling of error conditions — protocol discriminator error, message too short, call reference errors, endpoint reference errors (missing, invalid format, procedure), message type or sequence errors, message length, general IE errors, mandatory IE errors (missing, content), non mandatory IE errors (unrecognised), signalling AAL lost, signalling AAL released, status enquiry procedure, receipt of STATUS message

- Notification procedure — release 1 procedure enhanced for leaf and root application.

The above procedures are applicable at both the basic user interface, and also for use at the interface with private B-ISDN, with the necessary enhancements as required for interworking with private networks.

Supplementary service interaction

In addition to procedures to support the basic call to establish a point-to-multipoint connection, the protocol also covers the necessary interactions and subsequent modification of procedures for the Release 1 supplementary services to cater for the point-to-multipoint connection environment.

5.5 CAPABILITY SET 2 STEP 1 SIGNALLING

At the same time, work was progressing on the point-to-multipoint signalling mechanism studies were initiated to provide support for SG 13 Release 2 B-ISDN requirements. Since these requirements were fairly wide ranging, it was decided to produce the protocol in a series of steps. Not only does this ease the logistics and practicalities of producing the protocol, but in adopting a stepwise or modular approach the resulting mechanism is less complex. The aspects covered by the resulting Capability Set 2 Step 1 (CS2-1) DSS 2 signalling mechanism are:

- Additional traffic parameters
- Connection characteristics negotiation during establishment
- Connection modification
- Basic look-ahead
- Multiconnection.

Although not part of the Release 1 DSS 2 signalling mechanism, these aspects were considered to be still pertinent to the control of the bearer connection combined with the call, as well as in a separated environment. Hence, CS 2-1 is still considered to be part of the initial phase rather than the advanced target solution for B-ISDN signalling. Therefore, the protocol for CS2-1 is again based on the DSS 2 Release 1. However, while this proved to be the case for the additional parameters, negotiation, modification and look ahead, the combined call/connection approach was not appropriate for multiconnection; this is discussed further in Section 5.6.

Additional traffic parameters

The DSS 2 Release 1 protocol provided the necessary signalling capabilities to enable the establishment, in-call control and release of a point-to-point bearer connection in an ATM network. However, the characteristics of the bearer connection that could be signalled were limited to the Peak Cell Rate (PCR) and the required ATM adaptation layer characteristics. With regard to the former, it was realised that the full flexibility that the ATM environment could not be realised if only the PCR was signalled. Therefore, the incorporation of the full range of ATM traffic parameters being defined by SG 13 in DSS 2 was considered a matter of priority.

Scope of additional traffic parameters in DSS 2

Currently, studies have progressed such that parameters for

- statistical multiplexing, using the SCR parameter set (sustainable cell rate, maximum burst size), and
- tagging option, which enables the Usage Parameter Control (UPC) function to exercise 'traffic management'

are included in DSS 2, in addition to those specified in Release 1. In the near future, DSS 2 will be further enhanced to contain parameters to support the following connection characteristics:

- ATC (ATM Traffic Characteristic)
- ABR (Available Bit Rate)
- ABT (ATM Block Transfer)
- CDVT (Cell Delay Variation Tolerance)
- Global tagging
- Quality of Service.

Messages and information elements to support additional parameters

There are no additional messages to those contained in DSS 2 Release 1 required for the support of the additional traffic parameters. However, both the SETUP and the CONNECT have been enhanced to include the extended ATM Traffic Descriptor IE.

The ATM Traffic Descriptor IE has been extended to 30 octets in length (from the 20 octets as in Release 1) in order to support the additional parameters outlined above.

5.5 CAPABILITY SET 2 STEP 1 SIGNALLING

Procedures for additional parameters

The basic procedures contained in DSS 2 release 1 have been extended as follows to make use of the additional traffic parameters when establishing an ATM bearer connection:

- *Support of the Sustainable Cell Rate (SCR) set.* The mechanism operates on the parameter while adhering to cell loss priority, PCR limit, conditional tagging, Maximum Burst Size (MBS), Intrinsic Burst Tolerance, SCR and PCR relationship, SCR MBS and Cell Loss Priority (CLP), direction independence (all the above being dependant on the ATM layer traffic handling capabilities).

- *Local support of tagging.* At the originating interface the network responds to the 'tagging request' in the SETUP message, at the destination interface the user responds to the 'tagging supported' indication in the SETUP message and, if appropriate, includes a 'tagging request' in the CONNECT message to be sent to the originating user.

- *Specific error conditions.* If an ATM Traffic Descriptor IE is received with an invalid combination of parameters, it will be treated as a mandatory IE with content error as in DSS 2 Release 1.

DSS 2 supporting additional parameters can interwork with networks that do not support this capability, but only at the Release 1 level; it is not possible to interwork the additional parameters with N-ISDN.

The additional parameter capabilities have no impact on the supplementary services as specified for DSS 2 Release 1.

Negotiation

While the additional parameters included in DSS 2 make it possible to vary the characteristic of the connection being established, the Release 1 procedures still meant that the connection was fixed at establishment and for the duration of the call. Therefore, a major step forward in signalling was seen as the ability to negotiate regarding the ATM connection characteristics required at the time of establishment (Figure 5.8).

Scope and overview

The negotiation capability within DSS 2, when used in conjunction with the DSS 2 Release 1 signalling mechanism and the additional traffic parameters that have been specified, enables the characteristics of the

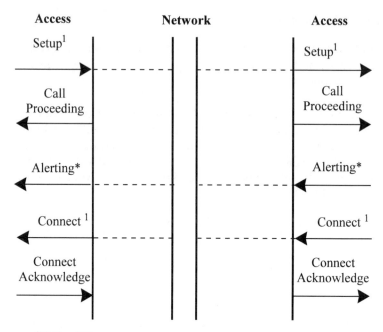

Figure 5.8 Negotiation of Additional Traffic Parameters Leading to Successful Connection Establishment

connection to be negotiated at the time of establishment. In addition to its application to the point-to-point connection, it can, when applied with the DSS 2 point-to-multipoint protocol, enable the characteristics of the connection established to the first leaf party to be negotiated.

The negotiation capabilities are only applicable during establishment catering for the following:

- negotiation of connection characteristics by providing alternative traffic descriptors,
- negotiation of cell rate traffic parameters using a minimum traffic descriptor.

With the Alternative Traffic Descriptor IE, parameters are handled as a single entity. When the Minimum Traffic Descriptor IE is used, a range of parameter values are allowed and handled independently. Traffic parameter negotiation is possible with the Alternative ATM Traffic Descriptor IE, whereas it is only possible to negotiate the Peak Cell rate when the Minimum ATM Traffic Descriptor IE is used.

Messages and information elements

There are no new messages specified to support DSS 2 negotiation. However, two DSS 2 Release 1 messages are used with enhancements, as follows:

- SETUP: either Alternative ATM Traffic Descriptor IE or Minimum ATM Traffic Descriptor are included (but not both);
- CONNECT: ATM Traffic Descriptor IE (included to indicate which traffic parameters have been successfully negotiated).

Two new information elements have been specified in addition to those contained in DSS 2 Release 1 to enable negotiation to be supported;

- Alternative ATM Traffic Descriptor: specifies the alternative ATM traffic descriptor to be used for negotiation of traffic parameter during connection establishment (can have any combination of parameters allowed for the ATM Traffic Descriptor for the specified bearer capability, alternative bandwidth must be less than originally requested).
- Minimum Acceptable ATM Traffic Descriptor: specifies the minimum acceptable ATM traffic parameters in the negotiation of traffic parameters at set-up (these are the lowest values that the user will accept).

Procedures supported

The following procedures are supported in DSS 2 negotiation:

- Negotiation of connection characteristics (originating interface): negotiation request, negotiation traffic parameter (minimum acceptable or alternative ATM traffic), negotiation acceptance.

- Negotiation of connection characteristics (destination interface): negotiation request, negotiation traffic parameter (minimum acceptable or alternative ATM traffic), negotiation confirmation.

If the negotiation is successful a CONNECT message is sent containing the ATM Traffic Descriptor agreed, if unsuccessful a RELEASE COMPLETE will be returned to originating interface.

Negotiation is uniform at both S_B/T_B (reference) and T_B (reference) interfaces and is terminated at the interworking points with other networks. There are no interactions with the Release 1 supplementary services.

Modification

With the ability to negotiate additional traffic parameters at connection establishment, the CS 2 enhancement was further advanced towards the fully flexible access signalling system that was required for B-ISDN if the full capabilities afforded by the ATM environment were to be realised. However, from the outset it was realised that, unlike the N-ISDN call dependant on a 'fixed' 64 kb/s switched environment, B-ISDN calls would be far more dynamic: the characteristics of the supporting connections are capable of being varied throughout the lifetime of the call to meet the end-users' demands. Therefore, the introduction of in-call modification of the connection was seen as being a crucial aspect of DSS 2.

Scope and overview

The Connection Modification capability within DSS 2, when used in conjunction with the DSS 2 Release 1 signalling mechanism and the additional traffic parameters that have been specified, enables the characteristics of the connection to be modified during the progress of the call to meet the end users' requirements. Currently, it is only the PCR that can be modified, but the mechanism is sufficiently generic to enable all the connection attributes to eventually be modified. Future developments will see negotiation also being introduced into the modification process. The ability to modify a connection is at present restricted to the connection owner and is only possible with point-to-point connections.

In addition, it is only possible to request modification to those parameters originally requested in a given direction. However, it is possible to request an increase in one direction with a decrease in the opposite direction in one modification request. To be modified connections must be established. The mechanism cannot be applied to connections that are in the process of being established or cleared. Clearing will always take priority over modification, hence if during the modification process clearing is invoked, the connection will be cleared down without the modification being completed.

When initiating a modification request to increase or decrease the PCR, the initiating user (connection owner) must

- be prepared to receive traffic with an ATM traffic parameter that is the greater of the unmodified or modified backward parameter for the connection being modified, and

- be prepared to transmit traffic with an ATM traffic parameter that is the lesser of the unmodified or modified forward parameter for the connection being modified.

5.5 CAPABILITY SET 2 STEP 1 SIGNALLING

It is also possible to modify in the given direction the PCR for $CLP = 0$ or $CLP = 0 + 1$ in the same modification request. All parameters must be decreased or all must be increased.

The modification process also allows the connection to continue to support the service application.

Procedures are also contained that allow for rejection of the modification request by the 'non-connection owner' and confirmation to the 'connection owner' and 'non-connection owner' that modification has been successful (see Figure 5.9).

Messages

The procedure with an enhancement to the DSS 2 Release 1 point-to-point mechanism is in fact an embedded mechanism. Therefore, there are new messages and Information Elements specified to support the requested modification of connections during the lifetime of the call.

The following new messages have been specified to support the modification procedure:

- MODIFY REQUEST: sent from initiating user to network and network to called user to request modification of connection,
- MODIFY ACKNOWLEDGE: sent to initiating user from network to indicate modification request is accepted,
- MODIFY REJECT: sent to initiating user from network to indicate modification request is not accepted,

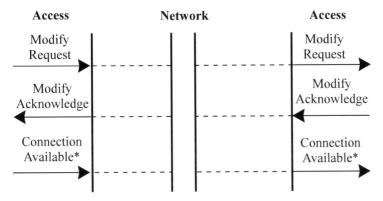

* Optional Message included only if requested in the broadband report type information element of the Modify acknowledge message

Figure 5.9 Successful Modification. Network signalling messages for support of the information contained in the DSS 2 CONNECTION AVAILABLE message. Not required if connection available information is not requested by the addressed user.

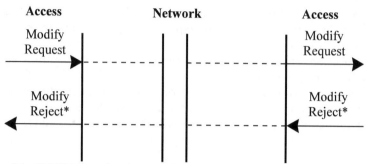

*the CAUSE information elements indicates why the modifcation was unsuccessful

Figure 5.10 Unsuccessful Modification—rejection by Called User

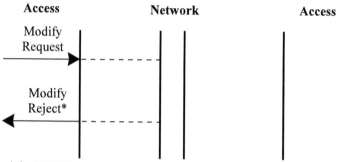

*the CAUSE information elements indicates why the modifcation was unsuccessful

Figure 5.11 Unsuccessful Modification—Rejection by Network

- CONNECTION AVAILABLE: sent from initiating user to network or network to 'non-connection owner' to confirm modification, if requested in MODIFY ACKNOWLEDGE.

Information elements

The modification mechanism in DSS 2 reuses many of the information elements that have been specified to support point-to-point connection control signalling. In addition, the following new IE has been specified for use in the MODIFY ACKNOWLEDGE message:

- Broadband Report Type: included in the MODIFY ACKNOWLEDGE message when the 'non-connection owner' requires confirmation of modification success from the 'connection owner'. Confirmation is sent using the CONNECTION AVAILABLE message. (This is considered to be a generic IE, and will probably have wider application.)

5.5 CAPABILITY SET 2 STEP 1 SIGNALLING

Procedures supported

The following procedures are supported in DSS 2 connection modification:

- procedures at requesting entity: modification request, modification acknowledgement, indication of modification rejection, response to STATUS while in modify request state, no response to modification request,
- procedures at responding entity: modification indication, modification acceptance, modification confirmation, modification rejection,
- transit entity conveyance of CONNECTION AVAILABLE transparently while in active state.

The above procedures are applicable at both the basic user interface and also for use at the interface with private B-ISDN. There is no interworking with other networks.

There are no identified interactions with the Release 1 supplementary services.

Look ahead

The basic Look-ahead capability was introduced into DSS 2 CS2-1 mainly to allow the network to determine the viability of establishing a connection. Although it is possible that in future this capability may be extended for user-to-user application, other developments may make this unnecessary. The Look-ahead capability in DSS 2 makes use of the Generic Functional Protocol (GFP) as specified for B-ISDN, which is based closely on the N-ISDN DSS 1 mechanism specified in Recommendation Q.932. The B-ISDN GFP using Connection Oriented Bearer Independent (COBI) techniques enables the activation and deactivation, invocation and operation, interrogation, status request, and, status notification, of a process/capability/ supplementary service in association with or outside an existing call. A brief outline of messages, IEs and procedures of GFP is contained in the following description of Look-ahead.

Overview and scope of Look-ahead

The basic DSS 2 Look-ahead allows the incoming side of the network (in response to a request from the calling user) the capability of looking ahead to the outgoing side of the network to, for example, ascertain the viability of supplying the network service requested by the 'calling user'. The decision as to whether the Look-ahead capability should be invoked is left

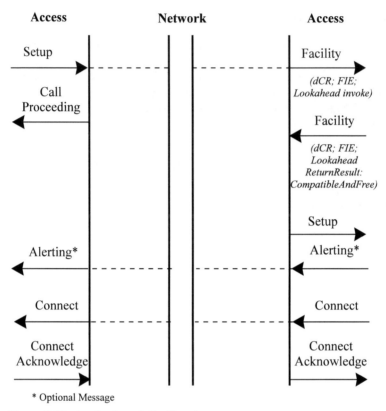

Figure 5.12 Look-ahead: Positive Answer

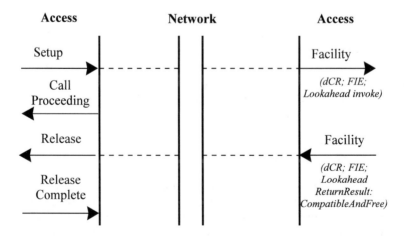

Figure 5.13 Look-ahead: Negative Answer

5.5 CAPABILITY SET 2 STEP 1 SIGNALLING

to the incoming gateway exchange. It is also possible for a private B-ISDN to indicate to the public B-ISDN that basic Look-ahead is required.

Further examples of how Look-ahead in DSS 2 can be applied are to check that the terminal addressed is compatible/incompatible, and that the terminal addressed is free or busy, and can be used prior to the establishment of the call connection. However, the application of DSS 2 Look-ahead is not part of the basic DSS 2 signalling service, which only defines the generic procedure.

Look-ahead messages

The procedure makes use of the point-to-point mechanism and the GFP, therefore messages specified in DSS 2 for Release 1 in Q.2931 and for DSS 2 GFP in Q.2932 are reused with enhancement to cater for Look-ahead. In addition, there is a new IE specified for Look-ahead together with Look-ahead operations and errors coded in ASN.1.

The following DSS 2 Release 1 messages and GFP message are reused with enhancement:

- SETUP: Network Look-ahead indicator IE to indicate capability used
- CALL PROCEEDING: As for Release 1, sent to calling party to indicate Look-ahead initiated by network
- FACILITY: Carries Look-ahead Invoke to called party and Look-ahead Return Result from called party.

Information elements

The Look-ahead mechanism in DSS 2 uses an IE that has been specified to support GFP signalling. In addition, the following new IE has been specified for use in the SETUP message (at T_B reference):

- Network Look Ahead is included in the SETUP (at T_B reference) message to indicate when a calling network has used the look-ahead capability in that call,
- Facility is used in the FACILITY message to convey the operations for Look Ahead at the destination interface.

Procedures supported

The following procedures are supported in DSS 2 Look-ahead at the S_B reference:

- normal operations: send Look-ahead invoke, start Look-ahead timer, enter Look-ahead state, stop timer/terminate procedure,

- exceptional procedures: stop timer/terminate procedure.

The following procedures are supported in DSS 2 Look-ahead at the T_B reference:

- originating T_B interface normal operations: send Look-ahead invoke, start Look-ahead timer (Tb), forward result to application, stop timer Tb/terminate procedure,

- originating public network normal operations: forward invocation to User B, enter Look-ahead state, send result, terminate procedure,

- exceptional procedures public network: stop timer/terminate procedure, for AAL-EST receipt abort procedure, for AAL-REL receipt ignore and remain in current state,

- destination public network normal operations: send Look-ahead invoke to private network, start Look-ahead timer (Tb), enter Look-ahead state, forward result, stop timer Tb/terminate procedure,

- destination T_B interface normal operations: forward invocation to User B, enter Look-ahead invoked state, send result to public network, terminate procedure,

- exceptional procedures public network: stop timer Tb/terminate procedure, for AAL-REL receipt abort procedure, for AAL-EST receipt ignore and remain in current state.

Using the GFP precludes interworking with non-ISDNs, frame relay and Packet Switched Public Data Network (PSPDNs).

5.6 MULTICONNECTION SIGNALLING

As with the other signalling recommendations constituting Capability Set 2 step 1 in ITU-T SG 11, it was intended that the signalling protocol to support multiconnection should be based on the CS 1 protocol Recommendation Q.2931. However, initial studies identified that the monolithic structure of Q.2931 prevented the fully required capability of signalling

5.6 MULTICONNECTION SIGNALLING

support for a point-to-point multiconnection call from being realised. Using Q.2931, it is not possible to add or delete connections to a call as required, since the call and connection are bound together.

Therefore, the protocol required for support of the multiconnection call requires a separation of the call and bearer control functionality to enable the degree of flexibility in control required to be achieved. To achieve simplicity and efficiency in the control mechanism for multiconnections, separate protocols for call and connection control are considered to be the best approach. This approach also allows the possibility of prenegotiation of the connections to be added, and a separate protocol is specified for this purpose.

Connection control

It has been possible to adapt Q.2931 to provide a connection control protocol, and this is currently under study in ITU-T SG 11. Use is made of many of the messages developed for Q.2931, and also the IEs used in the CS 1 DSS 2 protocol. However, since this is a connection control mechanism rather than a call and connection control mechanism, any IE, message or procedure that is call control related has been stripped out. Therefore, compatibility with Q.2931 is not possible.

Call control

Study of a mechanism for call control in the separated environment has also been undertaken in ITU-T SG 11 using Q.2931 as the basis. While this appeared to be an adequate solution for the simple point-to-point multi-connection call, it became apparent that it would not satisfy the more stringent requirements for the multipoint multiconnection environment. Since it is obvious that such a multipoint/multiconnection capability will be required in future networks, a number of organisations, including the Computer Manufacturers Association (ECMA) and the European Telecommunications Standards Institute (ETSI), decided that rather than use a short-term expedient solution (Q.2931-based), a full solution is required. More detailed information on this control protocol is contained in Chapter 15.

Pre-negotiation

In addition to call and connection control, it was realised that with separation of call and bearer connection it was now possible to negotiate ahead of establishment the characteristics of a connection that was to be associated with a call (except in the case of simultaneous call and connection establishment). This mechanism, known as *prenegotiation*, allows end users to check compatibility with the network, as well as compatibility and availability at the remote user with regard to the one or more connections they intend to establish within the lifetime of the call. The protocol is based on the Generic Functional Protocol, and the operations involved fulfil the requirements of prenegotiation without the reservation of any connection oriented resources within the network.

5.7 CONCLUSION

This chapter has identified how access signalling protocols have been developed from their N-ISDN predecessor to enable the support of the initial capabilities required for point-to-point connections. Their enhancement to enable the further network requirements for the support of the Capability Set 2 functions has also been covered.

These initial phases of the B-ISDN access signalling protocol DSS 2 have been the valuable basis not only allowing operators and manufacturers to provide initial broadband control, but also as a starting point for more advanced studies. However, with the need to move into the multipoint, multiconnection network control environment the scope of the initial solution to B-ISDN access signalling has reached its limit. Therefore, the studies into the separated call and bearer control environment are now vital if a viable solution to the control of advanced networks and the required access signalling support is to be realised.

REFERENCES

Law B, Signalling in the ATM Network, *BT Technol. J.* Vol. 12, No. 3, 1994.

STANDARDS

Q.920 *ISDN user-network interface data link layer—General aspects*, ITU-T (03/93).

Q.921 *ISDN user-network interface data link layer specification*, ITU-T (09/97).
Q.930 *ISDN user-network interface layer 3—General aspects*, ITU-T (03/93).
Q.931 *ISDN user-network interface layer 3 specification*, ITU-T (09/97).
Q.1400 *Architecture framework for the development of signalling and O&AM protocols using OSI concepts*, ITU-T (03/93).
Q.2931 *Digital Subscriber Signalling System No. 2 (DSS 2)—User-Network Interface (UNI) layer 3 specification for basic call/connection control*, ITU-T (02/95).
Q.2932.1 *Digital Subscriber Signalling System No. 2 (DSS 2)—Generic functional protocol: Core functions*, ITU-T (07/96).
Q.2961.1 *Additional signalling capabilities to support traffic parameters for the tagging option and the sustainable cell rate parameter set*, ITU-T (10/95).
Q.2962 *Digital Subscriber Signalling System No. 2—Connection characteristics negotiation during call/connection phase*, ITU-T (05/98).
Q.2963.1 *Digital Subscriber Signalling System No. 2—Connection modification: Peak cell rate modification by the connection owner*, ITU-T (07/96).
Q.2964 *Digital Subscriber Signalling System No. 2: Basic look ahead*, ITU-T (07/96).
Q.2971 *Digital subscriber signalling system No. 2 (DSS 2)—user-network interface (UNI) layer 3 specification for point to multipoint call/connection control*, ITU-T (10/95).

QUESTIONS

1. Why was a two-stage approach taken when specifying B-ISDN signalling?
2. There is a significant difference between the release procedure used in N-ISDN access signalling and B-ISDN access signalling; what is it?
3. Can you name three of the procedures that are supported in point-to-multipont signalling?
4. What are the two Information Elements that have been specified to support negotiation in DSS2 capability set 2 signalling.
5. What limitation of Q.2931 prevents it from providing full signalling support for multiconnection calls.

6

The ATM Forum Signalling Protocols and their Interworking

Nick Cooper

6.1 INTRODUCTION

The ATM Forum's origins, organisation and working method

The ATM Forum was formed in 1991[1]. It is an international non-profit organisation aimed at accelerating the use of ATM products and services through a rapid convergence of interoperability specifications. It has grown rapidly to its present 900 member companies.

The ATM Forum produces its specifications through debate and discussions within Working Groups, which report to a Technical Committee overseeing all technical activity. The ATM Forum Technical Committee and Working Groups meet for a week at a time—currently five times per year. Member organisations may submit 'contributions' for consideration by the Working Groups at these meetings. Such contributions will typically raise technical issues and propose solutions. If agreeable to the meeting, suitable text will be prepared for eventual inclusion in a specification.

The Forum's approach is significant. Being 'contribution-driven', it reflects the imperatives of those members who contribute technical proposals for inclusion during the specification's 'production cycle'. This is in contrast to the approach of the ITU-T, which drives its work with a set of questions

---

[1] ATM Forum home page: www.atmforum.com/

at the outset of each study period. There is always the risk with the former approach that the capabilities of a given specification are not as uniformly balanced as might be possible. On the other hand, the process should yield an early specification highly relevant to market needs.

Terminology

Traditionally, the terms 'Private' and 'Public' have been used within standards bodies to distinguish a public carrier network—often an incumbent national PTT—from independent, 'in-house' networks. In recent years, this clear separation has become blurred by changes in both regulation and the marketplace. After considerable discussion, the ATM Forum has chosen to use the terms 'Private Network' and 'ATM Service Provider Network' (in place of 'Public Network') to more accurately reflect the current situation. These terms appear in recent documents, though the reader will encounter the public/private nomenclature in earlier specifications.

6.2 THE ATM FORUM AND SIGNALLING

Recognising that signalling system requirements differ at different points in the network, the ATM Forum established three separate Working Groups to deal separately with each specification, together with a fourth to explore interworking issues. This structure, however, was superseded in Autumn 1997 by a re-structuring along functional rather than interface lines.

Original structure

The Signalling Working Group

This has dealt with the production of the User Network Interface (UNI) signalling specifications. It should be noted here that the UNI specification itself covers many aspects of the interface (such as the physical and adaptation layers), of which signalling itself forms only a part (albeit a major one). Because signalling is a feature of many topics, its advice has been sought through joint meetings with other Working Groups where signalling matters arise.

The Broadband InterCarrier Interface (B-ICI) Working Group

This has addressed the needs of interworking between networks and has been responsible for the B-ICI specification. Its scope, however, has concentrated on connection between service provider networks. As will be seen later, the issue of internetwork interfaces, and operation in general, has become a major issue, and has not been handled completely within this Working Group.

The Private Network Network Interface (P-NNI) Working Group

This Working Group has developed specifications for signalling both within and between private networks. It has been the major focus for routing expertise within the Forum, which has been reflected in its approach to the design of its P-NNI protocol.

Interworking among ATM Networks (IAN) Working Group

A fourth Working Group was created during 1997, though strictly it was a joint meeting of the above three Working Groups. This was the Interworking Among ATM Networks Working Group, which dealt with a number of issues raised by the growing need to interwork both public and private networks in a scaleable and coherent manner.

Revised structure

The ATM Forum restructured these groups late in 1997, to divide the work along functional rather than interface lines. The result was just two groups—the control signalling group and the routing and addressing group.

6.3 THE SIGNALLING SPECIFICATIONS IN MORE DETAIL

It has never been the ATM Forum's intention to duplicate work that is already under way; hence, whenever possible, it draws on the output of other standards bodies. However, in practice, it has often been found necessary to add to, or amend the functionality of, existing specifications to support particular requirements.

These considerations are particularly relevant in signalling, where much detailed work and output has already been produced. The Working Groups mentioned above have in each case adopted an ITU-T recommendation as the basis of their specifications. However, this has still lead to much activity, and additional work where the existing recommendation has not met a perceived need.

In most cases, this approach has resulted in specifications that refer back to earlier specifications and recommendations, either noting additional functionality or pointing out differences.

The Anchorage Accord

A major issue for implementers and users is 'backward compatibility'. The specification designer is implored to ensure that successive versions of his specification will interwork successfully with earlier versions. This is of course a major confidence factor for implementers and purchasers, who would otherwise need to make existing equipment redundant, when upgrading.

To recognise this need, the ATM Forum created the 'Anchorage Accord' at its Anchorage meeting in April 1996. This identified a set of some 60 specifications that it had produced that would form a foundation on which to build an ATM network. Subsequent specifications would be designed to be interoperable with those declared in the Anchorage Accord[2]. The main signalling specifications included in this set are:

- B-ISDN Inter Carrier Interface Specification v2.0
- Addendum to B-ISDN Inter Carrier Interface Specification v2.0
- Interim Inter-Switch Signalling Protocol (IISP) Specification v1.0
- Private Network-Network Interface Specification v1.0
- Addendum to Private Network-Network Interface Specification v1.0 for ABR Parameter Negotiating
- ATM User-Network Interface (UNI) Signalling Specification v4.0
- Addendum to UNI Signalling 4.0 for ABR Parameter Negotiation
- User to Network Interface Specification v3.1

---

[2] ATM Forum Anchorage Accord webpage: www.atmforum.com/atmforum/specs/anchorage.html

The User Network Interface

The Signalling Working Group have completed a number of specifications relating to the UNI, all based on the concepts embodied in the ITU-T equivalent UNI standard, Q.2931. Of greatest relevance are UNI version 3.1 and UNI version 4.0 (signalling). An earlier ATM Forum standard UNI v3.0 is incompatible with Q.2931 primarily, because they use different Signalling ATM Adaptation Layers (SAAL).

Two separate UNIs are identified—A private UNI and a public UNI. The private UNI exists between a private ATM Network and an End point terminal. The public UNI is used to connect a public ATM network to either an End point equipment or to a private ATM network. In practice, these two UNIs are very similar.

ATM Forum UNI Specification v3.1

UNI v3.1 includes definitions of the physical, ATM and management layers of the UNI. Section 5 alone deals with signalling. UNI v3.1 is broadly a subset of Q.2931, though it includes a number of important enhancements felt necessary for the early deployment and interworking of ATM equipment. The main additions are support for:

- point-to-multipoint connections
- additional traffic parameters, and
- private network addressing.

UNI v3.1 also differs from Q.2931 in the range of message types it supports. Some Q.2931 messages are not supported (ALERTING, PROGRESS, INFORMATION, NOTIFY), while others have been added to support point-to-multipoint working.

ATM Forum UNI Signalling Specification v4.0

The UNI Signalling Specification v4.0 was issued in July 1996. Unlike UNI v3.1, it covers purely the signalling aspects of the interface. Again, it is based on Q.2931 and is structured to identify differences between itself and Q.2931, rather than being complete in itself.

UNI Signalling v4.0 supports the following capabilities which are not covered by UNI signalling v3.1:

- Support for the Available Bit Rate (ABR) service
- Parameterised Quality of Service

- Switched Virtual Paths
- Narrowband services
- Leaf Initiated Join
- Anycast capability
- Frame Discard
- Proxy signalling
- a number of supplementary services (such as both calling and called party number information presentation and restriction, direct dialling in, user to user signalling).

The Private Network Environment and P-NNI

The ATM Forum have developed P-NNI v1.0 (published March 1996), and v2.0 is still under discussion. P-NNI offers both a control signalling protocol and a routing protocol (the next chapter provides more details on the PNNI protocol). An early specification for an Interim Interswitch Signalling Protocol (IISP) provided a limited capability, but was superseded by the issue of P-NNI v1.0.

The control signalling part of P-NNI is based upon the UNI signalling protocol, which is itself based on Q.2931. Whilst this supports point-to-point and point-to-multipoint connections, mechanisms have been added to support source routing, crankback procedures and alternate routing of call set-up requests in the event of set-up failure. P-NNI v1.0 does not support some features such as leaf initiated join.

The routing protocol is based on the link-state routing technique. It is designed to distribute topology information between switches, which is then used to compute paths through the network.

The InterCarrier interface

Early work in the ATM Forum was based upon the emerging Broadband ISDN User Part (B-ISUP), which forms part of Signalling System No 7 developed by the ITU-T. The model initially assumed by the ATM Forum in its work is one where carrier networks run B-ISUP as an internal network protocol between network nodes.

In the ITU-T view a version of this same protocol then operates between two such networks. However, the ATM Forum has created the Broadband Inter-Carrier Interface (B-ICI) protocol based on the internetwork B-ISUP protocol to fulfil this role (see Figure 6.1).

6.3 THE SIGNALLING SPECIFICATIONS IN MORE DETAIL

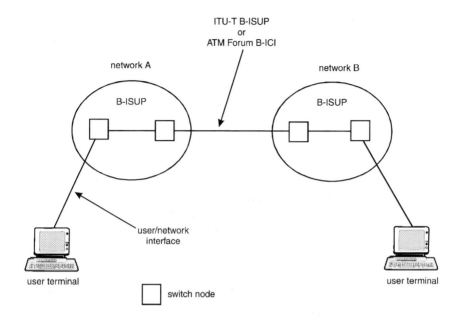

Figure 6.1 Inter-Carrier Interfaces to Interconnect B-ISUP-Based Networks

More recently, the enthusiastic take-up of the P-NNI protocol within the private network environment has led to some proposals within the industry to employ P-NNI within carrier or public networks in place of B-ISUP, mainly because of its perceived earlier availability. In this case, B-ISUP or B-ICI would not be an appropriate interfacing protocol.

This led to the consideration of an ATM InterNetwork Interface (AINI), to interconnect two networks where one runs B-ISUP internally and the other runs PNNI. It might also be used to interconnect two networks both running PNNI, but where full routing interworking is not desired, as shown in Figure 6.2.

Two full versions (1.0 and 2.0) of B-ICI have been produced, as well as an addendum to v2.0, otherwise known as v2.1. The B-ICI group's method of working has, in general, been to introduce features into the interworking specification, once they have been included in the UNI specification. Therefore, work within the B-ICI Working Group has tended to lag behind that of the Signalling Working Group for sound reasons. Hence, the latest version of B-ICI has been designed to support the features of UNI v3.1.

Being derived from Signalling System number 7, rather than Q.2931, B-ICI uses a slightly different terminology to both UNI and P-NNI signalling. Its signalling messages contain 'parameters' rather than Information Elements, and the call control messages are named differently in many cases.

B-ICI v1.0 was published in August 1993. From a signalling point of view, it is of limited significance since it supports only Permanent Virtual

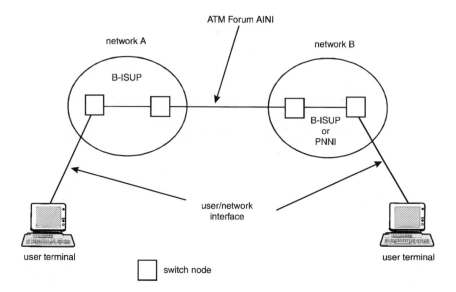

Figure 6.2 Location of the ATM Forum ATM Internetwork Interface (AINI)

Circuits. B-ICI v2.0, completed in December 1995, introduces Switched Virtual Circuits and much associated functionality. As with UNI v3.1, B-ICI v2.0 contains a full interface definition, hence only Section 7 details the signalling capability. Support is offered for both point-to-point connections and unidirectional point-to-multipoint connections.

B-ICI v2.1, an addendum to v2.0 was approved in December 1996 adding support for

- Variable Bit Rate service (VBR)
- Network Call Correlation Identifier (NCCI)
- ATM End System Addresses (AESAs).

In the Autumn of 1997 there were no specific plans to produce a further version of B-ICI.

6.4 INTERWORKING ATM NETWORKS

As ATM matures and larger scale networks are envisaged, it will be increasingly important to make sure that the various signalling specifications can be used to allow easy interconnection between networks—both private and ATM service provider. A number of important areas need to be considered

6.4 INTERWORKING ATM NETWORKS

A range of addressing options exist within the various signalling specifications, some originally intended for the private network domain, some for the public domain. It will be important to ensure that these can be used effectively to achieve call set-ups throughout a concatenation of ATM networks.

Such networks may be connected in a number of different ways. These options need to be clearly understood so that routing and control information can be passed satisfactorily as and when required. The choice of interface protocol to connect together private and ATM service provider networks also needs to be considered. A particular case involves the interconnection of two ATM service provider networks when B-ISUP is, for whatever reason, not an option.

Interconnecting networks

The protocol chosen to interconnect networks must be robust, easily implemented and support as much desired functionality as possible. Where both networks operate similar protocols internally, a version of this same protocol may be the obvious choice for interconnection. This is because it will require the minimum of interworking translation between the internal network protocol and the internetwork protocol, and support the functionality required as simply as possible. Where, however, the two networks do not run similar internal protocols, some form of interworking will be called for within at least one of the edge nodes. This will require additional functionality with potentially extra cost, and perhaps, development time.

The internal protocols most likely to be encountered in ATM networks are PNNI and B-ISUP. Depending on the combination of these encountered, the considerations outlined in the following sections may apply.

Interconnecting two networks running PNNI

PNNI was developed primarily for use within and between private networks. PNNI's approach to routing involves a logical grouping of the network into peer groups, each with a nominated peer group leader. The process is repeated hierarchically so that a logical tree structure of peer groups may be obtained. A key issue then becomes how exactly the intervening network(s) appears within this topological hierarchy. This will depend in part upon whether the two PNNI-based networks wish to share topology information.

Interconnecting networks where exchange of topology information is not permitted

If, for security or other reasons, exchange of topology information is not allowed between the networks, the use of the PNNI protocol is not suitable as an interworking protocol. An alternative would be to use the proposed AINI, whose control signalling functionality would be based upon that of PNNI.

Interconnecting networks where exchange of topology information is permitted

Where exchange of topology information is permitted between the interconnected 'end' networks, the intervening network(s) may be viewed with varying degrees of complexity by these networks. There are two main ways that the networks may be interconnected. The simplest is a direct connection, without any intervening network. The second case involves interconnect via one or more intervening networks, as shown in Figure 6.3. The significance of this becomes apparent when one considers the extent to which routing information is passed between the two networks.

At its simplest, the intervening network may appear as a simple link. At its most complex it may be viewed as a network of nodes in its own right, such nodes not necessarily corresponding to the physical ones. These network 'projections' are further described later.

Four possible 'projections' are identified which can be classified according to how the transit network processes the various items of signalling and routing information. These are summarised as Network Topology Abstractions (see Table 6.1).

Functionality of the intervening network

In addition to the way in which the intervening network is projected, the extent of support it offers is also considered. At least four scenarios can be

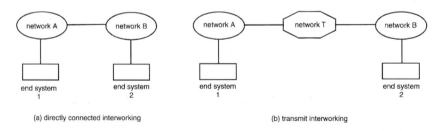

Figure 6.3 Interworking Configurations

6.4 INTERWORKING ATM NETWORKS

Table 6.1 Network Topology Abstractions

Projection type	Network Topology Abstraction
Link Projection	a single link between the two networks (see Figure 6.4)
Single Node Projection	a single P-NNI node between the two networks (see Figure 6.5)
Multiple Node Projection	several P-NNI nodes between the two networks. The abstraction may not map one-to-one onto the physical nodes interconnection the two networks (see Figure 6.6)
Node for Node Projection	This is a literal one-to-one projection and may be of limited applicability

Table 6.2 P-NNI peering scenarios

	Protocol support within intervening network		
Scenario	Basic signalling	P-NNI signalling	P-NNI routing
1	×	×	×
2	✓	×	×
3	✓	×	✓
4	✓	✓	✓

identified (see Table 6.2); they differ according to whether and how the intervening network handles the following information:

- basic signalling, i.e. the basic call control without the P-NNI-specific items
- P-NNI signalling—the specific signalling items associated with P-NNI (such as DTLs)
- P-NNI routing protocol.

Scenario 1 is effectively a tunnelling solution where no signalling support is offered by the intervening network.

Interconnecting networks—both B-ISUP based

Where both networks to be interconnected use B-ISUP as an internal protocol, then the ATM Forum's B-ISI protocol is appropriate for this purpose. Being itself based on B-ISUP, this approach will support maximum functionality across the network boundary.

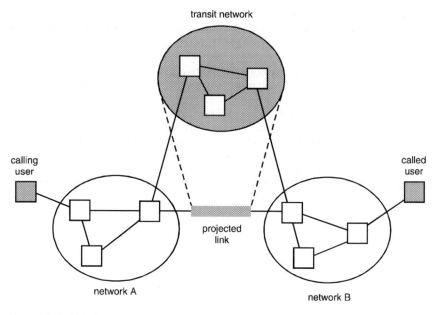

Figure 6.4 Link Projection

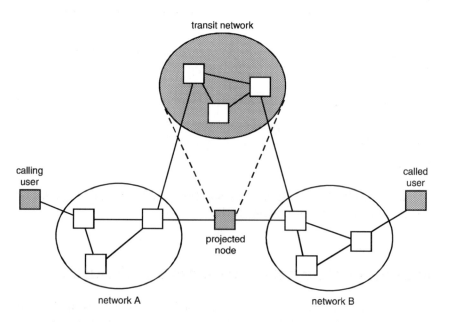

Figure 6.5 Single Node Projection

6.5 ADDRESSING

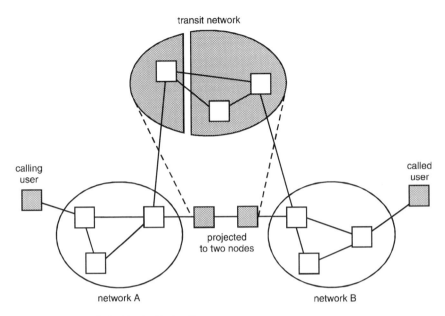

Figure 6.6 Multiple Node Projection

Interconnecting networks—a mix of P-NNI and B-ISUP based

This is the situation which suggested the need for the development of the ATM internetwork interface. The interface design would need to support as much functionality as possible consistent with compatibility of the two protocols. PNNI is being widely implemented in switches because of their deployment in private networks. Hence, it has been suggested that, to avoid the burden of such switches having to implement a second (B-ISUP) protocol stack, the AINI be based on the existing PNNI protocol, but without the routing protocol aspects of PNNI, since these would not be required.

6.5 ADDRESSING

Addressing has become of particular importance within the ATM Forum as the interworking of public and private networks has developed. Historically, the various B-ICI, P-NNI and UNI signalling protocols each developed their own statements on allowable addressing structures and methods. These reflected the needs perceived at the time by the specification developers, but did not offer guidance on how they should be used in an interworking scenario.

ATM Forum protocols support a number of address formats. The main division is between 'native' E.164 numbers and ATM End System Addresses (AESAs). The former have evolved out of the telephony environment via the ITU-T whilst AESAs have been primarily data oriented.

Native E.164 addresses

A native E.164 address may be up to 15 digits in length, and comprise a country code, a national destination code, and a subscriber number. Whilst Q.2931 and B-ISUP recommendations allow other 'type of number' formats, the ATM Forum allows only the international format.

AESAs

AESAs are derived from the ISO Network Service Access Point addresses. There are currently three types of AESA supported by the ATM Forum: Data Country Code (DCC); International Code Designator (ICD); and E.164-type; each being 20 octets in length. The various types each comprise an Initial Domain Part (IDP) and a Domain Specific Part (DSP). The IDP defines which type of address it is by specifying an administration authority responsible for assigning values to the remaining part of the address (the DSP).

A particularly important feature of addressing for scaleability to large networks is that addresses be aggregatable in a sensible hierarchical manner. If not, a flat, non-hierarchical scheme will lead to large routing tables within ATM switches.

Consequential impact of addressing philosophy

Some features and supplementary services which utilise addressing require careful attention to ensure that the interworking is facilitated where different address types are employed. An example is the Closed User Group (CUG) capability, which defines a group of users and the permitted connectivity both within and outside the group. This uses addressing information as part of the identification mechanism. A further example—the Network Call Correlation Identifier (NCCI)—is specified in B-ICI version 2.1. These are used to uniquely identify a specific call over a period of time, for purposes such as charging.

The ATM Forum recognises the importance of ensuring that practical interworking solutions for addressing are well understood, devoting considerable time to the issue.

6.6 OTHER ISSUES

A number of key issues, for example addressing and the InterCarrier interface, have already been covered. There are as well some other areas relevant to the development of ATM signalling standards.

Support for supplementary services

Support for supplementary services—such as are defined for voice services and narrowband ISDN—will be key for the full benefit offered by broadband networks to be realised. Where these are carried across networks using a range of protocols, it will be important to ensure that the protocol interworking provides a consistent service as viewed by the user, regardless of the network implementation.

Multipoint-to-point connections

Point-to-point connections are naturally the simplest to implement. Point-to-multipoint have followed, whilst multipoint-to-point remain under discussion. Among the topics that need to be covered are the aspects of merging streams of cells, as well as issues associated with routing and combining different qualities of service.

Mobile and wireless ATM

Signalling and routing protocols will play an important part in the development of effective mobile and wireless applications of ATM. In particular, the relevant signalling groups have been meeting jointly with other groups to ensure that, for example, routing expertise developed within the P-NNI specification work can be exploited in the design of call routing and hand-off procedures in the mobile environment.

Interaction with Internetworking Protocol (IP)

IP Routers may be interconnected over an ATM network by means of a simple network of Permanent Virtual Circuits (PVCs). This potentially makes inefficient use of the ATM network bandwidth, and requires establishment by the network management capability. A particular area of interest involves exploring more efficient and interactive methods of supporting IP traffic within an ATM environment. Two such methods have been suggested—P-NNI Augmented Routing (PAR) and Integrated P-NNI (I-PNNI).

PAR involves routers with ATM interfaces advertising their status into the P-NNI network, receiving similar information from other such routers, and using the information to establish SVCs across the P-NNI network. These will then be used to interconnect routers more efficiently.

Whilst PAR maintains a separation between the P-NNI and IP routing protocols, I-P-NNI seeks to use a single routing protocol—P-NNI—both for routers and ATM nodes. The ATM network routing information is effectively extended beyond the ATM network itself.

6.7 CONCLUSION

The ATM Forum membership has put much effort into developing specifications to accelerate the roll-out process of ATM switch technology. While much of its output has been based on existing ITU-T recommendations, it has extended these in its P-NNI specification by the inclusion of routing. A major challenge for the future will be to ensure that the signalling systems can interwork appropriately to meet operator requirements for manageable and scaleable networks.

As has been suggested earlier, many of the key items mentioned above are were still under discussion at the time this book was written. Given the energy with which the ATM Forum pursues such issues, they are likely to have evolved considerably in the following months and years.

STANDARDS

ATM Forum Specification: ATM User-Network Interface Specification v3.1 (af-uni-0010.002) 1994.

ATM Forum Specification: UNI Signalling v4.0 (af-sig-0061.000) July 1996.

ATM Forum Specification: B-ICI v2.0 (integrated specification) (af-bici-0013.003) December 1995.

ATM Forum Specification: P-NNI v1.0 (af-pnni-0055.000) March 1996.

ATM Forum Specification: B-ICI v2.0 Addendum or v2.1 (af-bici-0068.000) November 1996.

ATM Forum Specification: Interim Inter-Switch Signalling Protocol (af-pnni-0026.000) December 1994.

ATM Forum Specification: Addendum to UNI Signalling v4.0 for ABR Parameter Negotiation (af-sig-0076.000) January 1997.

ATM Forum Specification: Addendum to P-NNI v1.0 for ABR Parameter Negotiation (af-pnni-0075.000) January 1997.

QUESTIONS

1. What are the key features of the ATM Forum's main signalling protocols?
2. What are the protocol options for interconnecting ATM networks, using ATM Forum protocols? How do they differ?
3. How do E.164 numbers differ from AESAs?
4. What was the rationale for the Anchorage Accord? What signalling capabilities did it include?

7

The ATM Forum's Private Network Network Interface

Jennifer Scott and Ian Jones

7.1 OVERVIEW OF P-NNI

To-date, a large proportion of Asynchronous Transfer Mode (ATM) connections, particularly in the wide area environment, have been of a permanent nature—Permanent Virtual Circuits (PVCs)—requiring management intervention for set up and tear down. However, Switched Virtual Circuits (SVCs) offer the ability to set up and tear down connections of a range of desired characteristics on demand, to a reachable end user. The Private Network Network Interface (P-NNI) is just one of the signalling protocols that has been defined to do this. Others are described elsewhere in this book.

P-NNI consists of two protocols, namely a signalling protocol and a routing protocol. The signalling protocol is based on ATM Forum User Network Interface (UNI) signalling, which in turn is related to ITU-T Q.2931. Modifications have been made to provide a Network Network Interface (NNI) rather than UNI, to make use of the routing protocol, and to provide certain other extra functionality. The P-NNI routing protocol uses similar concepts to the OSPF protocol used within IP networks. It allows QoS-based dynamic hierarchical source routing, and provides for a hierarchical and scaleable network.

P-NNI routing

In contrast to the 'traditional' approach within telecommunications networks of using static routes, P-NNI operates a dynamic source-based routing protocol. As the name implies, source-based routing allows the originating or source node to select the complete hierarchical path from itself to the destination, and all the intermediate nodes have to obey the source node routing instructions. To achieve reliable source-based routing, topology information related to address reachability and current resource capabilities of the nodes has to be distributed around the network on a regular basis. As observed above, P-NNI is intended to be a very scaleable protocol, and this scaleability is achieved primarily via hierarchy and summarisation. In theory, P-NNI can support over 100 levels of hierarchy, although it is difficult to envisage a scenario where more than a handful of levels would be necessary or desirable.

P-NNI topology construction—the lowest level

A P-NNI network is arranged into a series of Peer Groups (PGs). The highest level of the hierarchy will be represented by a single PG, whilst at the lowest level of the hierarchy, there will be many PGs. The members of the lowest level PG may correspond to actual ATM nodes, whereas the higher level PG members are 'logical' nodes or abstractions of the lower PGs. This notion of PGs enables P-NNI to be more scaleable.

Neighbouring nodes within a PG exchange different packet types to reliably distribute topology information. These packet types are:

- Hello packets
- PNNI Topology State Packets (PTSPs)
- Database Summary Packets
- PTSE Request Packet
- PTSE Acknowledgement packets.

'Hello' packets are exchanged between neighbouring nodes on a particular predefined VC, VCI = 18. Hello packets are exchanged on every Virtual Path (VP) contained within a physical link. These Hello packets serve a number of purposes. First, they allow the nodes to discover the identity of their neighbours, to determine whether those neighbours are actually members of the same PG, and to verify the status of the physical links between the nodes. Nodes which have links to other nodes in other PGs are known as *border* nodes. These border nodes are responsible for creating *uplinks*, which provide connectivity amongst PGs at higher levels of the hierarchy.

Once the physical link is established and operational, topology information regarding the current capabilities of the nodes and links, along with reachable destination addresses from within the PG, can be exchanged.

7.1 OVERVIEW OF P-NNI

This exchange of information is crucial if P-NNI is to perform reliable source-based routing. This information is flooded through the PG in P-NNI Topology State Elements (PTSEs) encapsulated in P-NNI Topology State Packets (PTSPs).

Topology information includes details about both nodal and link state parameters. These parameters include values of the bandwidth and Quality of Service available across links and nodes, as well as policy-based parameters. Topology information exchanges continue as long as the P-NNI systems operate. It should be pointed out that PTSEs are only exchanged within PGs, and not between different PGs. Only Hello packets are exchanged between different PGs (Figure 7.1).

Once a node has learnt of the existence of other neighbouring nodes within its PG, it starts a database exchange process to synchronise databases. The exchange process involves a master node sending Database Summary packets to a slave node. The packets contain summarised details of the PTSEs stored in a node's topology database. When one of the nodes receives a Database Summary Packet, it checks its database against those PTSEs listed in the Database Summary Packet. If the database does not contain a PTSE, or if it is out of date, the node requests that its neighbour sends it the relevant PTSE(s), making use of the PTSE Request and PTSE Acknowledgements packets.

Each PG elects a Peer Group Leader (PGL). At start-up, each node is configured with a variable, known as the PG Leader Priority, which relates to the node's 'likelihood of becoming PGL'. On the basis of this variable, the nodes within a PG elect the node with the highest PG Leader Priority value to become the PGL. It is the responsibility of the PGL to represents its PG further up the hierarchy.

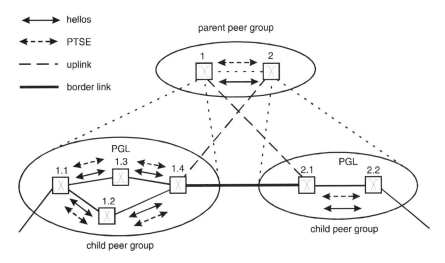

Figure 7.1 Exchange of Hello Packets in the P-NNI Hierarchy

A new PTSE will be generated to update the node's topology database whenever a 'significant' change occurs (for example, as a result of call setups/clear downs, topology changes, or network failures). The nature of a 'significant' change is suggested within the P-NNI specification.

Additionally, nodes check the age of the PTSEs stored in their database. This ageing function decrements the PTSE Remaining Lifetime to reflect how long they have been stored in the database. The ageing function also detects when the self-originating PTSE are to be refreshed, and when non-self-originated (i.e. received) PTSEs are to be deleted from the database as they have reached their expiry time. The expired PTSEs cannot be used for determining routing.

Building the hierarchy past the first level

The Peer Group Leader (PGL) represents the PG as a Logical Group Node (LGN) at the next level of hierarchy (i.e. its parent PG) (see Figure 7.2). The LGN represents a summarised view or abstraction of reachability and topology information pertaining to the child PG. It is important to remember that the LGN is not an extra physical node, it's only an abstraction of the child PG. The LGNs use the same process of exchanging information between themselves and their peers as happened at the lower level. The

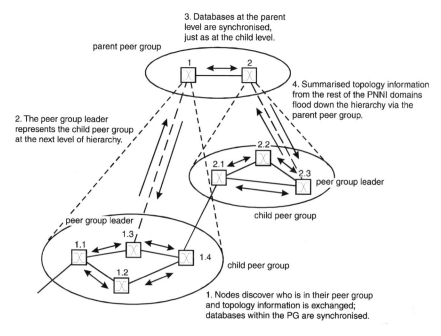

Figure 7.2 P-NNI Routing Hierarchy Construction in a 2-level Hierarchy

7.1 OVERVIEW OF P-NNI

LGNs in the Parent PG access their peers via the lower level border nodes creating uplinks to them. The nodes in the parent PG will establish SVCs between themselves for routing information exchange. This SVC is known as a RCC, and is used to exchange summarised addresses and aggregated topology information between the nodes in the parent PG. Once the databases are synchronised, PTSEs are flooded back down the hierarchy, to the child PGL, which then distributes them amongst the PG. This ensures that each node in the child PG not only has topology information about its own PG, but also about the parent PG too. In this way, each node now has reachability information as to how to reach every node encompassed by the parent PG.

A grandparent PG may be constructed 'above' a collection of parent PGs, and a great-grandparent, and so on, to extend the network to multiple levels of hierarchy.

Reachability and address summarisation

In addition to topology state information, address reachability information is also exchanged. To reduce the amount of information that needs to be distributed around the network, addresses which are reachable from a particular node are summarised. This summarisation corresponds to a 'common' address prefix. There may be a number of summarised addresses within a node, and additionally addresses which share no commonality with any other address on the node. These are known as *foreign addresses*, and are distributed intact. At the next hierarchical layer addresses are summarised further, where possible, and the information distributed around that PG. This continues right up to the highest PG. The summarised addresses are flooded down the hierarchy as before, so that each node has summary information about where to route calls with which address prefixes.

P-NNI is defined to make use of the ATM End System Address (AESA) scheme described in UNI 3.1. These addresses are based on the ISO NSAP (Network Service Access Point) scheme, and are 20 octets in length. There are three NSAP schemes described in UNI 3.1: they are the International Code Designator scheme; the Data Country Code scheme; and the NSAP E.164 scheme (in contrast to 'native' E.164). All three consist of a 13 byte prefix, a six byte End System Identifier (ESI) and a one byte selector. In P-NNI the 13 byte prefix ideally reflects the network topology. The ESI consists of some 'flat' address space to identify the destination UNI (a MAC address would be an example). P-NNI does not route based on the selector byte.

Each PG has an identity (the PGID) which corresponds to a prefix on an ATM NSAP type address of up to 13 bytes. The hierarchy can therefore be up to 105 levels deep. Each node also has a node ID, and an ATM address corresponding to the physical system implementing the node (since nodes above the lowest level of hierarchy are abstractions) which is used to set up the SVCs for exchange of routing information in PGs above the lowest level.

Clearly, an appropriate address structure within the network is an aid to conciseness and brevity of updates. While completely unrelated addresses can be supported at any node, they must be distributed individually and unsummarised potentially right up, down and across the network. Thus, if a chaotic address scheme is adopted, route table space is quickly filled and updates are all the lengthier.

More information on how to address the entities in a P-NNI network is given in P-NNI annex F.

Route computation

As ATM connections can vary enormously in their requirements, such as traffic types, bandwidth and Quality of Service (QoS), calculating a hierarchical source-based route (in contrast to performing hop-by-hop routing) to the destination which matches the connection requirements is a complex activity. Route computation in P-NNI can be based on a number of criteria, namely QoS, reachability and a Generic Call Admission Control (GCAC) algorithm, introduced in P-NNI, and described later in this section.

P-NNI selects a path to a destination which has the longest address prefix match. In a large P-NNI network there can be a large number of possible combinations of routes to the destination. It is the responsibility of the routing algorithm to find the an acceptable route through the network. To do this, the routing protocol needs to know the characteristics of the required connection, plus the capabilities of the nodes in the network. The connection characteristics can be obtained from the IEs within the Setup message, whereas the network resources are obtained the P-NNI topology information.

There are two types of route calculation that can be performed within P-NNI. The first is to use an algorithm which calculates all possible destinations from a node, based upon varying connection characteristics. This is known as a *pre-calculated route.* Pre-calculated routes are designed to reduce some of the processing overhead involved in calculating acceptable routes. The second type are on-demand route, these are used when the required setup values can't be met by any of the pre-calculated routes.

One of the factors ATM manufacturers seek to differentiate their products is Connection Admission Control (CAC). Thus, P-NNI cannot predict what the admission criteria will be at every node. Additionally, further down the DTL, the nodes do not correspond to actual physical switches, but rather are logical representations based on summarised information

7.1 OVERVIEW OF P-NNI

for which a completely accurate CAC algorithm which guarantees that the call will be admitted cannot be run. P-NNI therefore introduces the concept of Generic CAC (GCAC). Generic CAC is executed on receipt of a request to set up a connection to find which nodes are likely to admit the connection. As GCAC only predicts that a node is 'likely' to accept the call, it is not completely accurate, and the call may need to cranked back at some stage along the route.

The possible list of paths to the destination is first reduced to the links that are likely to accept the new connection, using GCAC. Path selection is via a routing algorithm and utilises required performance constraints (such as timing or loss criteria) configured within the network. Switches within a network can use different route selection algorithms.

Once the route has been calculated, it can then be converted into a Designated Transit List (DTL). These DTLs are explained later. Like CAC, route computation is not standardised, and can be used by switch vendors to differentiate their products.

P-NNI signalling

P-NNI signalling is designed to be compatible with UNI 3.1, but also provides functionality from UNI 4.0 signalling, including:

- Point-to-point and point-to-multipoint calls
- Parameterised Quality of Service (QoS)
- Anycast
- Signalling for ABR
- Carriage of Generic Identifier Transport
- Switched Virtual Path Connections
- Frame discard
- Negotiation of traffic parameters (PCR, SCR, MBS)
- Support for 64kbit Circuit Mode services.

Additionally, P-NNI builds on the capabilities of UNI 4.0 signalling to provide Soft Permanent Virtual Circuits (S-PVC) at both VP and VC level. Also, due to the source based routing of P-NNI, the signalling capabilities support the use of Designated Transit Lists (DTLs) and Crankback procedures. These additional capabilities are described below.

Call set-up and source routing

The DTL is formulated by processing the P-NNI routing topology information. A DTL is basically an array of node IDs, with a pointer which is used to indicate the current node ID. The DTL IE provides a route across the P-NNI domain from the ingress node to egress node, and is only carried within the SETUP message. The DTL is arranged hierarchically to provide a exact hop-by-hop route across the local PG and a hierarchical list of LGNs at each higher level within the P-NNI hierarchy, which must be traversed to reach the destination.

The DTL procedures are best explained using an example as shown in Figure 7.3.

1. Terminal A sends a SETUP containing called party number using UNI signalling, towards node 1.1.
2. Node 1.1 calculates the DTL from itself to the destination, based upon

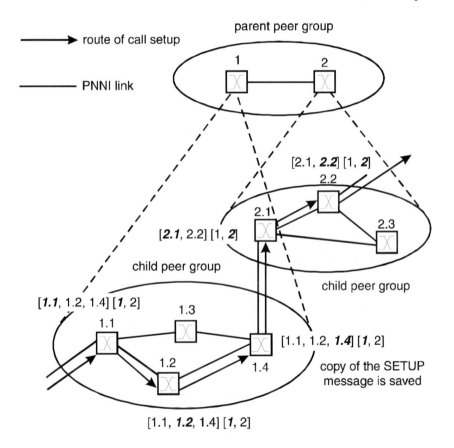

Figure 7.3 Call Setup in a P-NNI Network

7.1 OVERVIEW OF P-NNI

the called party number. Node 1.1 knows the complete path across its PG, which is included in the DTL as 1.1, 1.2 and 1.4. Node 1.3 was not chosen as, from its topology, information could not meet the desired characteristics of the connection. The path across the rest of the network has to be recorded using the next level PG, i.e. node 1 then node 2.
3. Once the SETUP message is received at node 1.4, the first level of hierarchy within the DTL is exhausted. The pointer is incremented and points to the next level of hierarchy, namely PG 2. From the Hello packets exchanged between border nodes (i.e. 1.4 and 2.1), node 1.4 knows how to route to node 2.1. A copy of the entire setup message, including the old DTL, is saved at egress to the previous PG in case of crankback.

Crankback

Crankback is a facility which allows, at connection establishment time, the network to reroute a call around a failed or congested node(s), instead of clearing the call back to the calling party. When a call cannot be processed according to the DTL, it is cleared back to the creator of that DTL (i.e. a previous node) with an indication of the problem. This indication is carried in the Crankback IE within the RELEASE or RELEASE COMPLETE message. Thus, the switch resources are released and no tromboning of the call occurs. The node which created the DTL now has the responsibility of finding an alternative route around the failed node/link and to the destination. The alternative route must support the original requested connection characteristics. If this node cannot find an alternative route, the call is cranked back again towards the source node, to the next previous node which created a DTL and this node tries to route the call. This process of crankback and rerouting continues until the source node is reached. If no alternative path can be found then, the call is released.

The types of causes for crankback include resource unavailability and network faults.

Soft Permanent Virtual Circuits

P-NNI allows for the setup of Permanent Virtual Circuits (PVCs) via the control plane. Such connections are known as Soft Permanent Virtual Circuits (S-PVCs), and may be of VC or VP nature. An S-PVC is very similar to a P-NNI Switched VC (SVC) in that it is source routed and makes use of P-NNI routing to find an acceptable route in real time. However, the two major differences are that S-PVCs terminate within the network, thus do not cross an UNI, and are initiated by management rather than the user.

Soft PVCs should look no different from the classical management controlled PVC common in ATM networks today, from the customer's

a) an SVC set-up

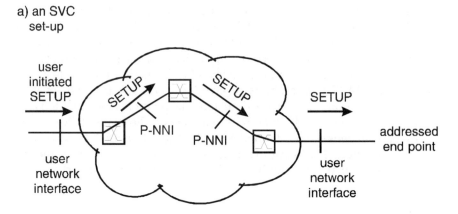

b) an S-PVC set-up

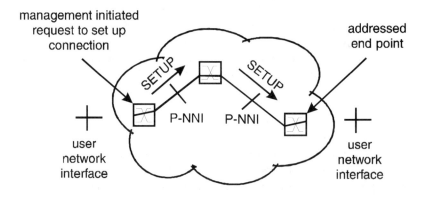

Figure 7.4 SVC vs. S-PVC

point of view. Soft PVCs can be configured to automatically reroute on network failure, and offer a standards based way to provide connectivity across a multivendor domain (Figure 7.4).

Two extra information elements have been defined to support Soft PVCs. These are the Calling Party Soft PVPC, or PVCC Information Element, and the Called Party Soft PVPC or PVCC Information Element. These Information Elements allow the setup message to identify the VPI/VCI combinations (or in the case of an S-PVPC, the VPI only) at the originating and terminating endpoints of the Soft PVC. Where the terminating endpoint is not specified, the allocated VPI/VCI is returned in the CONNECT message.

Using P-NNI in public networks

Notwithstanding its name, the *Private* Network Network Interface. P-NNI can be used in networks by public network operators (referred to in the ATM Forum as ATM Service Providers), as well as in 'enterprise' networks. There are a number of particular considerations in this case.

Topology information distribution

Many public network operators are unlikely to be enthusiastic about interconnecting directly with other operators using 'full' dynamically routed P-NNI. This would imply an interlinked P-NNI domain with associated agreements about PG levels and topology, and also implies topology state updates coming in and out of an operator's network domain. This scenario may have serious security implications, not just because it implies a certain visibility of an operator's network state, but because it could leave a network open to showers of routing update packets should an adjoining operator configure their network incorrectly or suffer a major outage. To overcome this problem P-NNI does allow for domains to be interconnected in a more traditional way, via exterior routes. Exterior routes are ones over which the P-NNI routing protocol is 'turned off', so that routing is static. Introduction of static routes into what is basically a dynamic routing domain requires care to prevent introducing routing loops, since dynamic topology changes elsewhere in the network will not be tracked by the manually configured static routes.

Where dynamic routing information is exchanged, it is possible to restrict the scope of advertisements so that certain reachability information is not distributed throughout the entire P-NNI domain.

Restricting P-NNI information

To help limit the advertisement of information, P-NNI has a number of capabilities designed to restrict the use of certain information. There are routing constraints which can be used to restrict the selection of certain paths, links or nodes to be used. In addition, summary addresses can be suppressed to stop the advertisement of addresses with prefixes. Suppressed summary addresses can be applied to both interior and exterior addresses. Finally, P-NNI supports point-to-multipoint (p-mp) connections, and a branching flag can be used to stops particular nodes from acting as a replication/branching point within (p-mp) calls.

Support for supplementary services

The ITU-T have defined a number of supplementary services (Q.2730/Q.2951.x) for use within a public network. These include number presentation and addressing services such as Calling Line Identification Presentation (CLIP), Direct Dialling In (DDI), sub-addressing (SUB) and Multiple Subscriber Number (MSN), security services, such as Closed User Group (CUG), and others. Many of these are supported in UNI 4.0. However, work to support some of these in P-NNI is still in progress.

Address support

P-NNI is defined to support ATM End System Addresses (AESAs) based on the ISO NSAP scheme. This is in contrast to the public telephony numbering scheme based on recommendation E.164 formulated by the ITU-T. Whether they choose to use them to address their customers or not, the majority of public network operators are likely to want to support 'native' E.164 style addresses, if only for interconnect and interoperability. Some equipment vendors may choose to implement P-NNI in such a way as to support native E.164 addresses directly—they could also be supported by mapping in and out of NSAP format as appropriate.

Support for intelligent network services

Intelligent Network (IN) type services (such as number translation) are typical of the sort of facility provided by a public network. However, private networks typically do not use the classic ITU style IN. Provision of IN-like services is possible with P-NNI but is outside the scope of this paper.

Interworking with ITU-T/Access signalling

P-NNI signalling, like UNI 4.0 signalling, is based around the ITU-T's broadband access signalling system Q.2931 (see Chapter 5), and refers back to it on a paragraph by paragraph basis for much of its functionality. This is not to say, however, that there are not some notable differences, many of which are covered in the comparison section below. This means that interworking issues need to be borne in mind during implementation.

Call correlation across multiple networks

To correlate information across different nodes within a network (for billing and reporting), or across multiple networks (e.g. for reciprocal billing arrangements), the Network Call Correlation Identifier may be

7.1 OVERVIEW OF P-NNI   **137**

used. This has only recently been defined in the standards bodies, and is not explicitly supported in P-NNI version 1, although it could be transported using the Generic Identifier Transport (GIT) Information Element, which is supported in P-NNI version 1. Generation and transport of unique NCCIs, particularly across multiple network domains, has formed the basis of discussions during formulation of P-NNI addenda.

Closed User Groups

Many large corporate business customers require enhanced security features over and above a basic any-to-any switched ATM service. The requirement might include a need to screen callers from outside their business from accessing their equipment and end-stations, or to prevent calls being placed outside the company. In its most basic form, such security can be provided by screening Calling and Called Party Numbers at connection setup time, either within the public network or on the Customer's Premises Equipment. This requires a matrix of calling and called party numbers to be held at each node, either for allowed calls, or for barred calls. This may be onerous to administer.

Closed User Group (CUG) defines a procedure which simplifies this process. Instead of having to check at each call setup whether a specific address is allowed to call another specific address, each user is assigned membership of a 'Closed User Group', with a particular unique ID of its own. At call setup a check is just made as to whether that caller is a member of the correct CUG, so at network ingress information about far end addresses need not be kept. Because the relationship between user and CUG is a many-to-one rather than a many-to-many relationship, CUG is easier to administer and is more scaleable than straight address screening tables. However, the signalling capability required to provide Closed User Groups is only recently being defined, partly because of the original thrust of P-NNI as a private network interface.

In the future, security features like Closed User Groups could be provided on ATM networks using Intelligent Network (IN) functionality.

Scaleability

Scaleability is clearly of paramount importance in a public network. As described above, P-NNI has been designed with the aim of being a very scaleable dynamic routing protocol. To achieve this scaleability, information about the network is aggregated. Of necessity, however, this aggregation must lead to some reduction in accuracy of information in proportion to the degree of abstraction. P-NNI is theoretically able to support a large number of peer group levels, likely to be well in excess of that required for a worldwide public network. However, practical limits imposed on aggregation are likely to be much lower.

7.2 COMPARISON WITH ITU-T NETWORK SIGNALLING

The most traditional choice for a public network operator is often to comply with international recommendations as published by the ITU-T. The ITU-T has defined the Broadband ISDN User Part (B-ISUP–Q.2761–4 inclusive) for layer 3 network signalling. In the first instance, this is based very much on its Narrowband equivalent, ISUP (Q.761-4 inclusive). For more details regarding B-ISUP, see Chapter 8, which provides more details on B-ISUP.

Protocol stacks

As can be seen from Figure 7.5, both stacks make use of the same lower layer ATM protocols. The SSCF (Service Specific Control Function), Service Specific Connection Oriented Protocol (SSCOP) and ATM Adaptation Layers (AAL) are usually combined in a single component, known as the *Signalling AAL*. The purpose of the Signalling AAL is to adapt the signalling messages into ATM cells, and *vice versa*. It is at the higher layers of the protocol stack, where the two differ significantly. Some of these differences are explained below.

Routing philosophy

The routing philosopies of the two protocols are very different, mainly due to the types of networks they were designed to operate in (i.e. P-NNI in private networks and B-ISUP in public networks). P-NNI consists of both a routing protocol and a signalling protocol, whereas B-ISUP is just a signalling protocol. In Signalling System No. 7 there is the concept of a signalling network. The maintenance of the signalling network, and the routing of signalling messages is undertaken by the Message Transfer Part Level 3 (MTP3). The call routing is the responsibility of B-ISUP, and CAC is performed by exchange applications, and are not the subject of standardisation as ITU-T leave the CAC and call routing algorithms to be determined by the network operator/switch vendor. While it is difficult to compare the call routing methods of PNNI and B-ISUP in detail, it is worthwhile discussing the routing of signalling in both protocols.

Unlike P-NNI, which uses hierarchical source-based routing, MTP-3b

7.2 COMPARISON WITH ITU-T NETWORK SIGNALLING 139

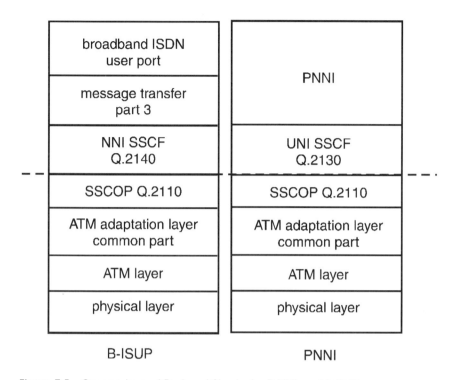

Figure 7.5 Comparison of Protocol Stacks for P-NNI and B-ISUP

uses hop-by-hop routing. In hop-by-hop routing, rather than the source determining the path towards the destination, it is the responsibility of each node along the path to route the signalling messages based its own routing knowledge. As the name implies, each node only knows how to route signalling messages to the next hop along the path towards the destination, rather than the complete route. With a hop-by-hop routing algorithm there is no need to distribute topology information around the network. Instead the routing information is configured into a routing database by management. The routing information normally consists of primary routes to be used in normal conditions and secondary routes to be used in failure conditions. Primary routes are normally the shortest path to the destination (i.e. least number of hops). Therefore B-ISUP and MTP-3b have no concept of Hello, PTSE or Database Summary packets and there is no database synchronisation or PGL election processes.

In either Broadband or Narrowband SS No.7 networks, two different types of signalling modes can be used to route signalling messages between nodes. These two modes are known as *associated* and *quasi-associated* modes of signalling. Quasi-associated signalling is a form of non-assigning signalling. P-NNI only supports the associated mode of signalling. Associated signalling is the more direct form of signalling between nodes. In this mode, the signalling messages are sent over a signalling link, which directly interconnects the signalling points. Quasi-associated signalling makes use of Signalling Transfer Points (STPs). One or more STPs are used to interconnect the signalling points. Chapter 8 contains diagrams of these signalling modes.

Whereas P-NNI can use dynamic routing, for the routing of signalling messages, MTP-3b employs a more static or pre-determined routing approach. Each node determines the outgoing link over which the signalling message will be sent based upon a Destination Point Code (DPC), and optionally a predetermined Signalling Link Selector (SLS), using a routing table stored either within each node or remotely. With this static routing method, the decision to select a signalling link is not affected by the signalling traffic loading, unless signalling congestion or failure occurs, on those outgoing links.

Recent studies (Zhu *et al.* 1995) have indicated that routing algorithms should have a close relationship with the node's CAC algorithm. This enables the routing algorithm to predict the probability of a node accepting the connection. In P-NNI , this is provided by the use of the GCAC in route computation, as described previously. SS No. 7 networks do not have this concept of GCAC, where information related to the status of the nodes probability of connection acceptance is flooded around the network.

One of the challenges facing network operators intending to implement source-based routing is to try and reduce the time delay in computing on-demand routes, plus manage the amount of topology related information being distributed amongst nodes with the network and to stop routes being selected without the awareness of the operator.

Behaviour under abnormal conditions

Due to the different backgrounds, P-NNI and B-ISUP differ in the level of resiliency they provide. As B-ISUP and MTP-3b were designed for use within a public network, they need to be robust to cope with failures and congestion within a network. One of the major parts of the MTP-3b functionality is to provide Signalling Network Management. This tries to ensure the integrity of the signalling network under failure or abnormal conditions. The Signalling Network Management is split into three sub-parts, known as Signalling Traffic Management, Signalling Route Management and Signalling Link Management. These are responsible for diverting signalling traffic from failed routes to operational routes, without message loss, managing signalling congestion, and determining the availability and status of signalling routes, along with controlling signalling links. Therefore, if a signalling link fails, once the call/connection has been established, provided there is another available signalling link, the signalling messages can be diverted and control of the call maintained.

Using B-ISUP and MTP functionality, the amount of signalling traffic transmitted within the control plane can be managed. If a node receives more signalling messages than it can process, then an indication is sent to the signalling user part causing this overload and instructed to back off until the congestion is cleared. This type of control plane congestion is not presently available in P-NNI.

P-NNI was originally developed to be used within private networks, and as such takes a different approach to availability and resilience. P-NNIv1 has a Crankback functionality, which can be used only at connection setup to route around congestion and/or failure along the selected route. As mentioned previously, P-NNI supports soft PVCs to provide resiliency for PVCs in the case of failure. P-NNI addenda, should contain procedures to provide fault tolerant routing. This would allow a failed connection to be automatically re-routed from the source node using a different DTL, without any intervention from the end users.

Quality of Service

The ITU-T has defined a series of QoS classes which are sets of values of delay and loss—held in the QoS Parameters IE. In contrast, the ATM Forum has gone one step further, in allowing individual specification of these parameters (seven in all)—held in the Extended QoS parameters IE. Unlike B-ISUP, which will only support QoS classes, P-NNI supports QoS classes and optionally Extended QoS parameters. The Extended QoS parameters was one of the capabilities carried forward from ATM-F UNI v4.0. They consist of both desired and cumulative Cell Delay Variation (CDV) and an acceptable Cell Loss Ratio (CLR) in the forward and backward directions. The cumulative QoS values allow each switch involved in a connection to add its expected QoS impairments to a running total. If the calling user's requested parameter values have not been exceeded when the connection reaches the called user, then the connection is established, otherwise the connection is released.

Addressing

B-ISUP currently supports native E.164 addresses and optionally NSAP E.164s. Work is under way to allow B-ISUP to support other types of NSAP formats.

Scaleability

It can be argued that both protocols scale to support very large networks. In a SS No.7. network, nodes are identified by Point Codes (PCs). Therefore, theoretically the size of a SS No.7 network is limited by the number of PCs available. The worldwide signalling network is split into two independent levels, namely the International and National levels. The format of the PCs is also split into two levels, i.e. International and National to reflect this. These PC values are administered on a worldwide basis by the ITU-T. SS 7-based networks have proved to be scaleable as they are used the world over today. As P-NNI is not based upon a traditional SS No.7 network, it does not make use of PCs. Instead, nodes within the network are identified by NSAP addresses. With P-NNI, the use of hierarchy levels allows for the aggregation of topology information and address summarisation, which enables P-NNI to support a large number of nodes. Although P-NNI has not yet been deployed on a worldwide basis, with careful use of hierarchy levels, for the aggregation of topology information and address summarisation, P-NNI should be able to support a large number of nodes.

Network intelligence and mobility

Again, due to the history of the two protocols, their support of aspects of Intelligence Networks (INs) and mobility differ. For the integration of IN and Broadband, B-ISUP will be used along with the services of Transaction Capabilities (TC) and Signalling Connection Control Part (SCCP). For the integration of mobile and broadband, it is likely that B-ISUP, plus TC and SCCP, will also be used, although no decision has yet been made. TC and SCCP are mainly used to transfer non-circuit related data between nodes in a SS No.7 network(s). Like B-ISUP, they make use of the transfer capabilities of MTP-3b. Non-circuit related information allows the network to retrieve data from another node/database without making a connection (i.e. connectionless transfer.) Examples include database access for number translation, mobile registration/location updates and terminal capability checking, such as Look-ahead.

As P-NNI was originally designed for use in a private network the support of IN and mobility was not seen as essential. If, however, public network operators wish to use P-NNI in their networks, they would seek to integrate P-NNI with their existing IN and mobility capabilities. As P-NNI doesn't specify the use of TCAP or SCCP functionality, a different approach needs to be found. Although P-NNI v1 does not support mobility, the ATM Forum Wireless ATM Working Group (W-ATM) may adopt P-NNI to be used in their mobility specifications. The W-ATM Specification is due for completion in 1999.

Interworking

As mentioned before, B-ISUP and P-NNI are very different protocols. To address this differing functionality, the ATM-F set up a Technical Committee known as Interworking amongst ATM Networks (IAN) at the beginning of 1997. The motivation for this activity is that it is seen as essential for the growth of ATM to have interworking between all networks, be they P-NNI or B-ISUP. The distinction between public and private networks is somewhat blurred/converging, as some public networks are/will using P-NNI for initial deployment of ATM SVCs.

The IAN Technical Committee are defining the interface between a number of broadband networks. This interface is known as the ATM Inter-Network Interface (AINI). The scope of this work is shown in Figure 7.6. It seems unlikely that AINI will be used to interconnect two B-ISUP networks where the natural choice would be to B-ISUP. There is still work to be done, as there are many aspects of the two protocols that may not allow for easy interworking, such as the support of CUG interlock codes and Network Call Correlation IDs (NCCIs).

P-NNI supports the necessary narrowband messages and IEs, allowing interworking with narrowband networks, although there are no explicit procedures within the P-NNI specification on how this is achieved.

BQ-SIG is the broadband equivalent of Q-SIG. Although Q-SIG has been

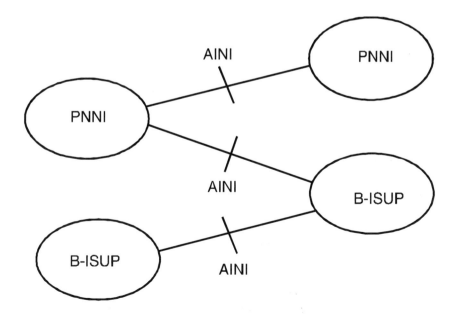

Figure 7.6 Scope of the ATM Inter Network Interface

defined for a number of years, BQ-SIG is still being defined by the European Computing Manufactures Association (ECMA). Both BQ-SIG and Q-SIG are intended to be used as inter-PBX signalling within private networks, and are based upon UNI signalling. ATM-F has recognised the need for this interworking, and P-NNIv2 will have procedures for BQ-SIG support.

7.3 EVOLUTION

The ATM Forum have developed a series of network signalling protocols based around the UNI signalling protocols. Initially, they defined IISP using UNI v3.1 signalling and static routing. After significant work they then released P-NNI v1, which was completed in 1996. An errata version of P-NNI v1 was completed in early 1997 to correct the errors found within P-NNI v1. Now work is progressing on P-NNI v2. Some of the proposed additional capabilities of version 2 are:

- Network Call Correlation Identifier (NCCI)
- Leaf Initiated Join (LIJ)
- BQ-SIG interworking (for private networks)
- Supplementary services
- Security

REFERENCES

Zhu T, Lui C and Mouftah H T, *Dynamic Routing for Multimedia Traffic over ATM Networks*, IEEE Press, 1995.

STANDARDS

ATM forum *Private Network Interface version 1* af-pnni-0055.00 (03/96)
ATM forum *ATM User Network Interface version 3.1* (09/94)
Q.708 *Numbering of international signalling point codes*, ITU-T (03/93).
Q.2761 *Functional description of the B-ISDN user part (B-ISUP) of Signalling System No. 7*, ITU-T (02/95).
Q.2762 *General functions of messages and signals of the B-ISDN user part (B-ISUP) of Signalling System No. 7*, ITU-T (02/95).
Q.2763 *Signalling system No. 7 B-ISDN user part (B-ISUP)—formats and codes*, ITU-T (02/95).

Q.2764 *Signalling system No. 7 B-ISDN user part (B-ISUP)—basic call procedures*, ITU-T (02/95).

QUESTIONS

1. What is the role and what are the responsibilities of a Peer Group Leader (PGL)?
2. What additional functionality is required for a border node?
3. How does topology aggregation affect routing performance and scaleability?
4. What are the main steps in P-NNI path selection?
5. What are the benefits (and pitfalls) of using a dynamic routing protocol for a public ATM network?

8

B-ISUP, ITU-T's Internodal Broadband Signalling Protocol

Ian Jones

8.1 INTRODUCTION

B-ISUP is an internodal signalling protocol, and as such can operate between nodes within either a national or international network. Like Narrowband ISUP (N-ISUP), it is based upon a common channel signalling concept. In common channel signalling, a dedicated signalling channel is used to transfer signalling messages between nodes. These signalling messages, related to the control plane, are used to establish, maintain and release user plane call/connections. The exchange of signalling messages creates an association between the control plane and user plane. Once this association has been created, a call/connection is said to be established and user plane data can be sent and received. When the call/connection is released the association between the control and user plane is then removed.

In an ATM environment, each Virtual Path (VP) can contain 65535 Virtual Channels (VCs). These VCs are divided into two categories: VC Identifier (VCI) values 0–31 are reserved for control and management plane functionality; whilst VCI values 32–65 535 are used for transferring user plane information. A single dedicated Signalling VC (SVC), using VCI = 5, is used to transfer signalling information between the nodes involved in the call/connection. The signalling VC is pre-configured by management (i.e. Permanent VC (PVC)), and is used to establish, maintain and release the user plane connections in real time. This concept is known as *on-demand* or *switched connections*. Depending on the configuration chosen, a SVC can either only control the user plane connections within its own VP, or alternatively, may in addition control user plane connections within other VPs. The first configuration is known as *associated signalling*, where the SVC is always associated with its VP, whilst the second is known as *non-associated signalling*. Therefore, a single signalling VC can control many thousands of user plane connections.

ATM connections are very flexible, allowing a wide range of their connections characteristics to be specified. Using signalling, end users can dynamically chose the bandwidth of their connections and, once established, have the freedom to then either increase or decrease this bandwidth. Also, a user can select from a range of transfer capabilities, including, constant, variable, block transfer or available bit rate services. A user can also put constraints on the Quality of Service (QoS) offered (i.e. how many cells can be lost, specify the transfer and variation delay of cells), and select a connection topology, be it point-to-point or point-to-multipoint, etc. Therefore, to control these ATM connections the signalling protocol has to be able to transfer this information in a flexible way, and the end user terminals, plus the switches have to process and understand this information.

8.2 OVERVIEW OF SS NO. 7 ARCHITECTURE

Figure 8.1 shows the Broadband SS No.7 architecture. As can be seen, the B-ISUP signalling procedures represent only one part of this overall architecture. Just like any other large complex system, the SS No.7 functionality is divided into a number of sub-systems, each designed to perform a specific task. The B-ISUP signalling procedures establish, maintain and release ATM connections. The Signalling Connection Control Part (SCCP), and Transaction Capability (TC), provide additional information required by the call/connection to be retrieved from either a local or remote database. The Message Transfer Part 3 for broadband (MTP3-b), plus the Signalling ATM Adaptation Layer (SAAL) and ATM layer, provide for a reliable transport mechanism for signalling information between nodes within a network(s).

8.3 B-ISUP SIGNALLING PROCEDURES

In the same way as N-ISUP (Q.761–4), the definition of the B-ISUP signalling protocol for basic call procedures has been divided into four parts, ITU-T Recommendations (Q.2761–4):

Q.2761: Defines a functional description or overview of B-ISUP.

8.3 B-ISUP SIGNALLING PROCEDURES

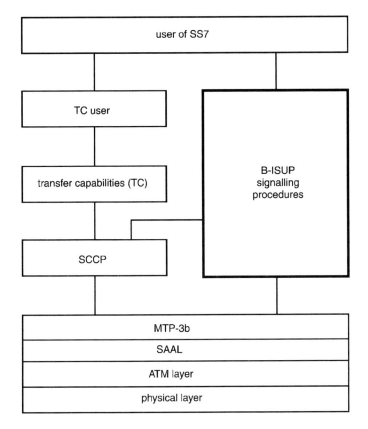

Figure 8.1 SS No.7 Architecture

Q.2762: Defines the messages and parameters applicable to B-ISUP.
Q.2763: Defines the specific format of the messages and parameters applicable to B-ISUP.
Q.2764: Defines the basic call/connection procedures or behaviour of B-ISUP.

B-ISUP architecture

In contrast to N-ISUP, the B-ISUP protocol architecture is based upon the concepts of the OSI Application Layer Structure (ALS). This architecture is shown in Figure 8.2. The ALS technique allows the signalling procedures to be defined in a very modular fashion, allowing easier evolution potential for future capabilities.

The term 'Exchange Application Process' is used to cover all the functionality found within a switch. The B-ISUP Nodal Functions and

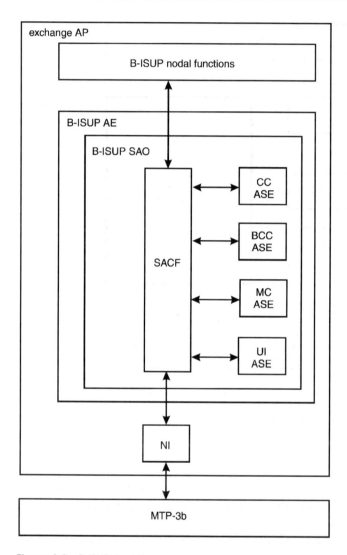

Figure 8.2 B-ISUP Architecture

the Network Interface (NI) entities are static, in the sense that they are not dynamically created/deleted whenever a new call/connection is received/released. Each Application Entity (AE) can contain one or more Single Association Objects (SAOs). These SAOs contain a Single Association Control Function (SACF) and a number of individual Application Service Elements (ASEs). The SACF acts as a co-ordinator, distributing signalling information to the correct entity, be it an ASE, the NI or an application.

8.3 B-ISUP SIGNALLING PROCEDURES

ASEs

One or more ASEs can be associated with a SACF. These ASEs are designed to perform a specific function. The B-ISUP signalling procedures consist of four basic ASEs, known as Call Control (CC), Bearer Connection Control (BCC), Maintenance Control (MC) and Unrecognised Information (UI).

The behaviour of a SAO is described using a formal description language known as the Specification and Description Language (SDL). SDL is based upon a finite state machine, where a state machine consists of a number of states. A change of state occurs due to an event. An event can be a message or primitive that has been sent or received, or the expiry of a timer. In B-ISUP, once the necessary signalling information has been exchanged between the nodes in the correct sequence, according to the signalling procedures, the state machine is said to be in the 'active state'. When all signalling nodes' state machines involved in the call/connection are in the active state, the call/connection is established and user data can be transferred.

The CC and BCC ASEs contain the procedures to establish, maintain and release an ATM call and a connection simultaneously. Although the CC and BCC ASEs are logically separated, presently B-ISUP is a combined call control and bearer control protocol. This can be seen from Table 8.1, where some messages are related to both the CC and BCC ASE. More information on call/bearer separation is described in a later chapter. CC and BCC ASE's consist of two parts, an incoming side and an outgoing side. The MC ASE handles the procedures for resource resetting, blocking, user part availability and consistency checking whilst the UI ASE deals with all unrecognised message and parameter handling. Additional B-ISUP capabilities can be incorporated by either modifying the existing ASEs, or by including a new ASE(s) within a SAO.

For each new call/connection, a new instance of a B-ISUP AE is created.

Table 8.1 ASE/Message relationship

Message	Related ASEs
Initial Address Message	CC, BCC
Initial Address Message Acknowledgement	BCC
Answer	CC, BCC[1]
Release	CC, BCC
Release Complete	BCC
Blocking	MC
Reset	MC
User Part Test	MC

[1]Depends upon message parameters.

These AEs are identified by a Signalling Identifier (SID). Like Point Codes, there are both Destination and Originating SIDs, (DSIDs and OSIDs). When a call/connection is released, the B-ISUP AE instance is deleted and the SID value is freed. The Network Interface (NI) function uses the SID parameter to distribute messages received from MTP-3b to the appropriate instance of B-ISUP AE. Figure 8.3, shows the B-ISUP AE instances that are created for a single call/connection scenario, involving three nodes as described in the Signalling Message Flows section.

Message structure

A B-ISUP signalling message is sent between nodes, encapsulated within the Signalling Information Field (SIF) carried by a MTP service primitive. The format or coding of the B-ISUP signalling messages is shown in Figure 8.4. The fields are shown coded as octet groups.

Signalling messages

The B-ISUP CS-1 signalling procedures supports 28 different signalling messages. The main signalling messages are described below:

IAM: Initial Address Message (IAM) is the first message exchanged between the originating and destination exchanges. It contains the Called Party Number parameter, plus the required ATM characteristics, such as

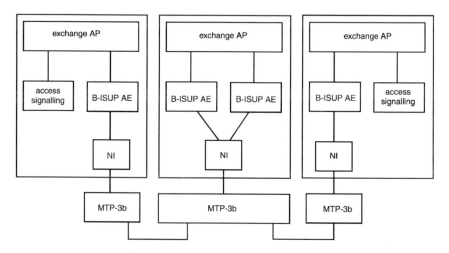

Figure 8.3 B-ISUP AEIs

8.3 B-ISUP SIGNALLING PROCEDURES

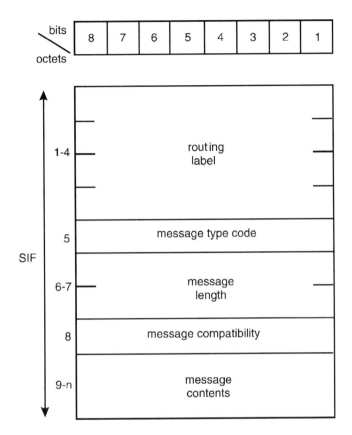

Figure 8.4 Relationship between the SIF and Associated Parameters.
Routing Label: specifies originating and destination point codes (OPC and DPC) plus the Signalling Link Selector (SLS); *Message Type*: identifies the type of message, IAM, IAA, ACM, etc.; *Message Length*: specifies the length of the message contents. The length of this will depend upon the number of parameters associated with each message; *Message Compatibility*: identifies what actions should be taken by the switch if the message is not understood (i.e. a CS-2 specific message being received by a CS-1 switch); *Message Content*: contains the parameters related to the message type

VPCI/VCI value, QoS class, ATM Traffic Descriptor, Broadband Bearer capability and AAL parameters.

IAA: the IAM Acknowledgement message (IAA) is used as an acknowledgement of the IAM, and indicates that the requested resources have been reserved for the requested call/connection.

IAR: the IAM Reject (IAR) message is sent in response to a IAM to indicate that the call/connection has not been accepted.

ACM: the Address Complete Message (ACM) is sent from the destination switch towards the originating switch. It is used to indicate that the destination switch has all the addressing information to reach the called party.

ANM: the ANswer Message (ANM) is sent from the destination switch towards the originating switch. It is used to indicate that the called party has excepted the call/connection.

REL: the RELease (REL) message can be sent from either the originating or destination switch. It is used to indicate that either party wishes to release the call/connection.

RLC: the ReLease Complete (RLC) message, is sent in response to a received REL. It is used to acknowledge that the far end has agreed to the release of the call/connection.

Parameter structure

Within the Message Content field, there will be a number of parameters which contain information associated to that particular message. The structure of the parameters is very similar to the message format, with type, length, compatibility and content fields. Some messages, have many more parameters associated with them than others. For example, the IAM message can contain over 20 parameters, whilst other messages contain only a single parameter. These parameters can be either a fixed or variable length, and can be either mandatory or optional. The structure and sequence of the parameters is defined using the Abstract Syntax Notation One (ASN.1) concept, as shown in ITU-T Rec. Q.2763.

When a switch receives a message it has to locate, identify and check all the parameters contained in the message content field, to see if the necessary information is present. Based on this received information, the B-ISUP procedures determine what action should be taken, such as to continue with the call/connection or reject the call/connection attempt.

QoS and ATC parameters

The ability to specify a level of Quality of Service (QoS) associated with a connection and for the network to guarantee that quality for the lifetime of the connection is one of the major benefits of ATM. Four classes have been defined by the ITU-T in recommendation I.356. The QoS class requested for the connection is signalled in the IAM, and once accepted can not be changed for the lifetime of the connection. If any of the node(s) along the route cannot support the required QoS class, the connection request will be rejected. The parameters which constitute the definition of QoS can be split into two basic categories:

- delays, which consist of cell transfer delays and cell variation delays, and
- errors, which consist of lost cells, erroneous cells, mis-inserted cells, and blocks of erroneous cells.

ATM can support a number of ATM Transfer Capabilities (ATCs in ITU-T, or the ATM Forum's equivalent, Service Categories). The ATCs supported by B-ISUP include Variable Bit Rate (VBR), Constant Bit Rate (CBR), ATM Block Transfer (ABT) and Available Bit Rate (ABR). VBR and CBR are the terms used by the ATM Forum, while ITU-T refer to the same characteristics as Statistical Bit Rate (SBR) and Deterministic Bit Rate (DBR)—as the ATM Forum's VBR and CBR is more intuitive, this term is mostly used throughout this book. VBR is sub-divided into three separate conformance definitions, known as VBR.1, VBR.2 and VBR.3. VBR.1 has time constraints imposed on the cell transfer and delay variation times, whereas VBR.2 does not. VBR.3 is also known as 'tagging', whereby rather than the ATM cell being discarded by the switch if it is non-conformant, the ATM Cell Loss Priority is changed from zero to one. A conformance definition describes the method used to determine conformant cells, against the connection traffic descriptor, and how those cells judged to be non-conformant are handled. Tagging and conformance are defined in ITU-T recommendation I.371.

These QoS classes along with their respective ATCs are shown in Table 8.2. Class 1 is defined as the highest QoS, whilst Class U the lowest.

Address parameters

B-ISUP supports the traditional E.164 numbering/addressing plan, plus additionally the ATM End System Address (AESA). The ASEA consists of an E.164 number plus a Network Service Access Point (NSAP) address. Routing within a B-ISUP network is currently based on the E.164 number contained in the called party number parameter, and the NSAP address is carried transparently through the network. Additional addressing schemes are likely to be included in future B-ISUP recommendations.

Table 8.2 Relationship between ATCs and QoS classes

ATC	Applicable QoS class
CBR, VBR.1, ABT	Class 1 (stringent class)
CBR, VBR.1, ABT	Class 2 (tolerant class)
VBR.2, VBR.3, ABR	Class 3 (bi-level class)
Any ATC	Class U

Signalled ATM parameters

To establish, maintain and release connections in an ATM environment, a number of ATM related parameters need to be signalled between the switches involved in the connection. The main ATM related parameters are described below:

ATM Traffic Descriptor: this parameter is used to select the required connection bandwidth for the required ATC in both the forward and backward directions. Different values for the forward and backward directions can be specified (i.e. asymmetric).
AAL: the ATM Adaptation Layer (AAL) parameter is used to select which type of AAL is to be used for a connection. Presently, the defined AAL types are 1, 3/4 and 5. Each AAL type has its own set of characteristics to be used with different ATM Transfer capabilities, such as CBR, VBR, ABR, etc.
Broadband Bearer Capability: this is used to select the Bearer Class and the ATM Transfer Capability (ATC). ITU-T Rec. Q2931 states which bearer classes are supported. The ATC identifies if the user's connection shall support CBR, VBR, ABR or ABT. The ATC type must be common across all the links making up the connection.
BB High/Low Level Information: both the BB-HLI and BB-LLI are used for compatibility checking between terminals, and as such are carried transparently across the network.
Connection Identifier: indicates which (VPCI/VCI) the user's connection is to be established on.
QoS: this is used to select which QoS class is required for the user's connection.
OAM: Operations, Administrations and Maintenance (OAM) flows are used for fault and performance management of ATM connections. The OAM F5 flow relates to the VC level, whilst the F4 flow relates to the VP level. ITU-T Recommendation I.610 contains detailed information.

ITU-T Recommendation Q.2721 lists the parameters supported by both N-ISUP and B-ISUP.

Signalling topology

The topology of SS No. 7 signalling networks can be split into two categories, known as *associated* and *non-associated modes*. The associated mode of signalling is the simplest of the two; the signalling path directly connects the two Signalling Points (SPs). The SPs are nodes A and B in the example shown in Figure 8.5. Quasi-associated signalling is a form of

8.3 B-ISUP SIGNALLING PROCEDURES

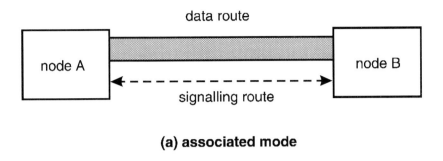

(a) associated mode

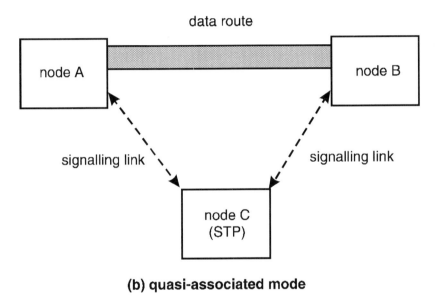

(b) quasi-associated mode

Figure 8.5 Signalling Topology Modes

non-associated mode signalling. In this mode, the signalling path and hence messages do not flow directly between the two signalling end points. Instead they are routed via an intermediate point, known as a Signalling Transfer Point (STP). STPs are packet switches which route signalling messages around the network. This is shown in Figure 8.5. Although quasi-associated is more complex than associated, it allows network operators to achieve greater flexibility and resilience. Depending on the network operators requirements, a signalling network can be configured using associated, quasi-associated, or a mixture of both signalling modes.

8.4 ROUTING

In an ideal network there would be no blocking, no failures of links or nodes, rerouting would never happen and the optimal route would always be chosen through the network. However, all networks can suffer from the above problems. Therefore the routing algorithms have to contend with these problems, plus the additional problems associated with ATM networks, where connections characteristics and the resources allocated to them, can vary enormously. The procedures used for routing B-ISUP signalling messages in a SS No.7 network is known as MTP-3b. MTP-3b is based upon MTP-3 functionality, with additional capabilities for ATM-based networks. MTP-3b functions in a very similar manner to ITU-T Rec. Q.704 MTP layer 3, with the exception of some MTP layer 3/layer 2 management processes. MTP-3b uses a hop-by-hop approach to route signalling messages across a network towards the destination. Unlike source-based routing, where the source or originating switch knows the complete route through a network towards the destination, switches using a hop-by-hop approach only know how to route to the next switch nearer to the destination. The hop-by-hop mechanism does not require topology related information to be flooded or aggregated throughout the network. Both types of routing algorithms can be used in an ATM environment, and both have their own advantages and disadvantages, as explained in the previous chapter.

MTP-3b introduction

Like MTP-3, MTP-3b operates in a connectionless manner. It is designed to reliably transfer B-ISUP signalling messages between nodes within a SS No.7 network. To achieve this, MTP-3b is designed to be able to reroute around failures, control congestion build up of messages and detect errors, such as messages received out of sequence. The major difference between the MTP-3 and MTP-3b is concerned with the maximum length of the Signalling Information Field (SIF) that can be transferred in a single message. With MTP-3b, this has been increased to 4091 octets to make more efficient use of the underlying SAAL capabilities, against the 272 octet limit for narrowband links. Therefore, provided the 272 octet limit is not exceeded, B-ISUP is equally able to utilise existing narrowband MTP-based networks. MTP-3b, like MTP3 is split into two main sections, known as Signalling Message Handling and Signalling Network Management.

8.4 ROUTING

Signalling Message Handling

This is responsible for routing the user part messages to their destination, and once there, to deliver them to the intended user part. The actual routing process is based on Point Codes (PC), which in the UK are 14-bits in length. Each PC uniquely identifies a node, be it in a national or international network. There are two types of PC within the routing label: Destination PC (DPC), which identifies the intended destination node; and a Originating PC (OPC), which identifies the originating node. ITU-T Rec. Q.708, has detailed information on PC formats. In addition, the Routing Label also contains the Signalling Link Selector (SLS). The SLS is used to evenly distribute the user part messages over a number of signalling links to help avoid congestion on any one link. The structure of the Routing Label is shown in Figure 8.6.

Receiving MSUs

Signalling information is transferred between MTP nodes by using Message Signal Units (MSU). Whenever MTP-3b receives a MSU from the SAAL, it is examined by the discrimination function. This examines the Routing Label within the Signalling Information Field (SIF) to determine the DPC value. If the DPC doesn't match the DPC value of the node, then the node

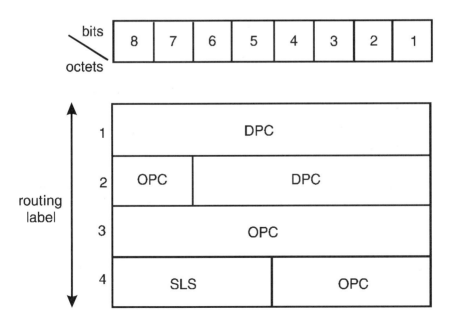

Figure 8.6 Routing Label

uses the route management function to consult its routing database to select the next link towards the destination. If the DPC matches that of the node, then the message has reached its destination and is passed to the Distribution Function. The Distribution Function then examines the Service Information Octet (SIO) within the MSU to determine which user part or network management functionality the message is intend for.

Sending MSUs

When MTP-3b receives a message from a higher layer, the process is the same as relaying a message onto another node. Therefore, the routing process uses the DPC to access the routing tables to decide which route is to be taken and the SLS determines which signalling link the messages are to be transmitted over.

Signalling Network Management

The Signalling Network Management maintains the integrity of the signalling network in the event of failure. It consists of three parts: Signalling Traffic Management, Signalling Link Management and Signalling Route Management:

- **Link Management:** this controls a Signalling Link Set. A link set contains one or more signalling links between two points. The link management covers link and link set activation, restoration and deactivation. MTP-3b link management, is identical to that of MTP-3 with the exception of link changeover procedures which employ different management messages and triggers appropriate to the SAAL.

- **Traffic Management:** this is used to divert signalling messages from one route to another without any loss, duplication or mis-sequencing, and also to divert back to the original route when required.

- **Route Management:** this is concerned with the status and availability of signalling routes. Information is distributed around the network to block or unblock particular routes from being used, such as when a node has failed and all routes to that node can be blocked.

Figure 8.7 shows how the signalling messaging handling and signalling network management functions interact with one another.

8.4 ROUTING

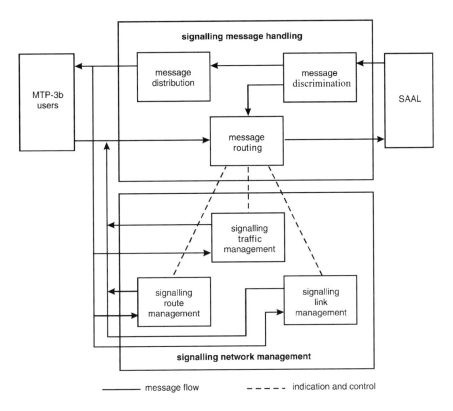

Figure 8.7 MTP3-b Overview

Signalling message flows

Figure 8.8 shows an example of the UNI and NNI signalling message flows for *en bloc*, manually answered terminals. *En bloc* sending means that all the addressing information is sent at once, instead of in stages, as in overlap sending. Examples of many other possible scenarios are shown in ITU-T Recommendation Q.2650.

The following gives a brief explanation of the main events in establishing and releasing an ATM call/connection. The paragraph numbering corresponds to the diagram numbering:

1. B-ISUP has the concept of assigning and non-assigning procedures. For all the VPCI values which the two adjacent switches control, one will be the assigning switch for all the even numbered values, while the other will be the assigning switch for all the odd numbered values. This procedure stops contention over VPCI/VCI values and bandwidth

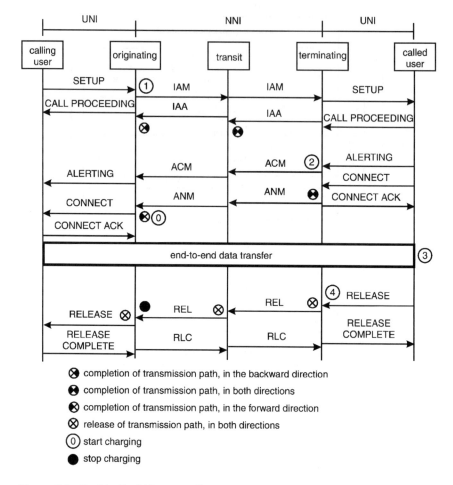

Figure 8.8 End-to-End Message Flows

between adjacent switches. The assigning switch is responsible for selecting a VPCI/VCI and reserving the required connection bandwidth on that VPCI/VCI. The Information Elements (IEs) contained in the UNI SETUP message are processed by the originating switch. Assuming this switch is the assigning switch, and the Connection Admission Control's (CAC) acceptance of the call/connection, the DPC is calculated from the called party number and a VPCI/VCI value chosen. Selection of the route can depend upon the called party number, plus ATM Cell Rate, Broadband Bearer Capability, QoS and associated delay parameters. The switch will create a new instance of the B-ISUP AE and the relevant IEs within the SETUP message, plus the necessary NNI parameters are transferred within the IAM towards the terminating switch. The actions performed by the two other switches will depend upon them acting as

the assigning or non-assigning switches. Both of these switches will create an instance of the B-ISUP AE, providing they can accept the call/connection. The IAA is sent back as an acknowledgement that the IAM has been accepted.
2. When the terminating switch determines the called party is being alerted, it concludes that all the necessary address information is complete, and sends an ACM back towards the originating switch. If the called party responds with a CONNECT, identifying the call/connection has been accepted an ANM is sent back towards the originating switch. These B-ISUP messages are routed back to the appropriate B-ISUP AE instance, within each switch along the route by the combination of the DPC, SIO and SID identifiers.
3. Once the connection has been established, charging is initiated and user plane information can be exchanged between the two end users. With connection-orientated networks, a substantial amount of processing has to be carried out to establish the connection. If, subsequently, the connection is only maintained for a short period of time, or if only a small amount of user data needs to be transferred, this represents an inefficient use of resources. Policing of the user plane ATM cells is employed by the switches to ensure the traffic contact agreed at connection set up time via the signalling messages is not violated. Any user plane cells judged to be in breach of this contract are either discarded or tagged (i.e. given a lower priority).
4. B-ISDN access signalling, Q.2931, supports a two phase release procedures instead of the three phase employed by ISDN access signalling, Q.931. Either the calling or called party can release the connection. In this example, the called party initiates the release. Once the RLC message has been received and processed, all resources associated with the call/connection are freed, including the VPCI/VCI value. In addition, charging is terminated and the BISUP AE instances associated with the connection along the route are deleted.

8.5 SCCP AND TC

MTP-3b was developed to allow signalling messages of a greater length to be transferred via the SAAL links with potentially much higher data rates than the MTP level 2. To make use of the new MTP-3b capabilities, SCCP has been modified so that it allows for the transportation of up to 3952 octets of data without invoking any segmentation procedures. Segmentation is used when data is too large to be carried by an individual message, therefore the data is to be split or segmented into a number of parts and carried in multiple messages.

B-ISUP uses the functionality provided by TC and SCCP mainly for non-call related applications. Typical applications include, services such a Look-ahead (Q.2724.1), which allows a network to check the called party terminal availability and compatibility without any commitment of network resources (i.e. establishing a user plane connection). In addition, Closed User Group (CUG) (Q.2735) and Multiconnection (Q.2722.2) can also make use the TC and SCCP functionality. With the integration of B-ISUP and Intelligent Network Application Part (INAP) protocols planned for ITU-T CS-3, plus the development of broadband mobile networks and terminals, the functionality offered by TC and SCCP will become more important. For more information regarding TC and SCCP functionality, see recommendations Q.771 and Q.711, respectively.

8.6 SAAL

In an ATM environment, MTP layers 1 and 2 are replaced by the ATM layer and the Signalling ATM Adaptation Layer (SAAL), respectively. This is shown in Figure 8.9. As its name implies, the SAAL is the adaptation layer for the control plane (i.e. signalling information). The SAAL consists of two parts, the common part and the service specific part. The common part is based upon AAL5, which is also one of the user plane AAL types. The purpose of AAL5 within the SAAL is to segment the signalling messages into ATM cells of 53 bytes, ready for transportation by the ATM layer, and also to re-assemble the ATM cells back into complete signalling messages. The service specific part consists of the Service Specific Co-ordination Function (SSCF) and the Service Specific Connection Oriented Protocol (SCCOP). These are used to transfer variable length packets, known as Service Data Units (SDUs), which contain signalling information. These SDUs can be transmitted between nodes using either an assured (i.e. using error detection, flow control) or an unassured manner.

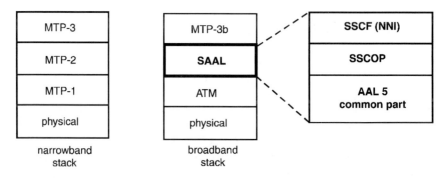

Figure 8.9 Narrowband and Broadband Protocol Stacks

8.7 INTERWORKING

Access signalling

The four main ATM UNI or access signalling protocols are the ITU-T's Q.2931, ECMA's B-QSIG, plus the ATM-Forum's UNI v3.1 and UNI Signalling v4.0. As B-ISUP and Q.2931 are defined by the same organisation, the ITU-T, interworking between the two is considered paramount, and both groups work together to achieve this. ITU-T recommendation Q.2650 defines the necessary procedures for NNI and UNI integration. Interworking protocols from different organisations is always a problem, as different organisations have different requirements. Interworking with UNI v3.1 causes problems, as UNI v3.1 only supports automatically answered terminals (i.e. it has no ALERTING message). UNI v4.0 supports the ALERTING message, plus other additional signalling capabilities. Some of the capabilities include individual QoS parameters instead of just QoS classes, Anycast, Available Bit Rate (ABR), Leaf Initiated Join (LIJ), Generic Id Transport (GIT), Proxy signalling concept, point-to-multipoint connections and VP switching. B-ISUP CS-2 either presently supports or is planning to support these additional UNI v4.0 capabilities, except for Proxy signalling, which is UNI specific, QoS parameters and Anycast.

P-NNI

The interworking between B-ISUP and the ATM-F's Private Network Node Interface (P-NNI) is seen as very important, and work is being carried out within the ATM-F to addresses this issue, as explained in the previous chapter.

B-ICI

Broadband Inter-Carrier Interface (B-ICI) was defined by the ATM-F to be used for interconnecting operators networks. B-ICI v1.0 was based upon Permanent VCs (PVCs), and thus had no signalling capabilities. B-ICI v2.0 is based upon B-ISUP, using a subset of MTP-3b allowing for only associated mode of signalling (i.e. no support STPs). Neither the ITU-T or ATM-F have published an interworking specification regarding B-ISUP and B-ICI, although a reasonable level of interworking should be possible as the two protocols are similar.

N-ISUP

B-ISUP supports the necessary procedures, messages and parameters to achieve interworking with N-ISUP. Various B-ISUP and N-ISUP interworking scenarios are defined in ITU-T Rec. Q.2660. Interworking at layer 2, between MTP-3 and MTP-3b, can be achieved so long as the information carried by MTP-3b is limited to 256 octets in length.

8.8 CONCLUSION

This chapter has given the reader an overview of the B-ISUP signalling protocol capabilities and its relationship to the traditional SS No.7 architecture, used by today's N-ISUP networks.

Since 1995 when B-ISUP CS-1 was originally approved, significant progress has been accomplished by the ITU-T in extending the capabilities of B-ISUP. In the near future, B-ISUP capabilities will be extended further as it incorporates concepts defined within the Internet Engineering Task Force (IETF) to further IP and ATM integration, and to support capabilities defined within the ATM-F to insure signalling interworking. In addition, B-ISUP will be further modified to enable it to support call and bearer separation, to enable multi-connection and third generation mobile services to be more efficiently supported.

It is expected in the next few years to witness more network operators, both within Europe and North America, deploying SVC services. This is due to the maturity of ATM signalling standards, the large number of vendors supporting signalling, and to enable operators to offer more flexible and scalable services to their customers. Indeed, some North American operators already offer an ATM SVC service to their customers. It is expected that these ATM signalling networks will either support end-to-end SVC services or S-PVC services to provide added resiliency for PVC services, in the event of a node or link failure. It will be interesting to see which signalling protocol, be it B-ISUP, P-NNI or AINI, that operators chose to implement, as they each have their strengths and weakness, depending on the type of services that are to be supported.

STANDARDS

Q.711 *Functional description of the signalling connection control part*, ITU-T (07/96).

Q.761 *Signalling System No. 7 ISDN User Part functional description'* ITU-T (09/97).

Q.762 *Signalling System No. 7 ISDN User Part general functions of messages and signals*' ITU-T (09/97).
Q.763 *Signalling System No. 7 ISDN User Part formats and codes*' ITU-T (09/97).
Q.764 *Signalling System No. 7 ISDN User Part signalling procedures*' ITU-T (09/97).
Q.771 *Functional description of transaction capabilities*' ITU-T (06/97).
X.207 *Information Technology—Open Systems Interconnection—application layer structure*' ITU-T (11/93).
Z.100 *CCITT Specification and Description Language (SDL)*' ITU-T (03/93).
AF-B-ICI-0013 *BISDN Inter Carrier Interface (B-ICI) Specification Version 2.0*' the ATM Forum (12/95).
AF-P-NNI-0055 *Private Network-Node Interface Version 1.0*' the ATM Forum (03/96).
AF-UNI-0010 *ATM User-Network Interface (UNI) Specification Version 3.1*' the ATM Forum (09/94).
AF-UNI-0011 *ATM User-Network Interface (UNI) Signalling Specification Version 4.0*' the ATM Forum (07/96).
I.610 *B-ISDN Operation and Maintenance Principles and Functions*' ITU-T (11/95).
Q.2650 *Interworking between signalling system No. 7—broadband ISDN user part (B-ISUP) and digital subscriber signalling system No. 2 (DSS 2)*' ITU-T (02/95).
266 *Broadband Private Integrated Services Network (B-PISN)—Inter-Exchange Signalling Protocol Basic Call/Connection Control (B-QSIG-BC)*' ECMA (1997).

QUESTIONS

1. What is the purpose of the four main ASEs defined for B-ISUP? Why does the B-ISUP architecture keeps these ASEs separate?
2. When a destination switch receives a B-ISUP message, how does it determine which B-ISUP AE the incoming message is intended for?
3. Why does the IAM normally take an ATM switch longer to process than other B-ISUP messages?
4. If an unrecognised B-ISUP message is received at an ATM switch, what actions could the switch take, and which parameter in the message might the switch examine?
5. How does an ATM switch correctly know how to decode B-ISUP messages and parameters containing variable lengths?
6. If a parameter needs to be repeated in a B-ISUP message, how is this coded?

9
The VB5 Interface

Mick Hale, Alex Gillespie and Keith James

9.1 THE BACKGROUND TO VB5

The introduction of fibre transmission and multiplexing technologies in the local loop necessitated the development of digital access interfaces to the local exchange for telephony. These interfaces were initially proprietary, but were eventually standardised in the early 1990s for telephony, Integrated Services Digital Network (ISDN) and leased lines. The European Telecommunications Standards Institute (ETSI) were first with their V5.1 and V5.2 interface specifications. The ITU-T Recommendations G.964 and G.965 followed, based on the ETSI specifications.

The current VB5 specifications draw on these earlier SNI specifications. Many of the higher level requirements are the same for broadband and narrowband networks, and consequently the nature of their interfaces.

It is unfortunate that the term Access Network (AN) is often used to describe two different things. In general, the term Access Network covers the 'local' network between the customer and the local exchange. The other definition relates to a specific integrated and managed network, between a Service Node (SN) and the customer's premises. This type of AN is defined in ITU-T Recommendation G.902 as 'an implementation comprising those entities which provide the required transport bearer capabilities for the provision of telecommunications services between a Service Node Interface (SNI) and each of the associated User Network Interfaces (UNI)'. This is the definition of an AN used in the VB5 standards, and throughout this paper.

The general Access Network architecture and boundaries defined in ITU-T Recommendation G.902 are shown in Figure 9.1.

In addition to the SNI, each equipment has a Q3 management interface. It is a principle in G.902 and carried into VB5 specifications, that time-critical management functions (e.g. user port blocking) are performed by the AN

THE VB5 INTERFACE

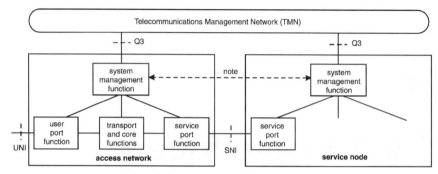

Figure 9.1 G.902 General Access Network Architecture and Boundaries

System Management Function (SMF) and the SN SMF, communicating over the SNI. In the VB5 specifications, a Real Time Management Coordination (RTMC) protocol is defined for this purpose. All other management communication is via the relevant AN or SN Q3 interface.

The SNI shown in Figure 9.1 corresponds to the 'V' reference point in the ISDN protocol reference model. The parts of the broadband ISDN protocol reference model significant to VB5 are shown in Figure 9.2.

Where the Access Network is of the integrated type described in G.902, controlled by a common management system (as opposed to disparate isolated transmission systems, for example), the VB reference point is then referred to as VB5, as detailed in ITU-T Recommendation I.414.

Two factors triggered the start of work on a VB5 interface, one being the thought that a broadband equivalent to the narrowband V5 interface would be necessary, and the other the liberalisation of telecommunications in Europe in 1998.

The work in ETSI on specifying the VB5 interface began with a Technical Report (ETR) in 1994. At the time that this work started, it was thought that there were probably already appropriate Asynchronous Transfer Mode (ATM) specifications available, or in the process of being produced. A VB5

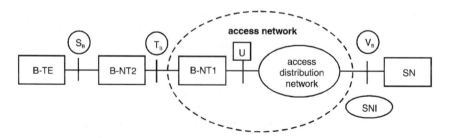

Figure 9.2 B-ISDN Access Protocol Reference Model

specification then need only consist of set of these specifications, with particular options mandated. However, once the ATM specification issues were studied, this proved not to be the case.

One significant conclusion of the report was that two VB5 specifications should be produced:

- VB5.1, a relatively simple Virtual Path (VP) cross-connect based version;
- VB5.2, a more sophisticated interface enabling connection-by-connection concentration in the AN under the control of the SN.

Each of these specifications would have a management specification associated with it to extend the AN and SN management models to include the VB5 specific aspects.

Many of the requirements identified in the report have been included in the current VB5.1 and VB5.2 specifications, but there are some significant changes, so the VB5 ETR is really now only of historical interest.

At the outset, the principle requirements of a VB5 interface were that it should

- be ATM based,
- be largely independent of the physical layer,
- be independent of user signalling protocols,
- provide the means for concentration in the AN,
- be scaleable,
- support B-ISDN and non-B-ISDN accesses,
- provide broadband/narrowband integration,
- enable efficient businesses operations and maintenance processes,
- support shared UNIs (i.e. one UNI may be connected to a number of different SNs via separate SNIs), and
- be specified with Open Network Provision (ONP) in mind.

9.2 THE VB5 INTERFACE

The VB5 interface is specified principally at the ATM layer. The only physical layer constraint at the VB5 reference point is a requirement to provide an embedded transmission path identification method (e.g. the path trace mechanism in Synchronous Digital Hierarchy (SDH) systems). Any path protection must also be done, at the moment, at the physical

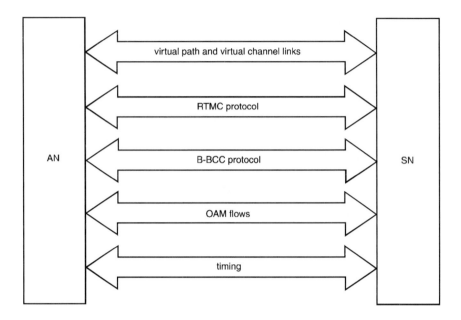

Figure 9.3 Functions at the Vb5 Interface

layer, though VB5 specifications will probably be updated to include VP level protection once a specification is stable. Figure 9.3 shows the functions at the VB5.1 and VB5.2 reference points. The Broadband Bearer Channel Connection (B-BCC) protocol is only used with VB5.2.

At the VP level, the normal ATM Operations Administration and Maintenance (OAM) procedures, specified in ITU-T Recommendation I.610, are required, e.g. generation of VP-AIS/VP-RDI (Alarm Indication Signal/Remote Defect Indication). The cell header format and encoding and the pre-assigned headers for use by the ATM layer used at the VB5 reference point use the Network to Network Interface (NNI) option of ITU-T Recommendation I.361, i.e. 12 bits of VP address space is available. It is assumed that if the Generic Flow Control (GFC) field at the UNI is used, it will be handled within the AN itself, and not transported over the SNI.

General application of the VB5 interface

The general application of the VB5 interface is shown in Figure 9.4. As illustrated in this figure, access networks and service nodes can have multiple VB5 interfaces. These can be interconnected via a transmission network whose management is entirely separate from either Q3(AN) or

9.2 THE VB5 INTERFACE

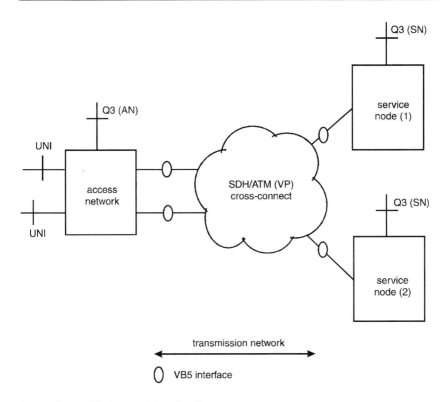

Figure 9.4 VB5 General Application

Q3(SN). The transmission network can be anything from station cabling to complex SDH or ATM networks. Since the VB5 interface is specified at the VP layer, it is possible to have multiple VB5 interfaces on the same physical link. The converse is also possible; a single VB5 interface may comprise multiple physical links.

The VB5 interface is specified to allow single physical UNIs to have access to multiple service nodes. In fact, the support of shared UNIs is a requirement from G.902 and meets a European Commission ONP requirement that a user, served by an access network, should have access to multiple service nodes so that the supply of service need not be a monopoly. UNIs can also consist of multiple physical links.

Since the VB5 interface is only specified at the VP layer some additional terms are required to identify specific groupings of VPs which comprise either a logical or physical interface (see Table 9.1). Identifiers related to these terms are used with certain VB5 specific protocol messages, as well as the management model.

The relationship between the additional VB5 specific logical and physical ports is shown in the VB5 functional model in Figure 9.5.

To identify VPs in multi-link physical interfaces at both UNI and SNI, an

Table 9.1 VB5 Specific Logical and Physical Ports

Logical/Physical port		Definition
LUP	Logical User Port	The set of VPs at the UNI or at a Virtual User Port (VUP) associated with one single VB5 reference point
PUP	Physical User Port	The physical layer functions related to a single transmission convergence function at the UNI
LSP	Logical Service Port	The set of Virtual Paths (VPs) at one VB5 reference point (i.e. associated with one and only one service node) carried on one or several transmission convergence functions.
PSP	Physical Service Port	The physical layer functions related to a single transmission convergence function at the VB5 interface

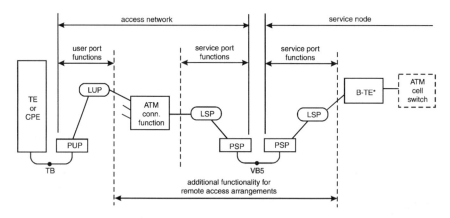

Figure 9.5 VB5 Functional Model

extended VP addressing scheme is used which introduces a Virtual Path Connection Identifier (VPCI). This has a range of 16 bits. When used at the NNI it allows up to 16 physical interfaces per SNI (4 bits/interface and 12 bits/VPC). When used at the UNI it potentially allows 256 physical interfaces (8 bits/interface and 8 bits/VPC).

The number of VPCIs at both the SNI and UNI can be extended, if necessary, by reducing the number of bits allocated to the VP range, and reusing those released to increase those available to identify the physical port. This requires co-ordinated provisioning in both AN and SN to ensure both entities are using the same addressing scheme.

The use of VPCIs is derived from the Digital Subscriber Signalling

9.2 THE VB5 INTERFACE

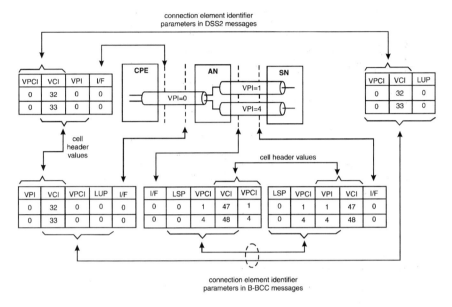

Figure 9.6 VB5 Connection Element Identification

system No. 2 (DSS2). At the UNI they are the same as those used in DSS2 signalling messages.

The relationship between the various connection element identifiers used in VB5, including the VPCI, is illustrated in Figure 9.6.

Narrowband support

VB5 interfaces provide for the integration of broadband and narrowband SNIs. This is achieved using a conventional ATM circuit emulation technique to encapsulate the narrowband V5 or V3 framed 2 048 kb/s serial data.

Non-B-ISDN accesses

In addition to the standardised broadband ATM interfaces, and legacy narrowband interfaces there are a number of other user interfaces that different network operators may require. These are supported by VB5 using a Virtual User Port (VUP) concept. Essentially, at some logical point in the AN a UNI is assumed but not realised physically. This part of the AN is visible to the VB5 specific management model and handled in a generic way. Additional adaptation is performed in the AN beyond the

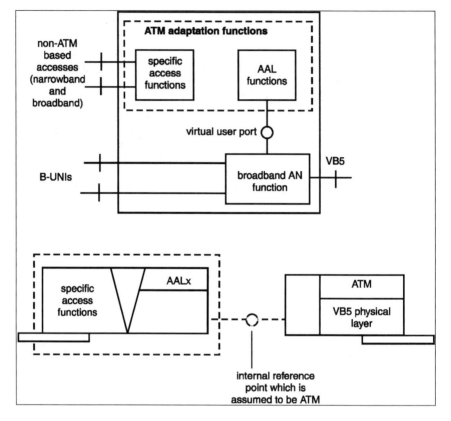

Figure 9.7 Virtual User Port

VUP, but this is not considered part of the VB5 management model (see Figure 9.7).

This technique allows the AN to support any user interface that can be adapted to ATM, over a VB5 reference point.

Connection types

A number of different VP and Virtual Circuit (VC) connection types are defined in VB5 specifications to identify the types of connections that can be supported via a VB5 reference point to an AN. These are classified by the termination point, point or multi-point type, and whether they are semi-permanent or switched.

VB5.1 can only support semi-permanent (VP and VC) connections directly. VB5.1 supports switched VCs as far as users are concerned, which are carried in semi-permanent VPs over the VB5 reference point. In this scenario a VP at the UNI maps directly to a VP at the SNI.

The AN can also provide unidirectional (SN to AN) point-to-multipoint connections via the VB5 reference point, i.e. the AN providing the cell replication function. The inverse multipoint-to-point connections or bi-directional multipoint connections cannot be supported over the VB5 interface directly. These connection types can be provided to end users via point-to-point links over the VB5 reference point. It is the SN which must provide any combining function.

VB5.2 with its B-BCC protocol can support switched VCs directly (either bidirectional point-to-point or unidirectional point-to-multipoint), in addition to all of the connection types supported by the VB5.1 interface.

9.3 THE REAL TIME MANAGEMENT CO-ORDINATION (RTMC) PROTOCOL

The main purpose of the RTMC protocol is to communicate real-time management information from the AN to the SN about the availability of AN resources which have an impact on the availability of service to the end users. In practice, the change in status of only two types of AN resource is communicated, VPCs and the LSP. Other AN resources, such as PUPs, for example, can be decomposed into sets of VPCs. Mechanisms are provided for the AN to indicate locked VPCs due to faults or administrative procedures. In the case of administratively locked VPCs, there are two modes available; one stopping all connections, the other allowing only test connections.

A further shut-down procedure allows the VPCs to be locked gracefully, by waiting until all switched connections on a given set of VPCs are released, before locking the resource.

In addition to VPC status procedures, there are some further housekeeping functions provided:

- VPCI consistency check: ensures alignment of AN and SN configurations.
- LSP ID check: confirms matching interface identities either side of the SNI.
- RTMC Reset: re-initialise the RTMC protocol entities.

ATM adaptation layer for the RTMC protocol

The RTMC protocol is carried in a VC in a reserved VP Connection (VPC) between the AN and SN. The VP Identifier (VPI) and VC Identifier (VCI) values of this channel are not fixed, and must be provisioned. The VPI range is unrestricted, but the VCI cannot use any of the reserved values between 0 and 31. It was originally intended to have reserved values for the VB5 protocols. Some manufacturers were concerned about accessibility of VCIs less than 32 using current hardware. It was also proving difficult to convince the relevant ETSI technical committee of the need to modify an existing standard to provide these values from a reserved set, so they were made provisionable in the VB5.1 standard.

ATM adaptation is performed using the Signalling AAL (SAAL), which comprises AAL5, SSCOP and SSCF. This is a specific ATM adaptation layer optimised for signalling applications also used in DSS2. Not all of the features available from these standards are required for the relatively simple RTMC protocol.

Port block/unblock and shut down procedures

The block/unblock procedure is used by the AN to inform the SN about the service relevance of the administrative actions taken in the AN, and to indicate service relevant fault conditions occurring in the AN.

This procedure is used in addition to the use of F4 OAM flows (e.g. VP-AIS) as described in ITU-T Rec. I.610. There are a number of reasons why it was not possible to use only the facilities provided in I.610, the principle one being some ambiguity as to the application of VP-AIS when resources are locked. The responsible standards group was unable to address these issues in the necessary timescales. There are other requirements related to resource locking which cannot be met by I.610 procedures which the RTMC provides, i.e.:

- Graceful 'shutting down' of resources.
- VP link status for user VPs terminated in the AN.
- Status of unidirectional and point-to-multipoint VPs.

The type of lock is indicated by reason codes in a block message. The following reason codes are used:

- Fully administratively locked (admFull): the resource is not available due to administrative actions in the AN.
- Partially administratively locked (admPart): the resource is not available for normal traffic due to administrative reasons in the AN, but available for test calls.
- Fault/Error condition (Err): the resource is not available due to a fault condition in the AN.

LSP ID check

The purpose of the LSP ID check is to confirm that the AN and SN RTMC protocol entities are communicating with the correct peer entity. The actual LSP identity is a provisioned value selected such that it is unique in any potential AN / SN combination.

The procedure is invoked automatically when the interface starts up, and can be invoked from either Q3(AN) or Q3(SN) when the interface is operational.

VPCI consistency check

The VPCI consistency check is a procedure inherited from the core network Broadband ISDN Signalling User Part (B-ISUP) of Signalling System No. 7. The purpose of the VPCI consistency check is to ensure that VPCIs provisioned in the AN and SN are consistent. Any differences between the VB5 procedure and B-ISUP are due partly to the asymmetrical nature of the SNI, and the fact that in VB5 the maintenance data is transported by a dedicated management protocol, the RTMC.

The procedure makes use of OAM loopback cells to check the physical path and the RTMC protocol to collect information in the SN about the reception of these cells in the AN. The message sequence chart in Figure 9.8 illustrates the process for a single VPCI check.

The process can only be started by the SN. The SN begins by sending a message to the AN requesting that the OAM monitor be attached to the VPC. Once the AN has acknowledged the request, the SN sends some loopback cells to the AN. When these are received back in the SN, it requests the AN for the result. The AN removes the loopback monitor and replies. If the AN has seen some loopback cells, it indicates success in the reply. The SN then moves to the next VPC.

The application of the VPCI consistency check is limited to VPCs, which terminate in the Access Network. The current specification does not make use of OAM segments, so it is possible only to use VPC end points. If end points exist outside the AN, for the test to be performed, the endpoint would have to be available (i.e. in the case of CPE, powered-up) and have a method of communicating the reception of loopback cells on specific VPs, which is not practical with existing standards.

The alternative, using VP OAM segments, was considered. The principle limitation is that since segments cannot overlap, once a segment is defined across the SNI for the VPCI consistency check, other segments crossing the SNI cannot then be used (i.e. from points inside the AN to the SN or beyond).

180 THE VB5 INTERFACE

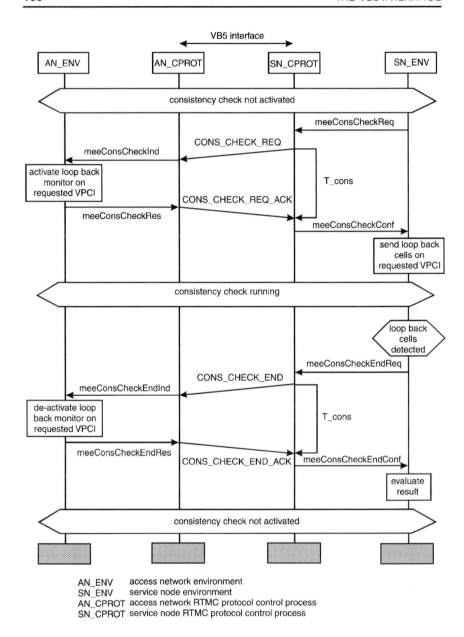

Figure 9.8 VPCI Consistency Check—Message Sequence Chart

Reset

The only resources which can be reset are VPCs. These can be reset individually or all the VPCs in the LSP can be reset with one message. The purpose of the reset is resynchronise the VB5 specific finite state machines relating to the RTMC protocol on both sides of the interface. To speed resumption of service, the initial state of the VPCs is set to unblocked. Any VPCs which were actually blocked in the AN are subsequently reblocked in the SN by messages from the AN.

9.4 THE BROADBAND BEARER CONNECTION CONTROL (B-BCC) PROTOCOL

The principle difference between VB5.1 and VB5.2 is the provision of a B-BCC protocol in addition to the RTMC protocol used with the VB5.1 interface. Any other differences between VB5.1 and VB5.2, such as the start-up procedure, occur only as a consequence of this additional protocol.

At the time of writing the VB5.2 interface is still in the process of being defined in both ETSI SPS3 and the ITU-T Study Group 13. It is therefore quite possible that some detailed aspects of the VB5.2 interface described here will have changed by the time either the ETSI standard or ITU-T recommendation is published.

The B-BCC allows the SN to control cross-connections in the AN, and hence enables call-by-call concentration in the AN. Such concentration allows a larger number of UNIs to be supported by an AN for a given amount of SNI bandwidth. The general principle is in line with ITU-T Rec. G.902. A similar technique is used in the narrowband V5.2 interface.

There are a number of advantages that concentration in the AN provides using this technique:

- The AN only has to handle one relatively simple signalling protocol, since only bearer channel control is required.
- A minimal number of signalling interfaces are required in the AN, compared to the large number required if user signalling was terminated in AN.
- No 'user profile' database is required in the AN.
- No routing information is required in the AN.
- The AN remains independent of any specific user/network signalling protocols.
- Utilisation of the SN physical interfaces is optimised, since traffic is concentrated by the AN.

The B-BCC protocol uses the same type of SAAL as the RTMC protocol but does not share the same channel. A common VB5 protocol discriminator is used with both RTMC and B-BCC protocols. The message codepoints of each protocol are therefore defined so as not to overlap. Figure 9.9 shows the general functional architecture of the AN from a protocol perspective.

B-BCC messages

The VB5.2 B-BCC protocol message set is summarised in Table 9.2. These messages are unique to VB5, but make use of information elements from DSS2, whenever practical, to reduce any interworking overhead. The initial concept of the B-BCC anticipated principally only the Allocation/De-allocation type messages. However, to interwork with DSS2 and B-ISUP, it is necessary to be able to modify connection parameters during call set-up and data transfer phases, hence the Update and Modify messages. Other

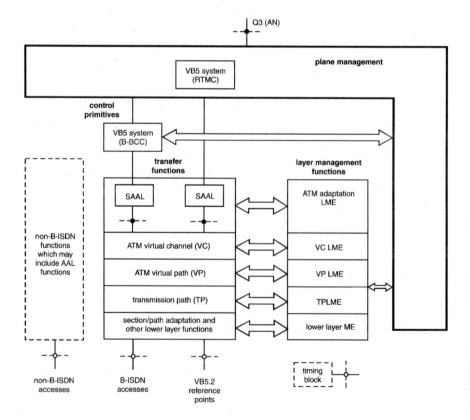

Figure 9.9 AN General Functional Architecture

9.4 BROADBAND BEARER CONNECTION CONTROL (B-BCC)

Table 9.2 B-BCC Protocol Messages

Message	Direction	Function
ALLOCATION	SN → AN	Allocate the AN part of a connection. Includes ATM transfer capabilities and QoS information
ALLOCATION COMPLETE	SN ← AN	Indicates successful AN connection establishment
ALLOCATION REJECT	SN ← AN	Indicates unsuccessful AN connection establishment
AUDIT	SN → AN	Check an AN connection
AUDIT COMPLETE	SN ← AN	Return data on an AN connection
DE-ALLOCATION	SN → AN	De-allocate (release) an AN connection
DE-ALLOCATION COMPLETE	SN ← AN	Result of De-allocation
MODIFY	SN → AN	Modify AN connection parameters
MODIFY COMPLETE	SN ← AN	Acknowledge successful modify request
MODIFY REJECT	SN ← AN	Indicate unsuccessful modify request
UPDATE	SN → AN	Update AN connection parameters
UPDATE COMPLETE	SN ← AN	Acknowledge successful update request
UPDATE REJECT	SN ← AN	Indicate unsuccessful update request

messages will be defined to handle interface start-up and resets, for example. At the time of writing, the details of these were still the subject of discussions within ETSI SPS3 and the ITU-T Study Group 13.

The B-BCC protocol is highly asymmetrical in that virtually all control rests with the SN. None of the connection related transactions can be initiated by the AN.

Connection and transaction identifiers

Messages are identified by unique transaction identifiers. These have relevance only for a single request/acknowledge transaction. This is the same as the mechanism used with the RTMC protocol. To avoid having to identify both ends of a connection with the AN end point resource identifiers (used originally to allocate the connection), a connection identifier is introduced. This is created by the SN, has the lifetime of a connection, and is first used in an Allocation message. It is subsequently released back into the pool of available references with the De-allocation message. All transactions after initial Allocation can use this method of referring to existing connections.

Connection establishment in the AN

The SN begins establishing a connection in the AN using the Allocation message. Table 9.3 lists all of the possible information elements which can be used with an Allocation message. The majority of these information elements are inherited from DSS2.

In general, it is not possible to determine the final connection parameters immediately when the Allocation message is sent, since other nodes in the connection may change the parameters as part of the negotiation during call set-up. An Update message is therefore provided, which allows the SN to revise the AN connection parameters, once it has received the updated parameters negotiated with other nodes.

It is possible, with local calls, for example, where no other nodes are involved, that an update message would not be required. A message sequence chart in Figure 9.10 shows the use of the B-BCC messages in connection establishment, as part of the description of interworking with DSS2 and B-ISUP.

Table 9.3 ALLOCATION message information elements (Note: Information elements shown as type 'M' are mandatory, whereas type 'O' are optional)

Information Element	Type	Comments
Connection reference number	M	Used over the lifetime of a connection
ATM traffic descriptor	M	
Broadband bearer capability	M	
OAM traffic descriptor	O	
Quality of service descriptor	O	
User port connection identifier	M	VB5 specific addressing
Service port connection identifier	M	VB5 specific addressing
ABR set-up parameters	M	
End-to-end transit delay	O	For further study
Cell delay variation tolerance	O	For further study
Alternative ATM traffic descriptor	O	
Minimum acceptable ATM traffic descriptor	O	

9.4 BROADBAND BEARER CONNECTION CONTROL (B-BCC)

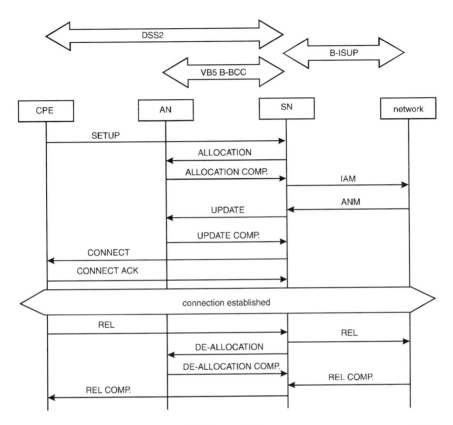

Figure 9.10 AN connection establishment/release—message sequence chart

Connection Admission Control

Connection Admission Control (CAC) for the AN part of the connection is performed principally by the AN. The SN has some information, such as the availability of certain AN resources via the RTMC protocol, and so could decide not to invoke the B-BCC in some circumstances. The normal case, though, would be for the SN to request a connection and the AN to acknowledge or reject the request, having performed its own CAC.

A related issue is the allocation of VPI and VCI values. If the AN is performing its own resource management and Connection Admission Control, there is some benefit in allowing the AN to allocate VPI and VCI values. To achieve a similar level of optimisation (if the SN allocated the resources) would require the SN to have detailed AN specific data. This, in turn, would have to be defined in a generic way to maintain an open interface. By keeping the data in the AN, manufacturers can optimise AN equipment without compromising the interoperability of the interface. However, a consequence is that the SN must be able to indicate more that just a single VPI/VCI value for connections in those situations where the AN can be offered a choice.

Interworking with DSS2 and B-ISUP

VB5 is not intended only to work with DSS2 and B-ISUP, however it is anticipated that the most significant application of VB5.2 will be in networks which use these user and network signalling protocols. It therefore makes sense to design the B-BCC such that interworking with DSS2 and B-ISUP is not unnecessarily complicated. The general design of the B-BCC is biased towards DSS2 rather than B-ISUP since architecturally the VB5 interface in the SN is more closely related to directly connected UNI accesses.

Figure 9.10 shows an outline connection establishment and release message sequence chart illustrating the use of Allocation, Update and De-allocation messages and their relationship with DSS2 and B-ISUP.

9.5 VB5 MANAGEMENT

The management (Q3) interfaces associated with access networks and service nodes which are linked with VB5 traffic interfaces need to support the configuration and fault and performance management functions associated with the VB5 interfaces. These Q3 interfaces are specified in terms of management models which are extensions of the ITU-T management model for ATM. In a management implementation, managed objects which conform to the classes defined for the model, can be manipulated to perform the various management functions. The management models for the AN and the SN differ slightly, and we start by looking at the model for the AN (see Figure 9.11).

Transmission Convergence (TC) functions both at UNIs and at VB5 interfaces are modelled by TC adaptor objects and connected VPs are modelled by contained VP Connection Termination Point (CTP) objects. End points of VP connections in the access network are represented by VP Trail Termination Point (TTP) objects, and connected VCs within these end-points are represented by contained VC CTP objects. Any end-points of VC connections are represented by VC TTP objects.

Logical user port objects group VPs at user ports according to whether or not the VPs terminate in the AN. It they do not terminate in the access network then the intermediate points of VP connections at a user port are identified and labelled with VPCI values (pointer 2). If they do terminate,

9.5 VB5 MANAGEMENT

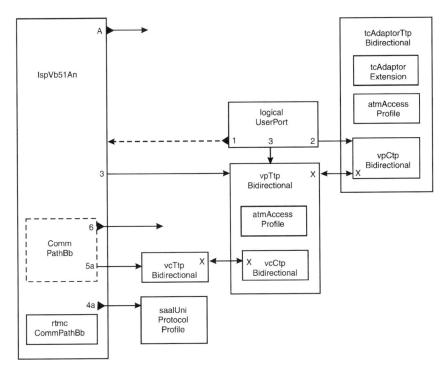

Figure 9.11 Object Relationships in the Access Network

then the end points are identified and allocated VPCI values (pointer 3 of logicalUserPort class).

When VP connections at the VB5 interface terminate in the AN, then the logical service port objects group their own VP connection end-points and label these with VPCI values (pointer 3 of lspVb51An class). Logical service port objects also contain an RTMC communications path object (a communications path subclass) to identify the end point of the VC connection for the RTMC protocol (pointer 5a) and its signalling adaptation profile (pointer 4a).

The modelling in the SN differs slightly from that in the AN because it takes account of user signalling (see Figure 9.12). If user VPs are only cross connected in the SN, then they are allocated VPCI labels by UNI access objects (pointer B) with a VP subclass used to allow the modelling of remote blocking. If they are terminated then it is the end points which are labelled (pointer 8), and again a subclass is used to allow modelling of remote blocking. This VPCI labelling in the SN assumes that the VP connections are directly related to user accesses. If VP connections are not directly related to a user access (for instance, if there is VC cross connection in the AN), then their terminations are identified and labelled by LSP objects (pointer 3 of lspVb51Sn class).

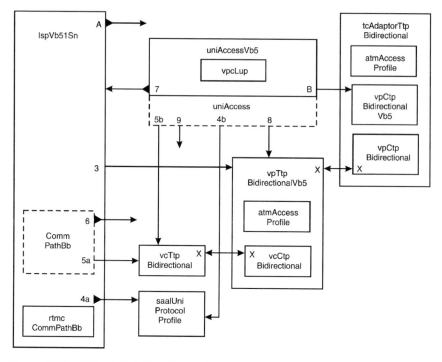

Figure 9.12 Object Relationships in the Service Node

The end points of VC connections for user signalling are identified, and profiles are assigned by UNI access objects in a similar way as for the RTMC protocol (pointers 5b and 4b). UNI access objects for indirect accesses also identify their logical service port (pointer 7), and contain VPC LUP objects to represent the remote blocking of any VP connections at the remote user port which are terminated in the AN.

The objects in the AN and SN models have data attributes (which the operations system can read and often write to), actions (which are requests sent by the operations system and which may trigger replies), and notifications (which can be sent to the operations system autonomously by the AN or SN). However, it is not appropriate to describe these here, and the interested reader is referred to the VB5 management specifications for further details of these, and how they are related to the information flows across VB5 interfaces.

9.6 SUMMARY

The availability of a standardised SNI, which the VB5 interface provides, enables open procurement of broadband access networks. This should

9.6 STANDARDS

advance their development and deployment, with a consequent impact on delivery of broadband communications. This is particularly significant for the more cost sensitive market sectors, such as residential and small business.

Progress in developing VB5 standards is now well advanced. In ETSI the VB5.1 interface specification will finish it's public enquiry phase in October 1997. In the ITU-T work originally done by ETSI on VB5.1 has been incorporated into the ITU-T Rec. G.VB51, which was about to be submitted for approval at the time of writing.

Work in ETSI and the ITU-T on the VB5.2 interface is running concurrently with good co-operation. It was anticipated that a stable draft, ready for submission into the approvals process, would be available by the end of 1997.

Management specifications will follow shortly after their respective interface specifications.

Other standards bodies have included references to VB5 in their documents, rather than attempting to derive their own standard for the SNI, in particular:

- The Digital Audio Visual Council (DAVIC) A4 reference point maps directly onto the ETSI/ITU SNI and uses VB5 protocols.
- The Full Services Access Network (FSAN) group (ATM PON 1997) have specified VB5 as the interface from the AN to the core network.
- The ATM Forum Residential Broadband group have adopted VB5 as one option for their Access Network Interface (ATM Forum 1997).

REFERENCES

Asynchronous Transfer Mode Passive Optical Network (ATM PON) Specification Issue 1, (2/7/97) Jan De Groote (Alcatel), Yoichi Maeda (NTT), Takatoshi Minami (Fujitsu), Alan Quayle (BT).

ATM Forum Technical Committee, *Residential Broadband Working Group: Baseline Text Document*, BTD-RBB-001.02, Chicago, Illinois, April 1997.

STANDARDS

G.902 *Framework Recommendation on functional access networks—Architecture and functions, access types, management and service node aspects* (11/95).

G.964 *V Interfaces at the digital local exchange (LE)—V5.1 interface (based on 2 048 kbit/s) for the support of access networks (AN)* (07/94).

G.965 *V Interfaces at the digital local exchange (LE)—V5.2 interface (based on 2 048 kbit/s) for the support of access networks (AN)* (3/95).

G.967.1 *VB5.1 reference point specification* (6/98).
G.967.2 *VB5.2 reference point specification* (2/99).
I.414 *Overview of Recommendations on layer 1 for ISDN and B-ISDN customer accesses* (09/97).
I.751 *Asynchronous Transfer Mode management of the network element view* (03/96).
EN 301-005 *V interfaces at the digital Service Node (SN); Interfaces at the VB5.1 reference point for the support of broadband or combined narrowband and broadband Access Networks (ANs)*, ETSI (1995).
EN 301-217 *V interfaces at the digital Service Node (SN); Interfaces at the VB5.2 reference point for the support of broadband or combined narrowband and broadband Access Networks (ANs)*, ETSI (1999).
Draft DEN/SPS-03049-1 Signalling Protocols and Switching (SPS) Management Interfaces associated with the VB5.1 Reference Point Part 1: Interface Specifications.
Draft DEN/SPS-03045-1 Signalling Protocols and Switching (SPS) Management Interfaces associated with the VB5.2 Reference Point Part 1: Interface Specifications.
ETS 300 324 *Signalling Protocols and Switching (SPS); V interfaces at the digital Local Exchange (LE); V5.1 interface for the support of Access Network (AN)*, ETSI (1994).
ETS 300 347 *Signalling Protocols and Switching (SPS); V interfaces at the digital Local Exchange (LE); V5.2 interface for the support of Access Network (AN)*, ETSI (1991).
ETR 257 *Signalling Protocols and Switching (SPS); V interfaces at the digital Service Node (SN); Identification of the applicability of existing protocol specifications for a VB5 reference point in an access arrangement with Access Networks*, ETSI (1995).
DAVIC Specification Part 02, Revision 3.0, *System Reference Models and Scenarios*, Technical Report, DAVIC (1997).

QUESTIONS

1. What is the difference between an SNI and a UNI?
2. Why do the VB5 interface specifications *not* specify the physical layer of the SNI?
3. What does the abbreviation RTMC stand for and why is it so important?
4. Does the VB5 interfaces support only ITU-T terminals?
5. Is there likely to be a need for another VB5 interface, such as a VB5.3?

10
The use of Session Control in DAVIC to provide Interactive Multimedia services

Richard Miles, Paul Reece and Laurent Boon

10.1 INTRODUCTION TO DAVIC

The Digital Audio-Visual Council (DAVIC) is a non-profit making association registered in Geneva. Its primary objective is to advance the success of emerging digital audio-visual applications by the timely production of specifications. Such specifications must be internationally agreed and based upon open interfaces and protocols that maximise interoperability across different countries and implementations. Initially, the applications addressed by DAVIC include the broadcast and interactive type in which there is a significant high-quality digital audio video component, such as Video-On-Demand (VOD) or home shopping.

The vision of the broadband future, where on-demand services and information is available at home, in the office and on the move is now well established. Such a vision relies upon the integration of many networks and associated services, signalling and control. The current lifecycle of multimedia applications and standards can be very short. Technical superiority does not necessarily guarantee universal adoption by all manufacturers, nor international standardisation. With this in mind, choosing the right building blocks for the broadband future vision is a risky business. Development of such a network is also extremely costly, which leads participants to seek collaboration in order to maximise interoperability and economies of scale. The goals of DAVIC are to identify, select, augment, develop and obtain the endorsement by formal standards bodies of specifications of interfaces, protocols and architectures of digital audio-visual applications and services. It was these goals that brought many of the key players in the fields of entertainment, information technology and telecommunications to open international collaboration to produce the DAVIC specifications.

Membership

Membership of DAVIC is open to any corporation or individual firm, partnership, governmental body or international organisation. By December 1995, DAVIC had membership of over 200 corporations representing more than 20 countries from all over the world and virtually all business communities with a stake in the emerging field of digital audio-visual applications and services.

The DAVIC 1.0 specification

The DAVIC 1.0 specification has been developed by participating DAVIC members on the basis of submissions from both members and non-members in response to two Calls for Proposals which were issued respectively in October 1994 and March 1995. The public specification was finalised and published in January 1996.

The DAVIC 1.0 specifications have been developed within a multi-industry environment, and must therefore be flexible enough to accommodate many distinct implementations based on individual requirements. The approach taken by DAVIC was to specify a set of tools which lend themselves to flexible implementation, rather than defining systems which tend to be application-specific. Compliance with this principal leads to substantial benefits in terms of interoperability and availability of technology as the tool is applicable to a wider field of the technology. Examples of DAVIC tools include Digital Storage Media—Command and Control (DSM-CC) based session control and Motion Pictures Expert Group (MPEG) encoded video streams.

10.2 THE DAVIC SYSTEM REFERENCE MODEL

Figure 10.1 shows a general representation of the system addressed by the DAVIC specifications. It comprises five entities: the Content Provider

10.2 THE DAVIC SYSTEM REFERENCE MODEL

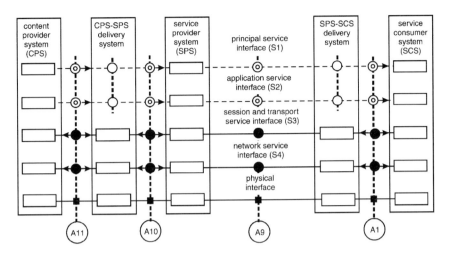

Figure 10.1 1—DAVIC System Reference Model

System (CPS); the Service Provider System (SPS) and the Service Consumer System (SCS); and the three entities being connected by two Delivery Systems (DS), the first connecting the content provider to the service provider, and the second the service provider to the service consumer. In principal, DAVIC specifications can address any part of the general system below. The general DAVIC reference model is multi-dimensional in the sense that multiple SPSs may connect through many delivery systems to many SCSs. Each entity is composed of sub-system objects, which may or may not exist in a specific implementation of the architecture. The links between entities represent logical information flow. Similar flows may also exist between sub-systems within an entity. The combination of entities, flows and sub-system objects can be used as a reference against which tools can be adopted and positioned within a particular implementation.

Reference points

Within the DAVIC system reference model, reference points define sets of logical interfaces through which peer objects transfer information. Within Figure 10.1, the following reference points are shown: A11 linking CPS and DS, A10 linking DS and SPS, A9 linking SPS and DS, A1 linking DS and SCS.

System entities

The main DAVIC 1.0 system entities are the Service Provider System (SPS), Delivery System (DS) and Service Consumer System (SCS).

Service Provider System

The service provider system is the primary means of accessing and manipulating the content that exists within, and flows through, a DAVIC system. The typical DAVIC 1.0 SPS could be a provider of a Video-on-Demand application.

Delivery System

Within DAVIC the term *delivery system* has a very broad meaning. It refers to virtually any means to deliver information from one entity to another entity in order to support services defined by DAVIC. Entities involved are, among others, content provider systems, service provider systems and service consumer systems. Figure 10.1 depicts the general DAVIC model, showing how the various entities are connected to the delivery system and through which reference points. Service providers can download content from the content providers and offer this content to end-consumers.

For DAVIC 1.0 only the delivery system between reference point A1 and A9 is considered. This delivery system is essential to bring the services offered by the service provider to the end-consumers. In future versions of the DAVIC specifications, the delivery system between the reference points A10 and A11 will be specified to ensure easy transfer of content between the content providers and the service providers.

Types of networked delivery systems range from telecommunication networks, such as co-axial cable or Asynchronous Transfer Mode (ATM) networks, to satellite broadcast systems, such as direct broadcast satellite and terrestrial broadcast systems.

Service Consumer System

The SCS entity represents the combination of Network Interface Unit (NIU), Set-Top Unit (STU) and human or machine (i.e. video recorder) service consumer. The SCS contains functionality to permit the human or machine service consumer to interact with the service provider system and exert session control over the delivery system.

Sub-system objects

Internal to each of the main functional entities are additional sub-system objects that represent functionality within a particular element. These objects may be considered as functionality required to support the information flows between system entities. The location of sub-system objects within the main DAVIC functional entities is shown in Figure 10.2.

Service provider system objects

The service provider system comprises four core service elements: content service, application service, service gateway and stream service. The content service element is responsible for the management of content, including loading and unloading content, between the server and the content provider as well as among the service elements. The application service element is the collection of DSM-CC interface definitions on which all applications build. The service gateway element is usually the first element the SCS observes, and as such acts a broker to the applications available within the SPS by providing the following functions:

- Organising the service provider domain where services register (or install), make their existence known and de-register when they are

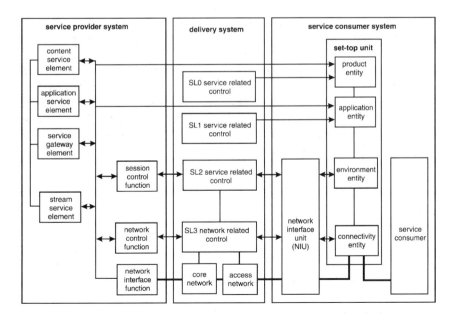

Figure 10.2 DAVIC Functional Entities

decommissioned. The service Gateway adopts a name context interface, which provides a navigable name graph (analogous with a directory structure) to organise the service domain.

- A means for the client to discover the existence of a service (i.e. a stream object for VOD). This is accomplished by browsing the service domain at the S2 level. The objective is often to locate a specific service, known through the name bound to the name graph, i.e. an object of kind 'stream' and id of 'batman'.

- A means by which a client activates and deactivates an instance of a service. Included in this process is session establishment and management through requests to the session control function (Session object) leading to appropriate S3 and S4 messages.

The Stream service element is a repository and source for streams. It provides the S2 interface through which the client device controls the media stream (play, pause, etc.).

Delivery system objects

The SL0, SL1 and SL2 service related control objects should be considered together as providing a brokering interface to the capabilities of the delivery system. One such use of these objects could be in providing DAVIC Level 1 gateway functionality, whereby the end user can browse available DAVIC service provider systems in a manner analogous with a Yellow Pages or the Yahoo service. In this particular case, the name graph would represent a navigable structure of service gateways with which a session may be established or transferred

The network related control entity provides control functions for network configuration, connection establishment and termination, and information routing. Within the delivery system, different sets of internal interfaces (not shown) may be used to encapsulate different core network and access network elements to provide interoperable, or compatible, delivery system services at the A1 and A9 reference points.

Service consumer system objects

The product entity is responsible for accepting content information and presenting the information to the end user of the SCS. Typically, this role would be the decoding of Multimedia and Hypermedia Expert Group (MHEG) objects and MPEG content information and presentation to a television set or other end user device.

The application entity represents and supports the end user application running on the SCS. The DAVIC 1.1 specification has chosen a Java virtual machine to provide this role.

10.2 THE DAVIC SYSTEM REFERENCE MODEL

The environment entity is responsible for establishing and terminating the environment, or session, in which an application will operate. This session management may make use of either the Connectivity entity to perform connection control or a proxy signalling capability.

Information flows

The DAVIC system reference model, shown in Figure 10.1, contains four distinct information flows, which are labelled S1 to S4. Two types of flow exist: content information and control information. Figure 10.3 shows these information flows.

The S1 flow is unidirectional, between content source (SPS) and content sink (SCS), and used for the transportation of content information, such as MPEG encoded video. The flow is in the user plane and, as such, is transparent to any intermediate control objects through which it passes.

The S2 flow is bidirectional between end-to-end application control functionality. This is not an information flow in the traditional telecommunications sense, but a Remote Procedure Call (RPC) plane based on an underlying Common object Request Broker Architecture (CORBA). As such, information flows are of the form of remote procedure calls upon objects with defined methods and interfaces, rather than the exchange of sequences of messages between state machines.

The S3 flow is a bidirectional control information flow between end

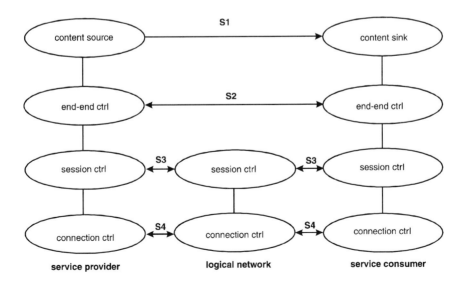

Figure 10.3 DAVIC Information Flows

users, such as SPS and SCS, and the delivery system. Control information is exchanged in the form of sequences of messages to exert session control over the network resource in response to method invocations at the S2 level.

The S4 flow is a bidirectional control information flow between end users, such as SPS and SCS, and the delivery system. Control information is exchanged in the form of sequences of messages to exert connection control over the network resource in response to session level requests at the S3 level.

10.3 DAVIC PROTOCOLS

MPEG-2 (S1)

The tool chosen by DAVIC to provide this S1 information flow is the ISO MPEG encoding standard for video material. In particular, DAVIC specify the use of MPEG-2 transport streams, which can comprise a number of individual MPEG video streams as well as private data (such as executable code).

The protocol stack defined by DAVIC for the delivery of S1 content information within an end-to-end ATM environment is shown in Figure 10.4.

DSM-CC user-to-user and session objects (S2)

The S2 level is underpinned by a CORBA layer, on which DSM-CC User to User (U-U) objects within the service provider system may be manipulated via remote procedure calls. This section aims to give the reader a summary understanding of the operation of DSM-CC U-U, concentrating on its relationship with the session layer. To give an overview of the DSM-CC U-U environment, a truncated version of the DSM-CC system reference model is presented in Figure 10.5:

The session managers are not at the lowest layer of the system; the

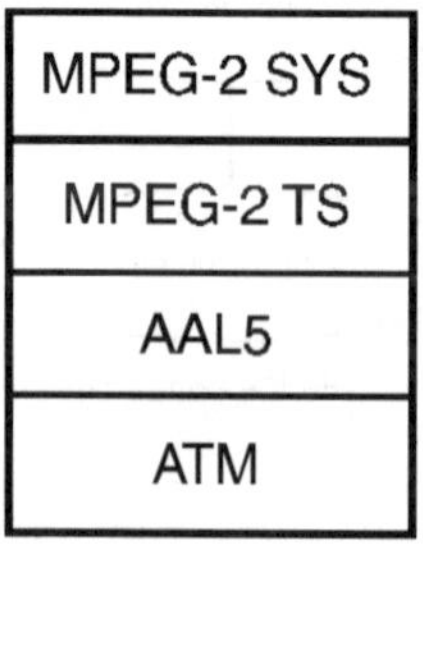

| MPEG-2 SYS |
| MPEG-2 TS |
| AAL5 |
| ATM |

Figure 10.4 S1 Protocol Stack

10.3 DAVIC PROTOCOLS

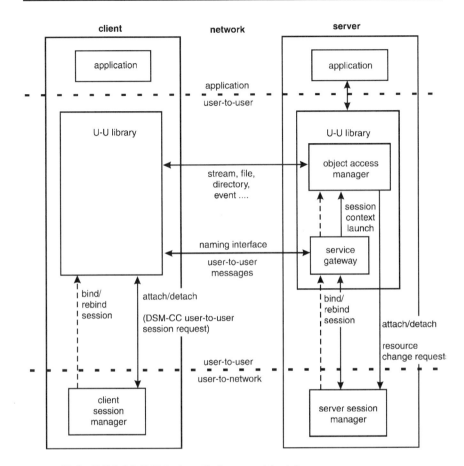

Figure 10.5 DSM-CC U-U System Reference Model

original reference model has been truncated. Below the server session managers there are the ATM session, resource and control layers, but these are outside the scope of this section.

In summary, the DSM-CC U-U protocol is used to navigate to and then control a piece of content on a local or remote server.

Not all of the arrows shown in Figure 10.5 represent U-U interfaces. According to the DSM-CC specification, Table 10.1 represents those high level interfaces implemented by the U-U protocol.

Each of the interfaces shown in Table 10.1 can be broken down into smaller component interfaces. The following sub-sections describe these component interfaces. The total grouping of component interfaces is called the *DSM-CC U-U library*; this can be seen in Figure 10.5. For simplicity, only the core DSM-CC U-U libraries will be considered. Finally, the session interface will be examined in greater detail, showing the interaction of DSM-CC U-U and the session managers (client and server).

Table 10.1 DSM-CC U-U High Level Interfaces

Peer 1	Peer 2
Client U-U Library	Server service gateway
Client U-U Library	Server object Access
Client Application	Client U-U Library
Server Application	Server U-U Libraries

Core U-U library

To explain these interfaces fully would be a detailed exercise, and would virtually involve a full replication of the DSM-CC specification. A high level understanding, though, is likely to help the reader comprehend the interaction between the DSM-CC U-U layer and the session managers (client and server).

- *Base interface*
  The base interface provides common operations for the deletion of DSM-CC object references and objects. All other core interfaces except Access and Directory interfaces inherit from this interface. Directory objects have their own closure and deletion procedures, for example:
  — a client has finished using an object reference to access a remote service, the client uses the base interface to close it;
  — a service object on a server is no longer required. The management of the server uses the base interface to delete it.

- *Access interface*
  The access interface provides common description and access control attributes. These include size, version, date, lock and permission attributes, for example:
  — a file object on a server contains important reference information on the services offered by that service provider. It is publicly available but should only be amended by the management of the server. The access interface is therefore used to place access restrictions on that object.

- *Stream primitives interface*
  Stream primitives are used to emulate VCR-like controls for manipulating MPEG continuous media streams, for example:
  — a client has navigated his/her way to the film *Batman*. To view the film in a controllable fashion, (s)he must be presented with a recognisable interface, that is, recognisable in terms of the clients application being able to make use of it. The interface can then be used to take control of an MPEG stream with 'play', 'stop', 'pause', etc.

- *File interface*
  The file interface supplies two operations: file read and file write for the manipulation of the content of files, for example:
    — a text review of the film *Men in Black* has been located by a client. To be able to view the contents of that file, a recognisable interface must be presented to the clients application. This is similar to the case of the Stream interface.

- *Binding iterator interface*
  This interface provides the methods to return more bindings from a context (directory), if there are any, for example:
    — a client has begun to list the contents of a directory and wishes to obtain the next name in that listing. Methods provided by the binding iterator are used to obtain those subsequent names, if they exist.

- *Directory interface*
  The directory interface provides a general name space for binding names to services or data, for example:
    — a service provider has a collection of service objects that must be ordered and associated in such a way that they are manageable and a client can gain easy access to them. The directory interface is used to perform this process of association and ordering;
    — a client presented with a service provider faces a directory structure that contains the object references for service objects that (s)he wishes to consume. To navigate this structure, the client uses the directory interface; it is also used to obtain the object reference that is required.

- *Session interface*
  The session interface enables a client to establish a session with a service gateway domain of services, for example:
    — a client wishes to connect to a service provider. It is the operations of the session object that are invoked to perform the connection;
    — a client wishes to resume watching the film *Batman*. When they last left the service provider that they were using to view the film they suspended their session. By passing the correct parameters across the session interface, the client application can reconnect to the service provider and the user can resume their viewing of the film.

- *Service gateway interface*
  This interface merely inherits from the directory and session interfaces, for example:
    — a service providers management requires a single root or point of reference to which all of its services can be attached. A single point of client access is also required. The service gateway interface provides this.

- *First interface*
  This interface enables an application to obtain its first objects, for example:
  — a client wishes to obtain the object reference of the primary service that has been returned as the result of attaching to a service provider. This is not the object reference for the service gateway but rather a service further down the directory structure. It is the first interface that provides the necessary operation.

The session interface

The session interface consists of two operations, 'attach' and 'detach'. These are invoked across the U-U/U-N interface, as shown in Figure 10.5. The session object itself resides on the U-N side of the interface.

The operation `attach()` takes three parameters: `serviceDomain`, `pathName` and `savedContext`. `serviceDomain` contains the NSAP address of the server that the client wishes to contact. `pathName` contains the path to a service that the user wishes to activate upon connection. This is a path though the directory structure of the service provider system. The DSM-CC specification states that either of these can be null if the other is set. `savedContext` is used to send information that is required by the server to resume an application from the point at which it was suspended. DSM-CC does not state the format that the information takes.

The operation call `attach()` returns one parameter: `ResolvedRefs`. This contains the object reference for the service gateway that has been connected to and if the `pathName` parameter was set one or several object references for a first service.

The operation `detach()` takes one parameter—`ASuspend`—and returns one—`SavedContext`. If `ASuspend` is set to `TRUE` then the state of the currently executing application will be returned in `savedContext`. It is this parameter that can then be passed in a subsequent `attach()` invocation.

DSM-CC states that `attach()` be invoked on the local session object and `detach()` be invoked on a remote session object. `Detach()` is invoked on the remote session object using the object reference for the service gateway connected to. This object reference was originally obtained when `attach()` was invoked.

Summary

DSM-CC U-U uses session objects to obtain and resume connections to service providers and the services found on them. Local session objects are used to obtain initial connections. Remote session objects are used to close those connections. Once a session has been established, the client application can use the U-U library to manipulate service objects on the service provider system that it has established a session with. In all of this, it is implicit that the session objects form the bridge between the application and transport layers.

DSM-CC user to network (S3)

The protocol specific to the S3 flow is the ISO MPEG session control protocol, DSM-CC User to network (U-N). The protocol stack for DSM-CC U-N is given in Figure 10.6.

DAVIC supports a sub-set of the DSM-CC U-N messages. These are the messages involved in client session set-up, server add and release resource, server client and network session release. DAVIC also specifies messages for session transfer should it be required in preference to subsequent release and set-up of the session to a new service domain.

Connection control—Q.2931 (S4)

The S4 flow is a bidirectional flow supporting call/connection control and resource control functions. The call/connection control functions consist of the capabilities to establish calls and connections in a B-ISDN network.

The protocol specific to the S4 flow is the ITU-T call/connection control protocol, Q.2931. This protocol supports the establishment and release of calls consisting of a single point-to-point connection in a B-ISDN network. The protocol stack for call/connection control is given in Figure 10.7.

DAVIC supports a subset of the Q.2931. These are Call Establishment (CALL PROCEEDING, CONNECT, CONNECT ACKNOWLEDGE and SET-UP), Call Clearing (RELEASE and RELEASE COMPLETE), Miscellaneous (STATUS and STATUS ENQUIRY) and Global Call references

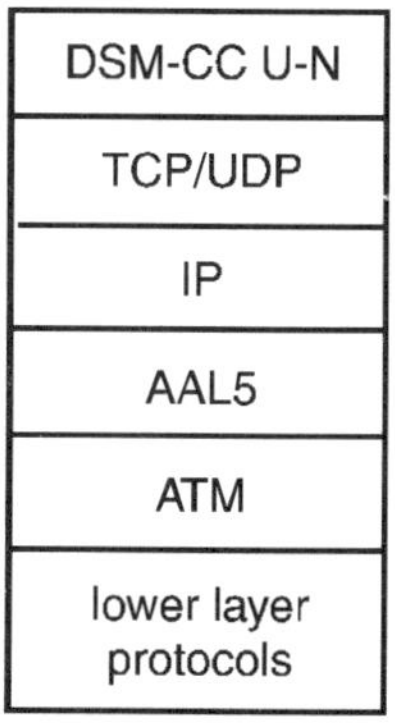

DSM-CC U-N
TCP/UDP
IP
AAL5
ATM
lower layer protocols

Figure 10.6 S3 Protocol Stack

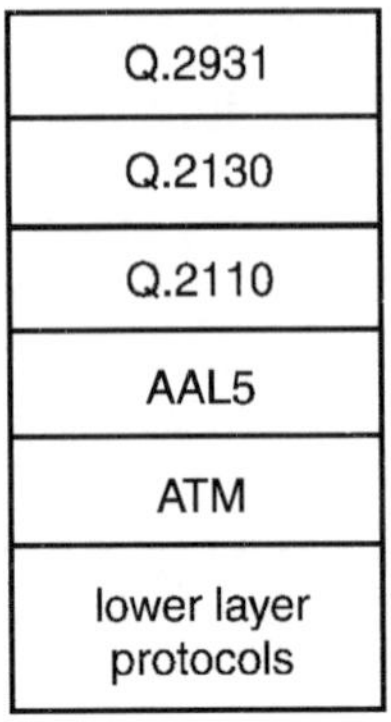

```
┌─────────────────┐
│     Q.2931      │
├─────────────────┤
│     Q.2130      │
├─────────────────┤
│     Q.2110      │
├─────────────────┤
│      AAL5       │
├─────────────────┤
│      ATM        │
├─────────────────┤
│   lower layer   │
│   protocols     │
└─────────────────┘
```

Figure 10.7 *S4 Protocol Stack*

Messages (RESTART and RESTART ACKNOWLEDGE). In addition to the messages used, DAVIC also specify the Call/Connection states and Information elements supported and coding of specific Information elements.

10.4 SESSION CONTROL

The DAVIC session is a means of encapsulating the set of network connections required for an instance of service usage. A session may require multiple connections at the S4 level, which are managed and controlled by a single S3 session. The application-programming interface to the session is offered at the S2 level by the session object, within the end user terminal (SCS and SPS). Thus, the session object exerts control over S3, which in turn subsequently controls S4.

The session control process leads to a sequence of information exchange both internal to a user system from S2 to S3, and S3 to S4 message parameter mappings and end-to-end between user and network across S3 and S4. The end result of session establishment is a set of S1 and S2 connections between SPS and SCS, through which the user browses the service domain and retrieves application content.

S2 to S3 parameter mapping

When the operations session attach() and session detach() are invoked on a local session object their parameters must be carried to the remote system in S3 messages. This is because there is no S2 connection established to the entity with which they wish to establish a session. The parameters that need to be carried in the case of session attach() are: serviceDomain; pathName; savedContext and resolvedRefs. For session

10.4 SESSION CONTROL

detach(): ASuspend and savedContext. These are described in section 3.

All of the U-N session establishment and release messages contain a field for what is described in the DSM-CC specification as User to User Data (uuData). It is into this field that the marshalled S2 parameters needing to be conveyed across the network are inserted. The uuData field is of variable length, and so can cope with the variable length parameters that are passed.

S3 to S4 parameter mapping

In the flows shown in Figures 10.8, 10.9 and 10.10, it can be seen that on receipt of certain DSM-CC U-N messages the server must establish and

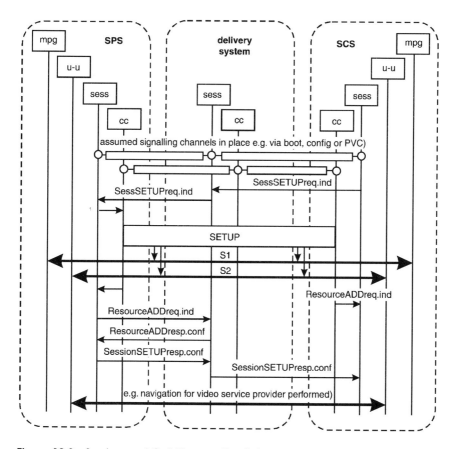

Figure 10.8 Session and Call/Connection Set-up

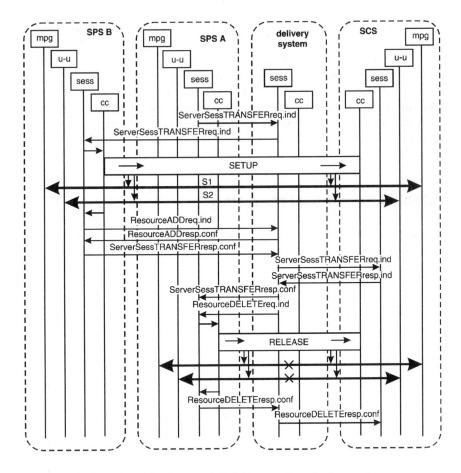

Figure 10.9 Transfer of Session and Call/Connections between Servers

release connections across the B-ISDN. These call/connections must be associated to the session, and must be established at the required bandwidth and QoS.

It is not enough that the server knows the local call reference of the call/connections it has established in association with a session; the STB must also know which session the call/connections are associated with, and this requires an end-to-end association. This association is made using a resource identifier, which is made up of a session identifier and a resource number. This information is to be carried by the generic information transport information element within Q.2931 messages. The recommendation was under discussion within the ITU-T when this identifier was needed, and the interim solution used the Broadband Higher Layer Identifier (BHL-I) Information element.

10.4 SESSION CONTROL

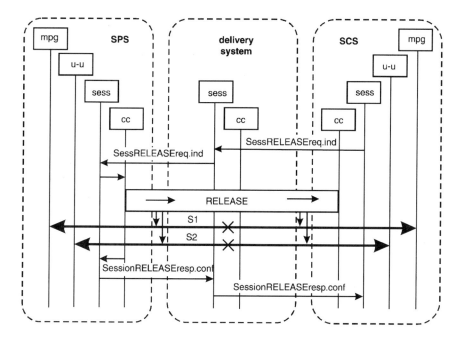

Figure 10.10 Session and Call/Connection Release

The DSM-CC Resource Types is used to indicate that two connections are required through use of the *AtmSvcConnection* resource. However, there is no explicit indication of bandwidth and QoS for the S1 or S2 connections. In implementations, detailed by Reece *et al.* and other articles within this publication, this has been overcome by explicitly defining a bandwidth and QoS for S1 and S2 connections. The Resource Number field is used to identify the S1 and S2 connections.

Dynamic operation

DAVIC use dynamic system modelling to specify how a DAVIC compliant system behaves with respect to session and connection configuration. This takes the forms of message flows between functional entities within a DAVIC system (see Figure 10.3).

A number of information flow scenarios are specified within the DAVIC recommendations; session and Call/Connection Set-up, Resource Add, Resource Delete, Transfer of session and Call/Connections between Servers, service Transfer, session and Call/Connection Release. For many of these there are also further scenarios defined depending on whether SCS-initiated, network-initiated or server-initiated.

This section focuses on what are envisaged to be the three main information flow scenarios: session and Call/Connection Set-up (SCS-initiated); transfer of session and Call/Connections between servers (server-initiated); and session and Call/Connection Release (SCS-initiated). The scenarios below concentrate on the session and call/connection flows.

Session establishment

The information flows in Figure 10.8 show how a SCS makes a request to the SPS to establish a session. The requested session requires two connections across the network; a connection for the S1 flow and a second connection for the S2 flow. The SPS establishes these connections within the context of the session, and confirmation that the requested session has been established is sent to the SCS.

Session transfer

The information flows in Figure 10.9 show how a SPS, SPS A, informs a second SPS, SPS B, that it has a session (and associated connections) it wishes to transfer (from SPS A to SPS B). Due to the limitations of the call/connection control protocol, it is not possible to simply transfer the connections to the new SPS. SPS B must set up two new connections in parallel to the two existing connections set-up by SPS A. Once these connections have been set up, SPS B can then notify the SCS that the session has been transferred (and the new connections can now be used). The SCS then notifies SPS A that the session has been transferred, and the two connections established by SPS A are no longer required and are released.

Session release

The information flows in Figure 10.10 show how a SCS makes a request to the SPS to release a session. The release of the session requires that the two connections associated with the session are also released. The SPS releases these connections, and confirmation that the session has been released is sent to the SCS.

The information flow scenarios are further supported in the DAVIC recommendations by functional entity actions.

10.5 DAVIC DEVELOPMENT

Recent DAVIC History

Since the publication of the DAVIC 1.0 specification in January 1996, work has been ongoing to widen its scope beyond Video-On-Demand applications and access network technologies. This has been achieved through the addition of new tools and the publication of new versions of the DAVIC specification. The DAVIC 1.1 specification, published in September 1996, primarily focused on enhancing the capabilities of the service consumer system, such as definitions of a network independent interface, Internet access tools and a virtual machine based upon Java.

Version 1.2 of the DAVIC specification was released in December 1996, and focused upon improving the security of content and communications, enhanced Internet access and High Definition Television (HDTV) quality video. DAVIC 1.2 provides a common specification that allows access from a SCS to high quality audio-visual services and also to the Internet.

DAVIC and the Internetworking Protocol (IP)

The Internet is the main source for interactive information today. The broadcast television industry is already taking advantage of the Internet (broadcasters have web-sites, links to web pages are shown in many television programs and commercials). The penetration and growth of IP-based applications has made it impossible for DAVIC to ignore the IP. The strategy for the DAVIC 1.3 specifications is now strongly focused on the convergence of DAVIC and the Internet.

The worldwide web provides a fantastically powerful and rich interactive environment based upon the Hypertext Mark-up Language (HTML) and Java code. The one single thing preventing the Internet from becoming the means of delivering all interactive and broadcast services is its lack of throughput, or bandwidth. DAVIC delivery systems have the bandwidth and low latency to deliver very high quality broadcast services as well as manageability, accountability and security included from day one.

The ideal DAVIC 1.3 system will see the adoption of Internet protocols that will enable existing Internet use to continue unchanged, with the same equipment and protocols, but within the framework provided by DAVIC members through the specification process.

10.6 CONCLUSIONS

At the time that the DAVIC 1.0 specifications were drafted, VOD was considered to be a great revenue earner and, as such, many of the leading players in the IT and telecommunications field were racing to test and launch VOD equipment and broadband network technologies. It was on this wave of enthusiasm for VOD that the DAVIC specifications were created. From these initial steps towards a public VOD service, one thing became clear—VOD was not a 'get-rich-quick' investment, requiring considerable up-front deployment of new technology, tight profit margins and a long time to wait before the service became profitable. Since the release of the 1.0 specification, enhanced digital broadcast has become the preferred medium to deliver interactive and broadcast services to the mass market. DAVIC have followed suit, and are continuing to play a leading role in drafting specifications for enhanced digital broadcasting through alignment with and input to the European Digital Video Broadcasting (DVB) specification.

DAVIC was successful in raising the awareness and interest in the use of session control as a mechanism to deliver interactive applications across a networked environment. Session control is certainly defined in a more rigorous and flexible manner by the Telecommunications Information networking Infrastructure (TINA) specifications. The DAVIC session is, by comparison, naïve, with user, communications and application attributes all tightly bound within the capabilities of DSM-CC. Inevitably the breadth of native DAVIC applications was limited by the original scope of DSM-CC; which was the client-server based retrieval of video-based applications.

In the early days of DAVIC, few would have predicted the growth of Internet use that we see today. The Internet is now the leading platform for the provision of mass-market interactive information services. As with VOD in 1994, the challenge facing the IT world today is adding quality of service to IP-based applications in the Internet and Intranet. There are many competing technologies from virtually every ATM switch and IP router manufacturer, much as we saw with proprietary solutions for VOD in 1994. At present a winning technology has yet to emerge from this battle. DAVIC have now positioned themselves with a systems integrator role, whereby existing and emerging best-of-breed IP standards and technologies can be chosen and integrated with existing DAVIC tools. The vision for DAVIC 1.3 is a network architecture that will allow existing IP-based applications to remain unchanged in their use of IP and the Internet, but enable them to make use of high bandwidth, low latency DAVIC infrastructure as required. Further reading on the activities of DAVIC is available from the DAVIC web-site.

REFERENCES

Reece P W *et al.*, An early implementation of a DAVIC V1.0 system—use of dynamic connections for interactive multimedia services, *BT Technol. J.* Vol. 16, No. 1, January 1998, pp. 114–126.

SPECIFICATIONS

DAVIC 1.0 Specifications.
DSM-CC Specifications (July 1996).

FURTHER READING

DAVIC:www.davic.org
DSMCC:ftp://playground.sun.com/pub/dsmcc/

QUESTIONS

1. Why is a service gateway element needed?
2. What are the two types of information flows in DAVIC, and how do they differ?
3. Why did DAVIC choose ATM rather than the internet as the base technology for its studies?

11

Design for Performance of Broadband Signalling and Services

Dave Morris and Ann Elvidge

11.1 INTRODUCTION

The aim of performance analysis is to look for areas where the signalling performance of a broadband network, as a whole, may be inadequate. The effect of inadequate performance, as perceived by the user of broadband services, is increased service delays and lower levels of service availability. Long service delays are caused by a greater demand for service than the network can handle. The requests for service are then queued in a buffer, which causes the perceived delay.

From the users' perspective, the key performance parameter is the service delay. The user is soon frustrated if, after accessing a menu screen to choose their service, the system response time rises to a level where it becomes slow to navigate the menu.

It is most likely that the component with the longest response time is also the most overloaded (unless the network components have been poorly dimensioned), and as such we should be looking, in the first instance, for the 'bottleneck' in the network. This will point to the component most likely to cause problems by adding the longest delay to the service. When this has been determined the load dependent delay can be estimated using simple queuing models. In this way, a first cut analysis of the major broadband signalling issues can be carried out pointing the way to better design and implementation.

11.2 ASSUMPTIONS

Information flows and functional to physical mapping

As yet it is unclear what information flows will occur during a broadband multimedia call. There has been some early work to sketch out information flows, but these have only dealt with information flows between core network functional entities. Our aim was to establish a measure of the quantity of signalling during a broadband call so the performance measures can be estimated at this early stage. The information flows are based on a DAVIC view (ISO/IEC 1995), which is one particular variant on the design of broadband networks.

The information flows are divided into four call processes: Initiation, Set-up, Transfer and Release. Initialisation occurs when the set top box is powered up. At the end of Initialisation the set top box has a network address and is able to send signalling messages to the network or a proxy signalling agent (if it requires one). During Set-up, a user-to-user information stream and an MPEG information stream is established between the set top box and the L1GW. During the Transfer phase, the set top box is disconnected from the L1GW and subsequently connected to the IMS server. A user-to-user information stream and an MPEG information stream is established between the set top box and the IMS server. Finally, a Release will occur, which tears down information streams between the set top box and the IMS server. Note that after the Release the set top box does not need to go through an Initialisation stage in order to set up a new broadband connection. This only has to be done after the set top box has been powered down. The information flows in Figures 11.1 to 11.4 show the functional entities of the network. (Note that information flows containing a proxy signalling agent are not shown.)

The information flow marked as 'User Data' is the point in a call where the set top box and the Level 1 Gateway or IMS server exchange DSM-CC user-to-user signalling messages and an MPEG video stream.

The names given to the information flows shown in Figures 11.1 to 11.4 are followed by a number in brackets. This number represents a best estimate of the actual number of messages involved in supporting the information flow, and it is included because, from a performance point of view, it is the number messages which are of importance. The estimate of the actual number of messages used is based on the format of the Q2931 protocol.

11.2 ASSUMPTIONS

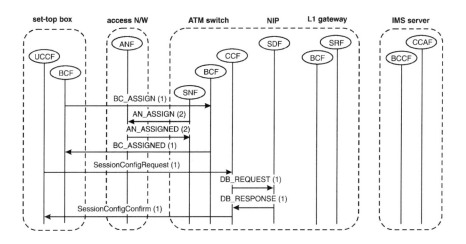

Figure 11.1 Session Initiation

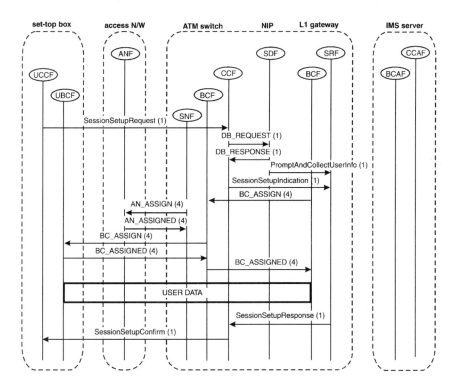

Figure 11.2 Session Set-up

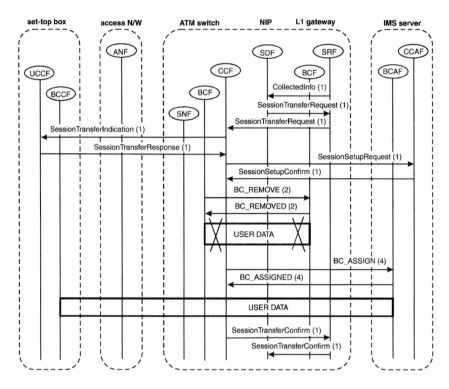

Figure 11.3 Session Transfer

Network architectures

At an early stage in broadband network design, there is no one definitive design for a broadband network architecture. Thus, from a performance point of view, rather than there being a single network scenario to assess, a number of variants need to be considered. In this study the number of network architectures has been reduced to two; the session manager and L1 gateway in a single network node; and the session manager, L1 gateway and proxy signalling agent in separate network nodes. The aim of these configurations is to test how much processing is required at a physical network node. Putting more functionality into a single node will generally decrease its performance, since each supported function requires a finite amount of processing. On the other hand, separating functionality increases demand for installed signalling and user bandwidth.

General assumptions made about the scenarios in the context of interactive multimedia services are:

- The session manager is located within the ATM switch.

11.2 ASSUMPTIONS

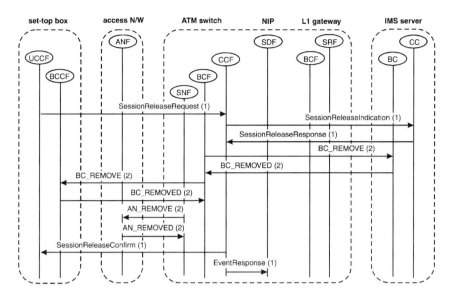

Figure 11.4 Session Release

- The ATM switch signals to some Network Intelligence Platform (NIP) using signalling similar to INAP.

- The L1GW is an Intelligent Peripheral (IP), which provides video dial tone and collects user selection information.

- The L1GW is not involved in the call beyond the call set-up phase.

- The NIP directs the call Set-up between the L1GW and the set top box, or the IMS server and the set top box, using signalling similar to INAP.

- The PSA is not located in the access network (i.e. at the VAP).

Figures 11.5 and 11.6 show the two network architectures considered.

Number of visits

The message sequence charts give us an indication of how busy various network components will be for a given network architecture and call procedure. This is done by counting the number of times each network component is used, or visited, for each call procedure. To maintain a consistent view a visit occurs when a signal is sent OR received at a network node. The reasoning here is that processing is required when a signalling message is sent or received. Tables 11.1 and 11.2 summarise the number of visits to each network component.

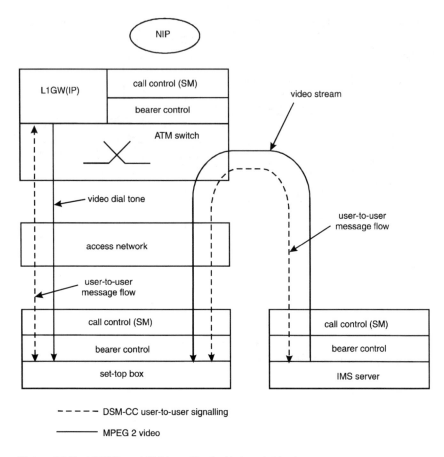

Figure 11.5 L1GW and SM in a Single Network Node

However, knowing the number of visits to each component is not enough, since we need to know what functional to physical mapping exists. Put another way, each signal sent or received has to be processed and where that processing is carried out can have an effect on system performance. Table 11.3 shows the number of visits each network node experiences given the two scenarios assumed.

The bottleneck

The bottleneck is the network component which is most highly utilised. This component will handle the lowest number of calls per second and will be the first to overload. To ascertain which component is the bottleneck we need to know the number of visits per call made to that component, the

11.2 ASSUMPTIONS

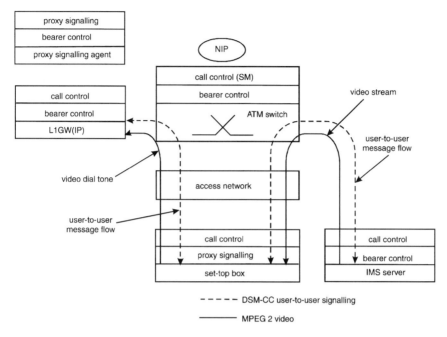

Figure 11.6 L1GW, PSA and SM All in Separate Network Nodes

Table 11.1 Number of Visits with NO Proxy Signalling Agent

	STB	SM	L1GW(IP)	NIP	IMS	AN
Initiation	4	10	0	2	0	4
Set-up	10	30	11	3	0	8
Transfer	2	18	9	3	10	0
Release	6	17	0	1	6	4
Total	22	75	20	9	16	16

Table 11.2 Number of Visits with Proxy Signalling Agent

	STB	SM	L1GW(IP)	NIP	IMS	PSA	AN
Initiation	6	16	0	2	0	4	4
Set-up	3	36	11	3	0	13	8
Transfer	2	18	9	3	10	0	0
Release	3	21	0	1	6	7	4
Total	14	91	20	9	16	24	16

Table 11.3 Number of Visits to each Physical Node. (I—Initialisation, S—Set-up, T—Transfer, R—Release)

	{L1GW,SM}					{L1GW},{PSA},{SM}				
	I	S	T	R	Total	I	S	T	R	Total
Switch Node	10	41	27	17	95	16	36	18	21	91
L1GW Node					0	0	11	9	0	20
STB Node	4	10	2	6	22	6	3	2	3	14
NIP Node	2	3	3	1	9	2	3	3	1	9
IMS Node	0	0	10	6	16	0	0	10	6	16
PSA Node					0	4	13	0	7	24
AN Node	4	8	0	4	16	4	8	0	4	16

average service time for each server and the number of servers that component has. (Note that for a queuing system the service time is the time it takes to service a message once it reaches the head of the queue.) The maximum capacity of that network component is represented as

$$\text{Maximum Call Handling Capacity} = \frac{N}{S \times V} \text{ [calls/s]}$$

where N is the number of servers, S is the average service time for each server, and V is the total number of visits (to all servers). The bottleneck is then by definition the component with the lowest maximum capacity.

The above is a theoretical capacity unachievable in real systems. The maximum call handling capacity is the capacity at which the response time rises to infinity, where the response time is the total of the queuing time and the service time. To achieve reasonable response times and to allow for fluctuations in traffic network, nodes are typically operated at a fraction of the maximum call handling capacity.

In fact, we do not actually have enough information at present to determine the bottleneck, because we do not know the average service time per visit for most of the network components; these components have yet to be designed and built. From a performance point of view this is a problem, because without knowing where the bottleneck is it is not possible to analyse the system as a whole. So we must make an assumption, which is that the ATM switch node will be the bottleneck. This seems reasonable given that the session manager which is assumed to reside on the switch experiences the most number of visits (see Tables 11.1 and 11.2), and that the complexity of the operations processed by the session manager are high since it is handling call and bearer control. However, we do have to keep in mind that this is an assumption and in fact any network component could be the bottleneck.

11.3 PERFORMANCE ISSUES

Table 11.4 Response Time per Visit

Independent network component	Response time per visit (ms)
STB	10
SW	Load dependent
PSA	10
L1GW	10
	25
NIP	
IMS	10
AN	10

The average service time per visit depends upon the make of the ATM switch and its processing characteristics. For the purposes of this study, we assume that the average service time per visit is 4 ms.

Load dependent response time

All components in a network will have a load dependent response time, but only the bottleneck will experience a dramatic rise in response time as the system reaches its maximum capacity. Response time assumptions are shown in Table 11.4.

The queuing model for the ATM switch was assumed to be M/D/4 with an additional fixed delay, following an analysis of narrowband switch behaviour for narrowband calls.

11.3 PERFORMANCE ISSUES

System capacity

The term *maximum system capacity* is defined in this report as the maximum rate at which the system can handle broadband calls. The bottleneck determines maximum system capacity, and so it is the capacity of the ATM switch node that we need to resolve. If we assume a complete call requires a Set-up, Transfer and Release call process, then we can estimate the load on the ATM switch node by noting the figures contained in Table 11.3, and remembering that the average service time per visit for the ATM switch is assumed to be 4 ms. To compute system capacity we need to specify the ratio of the number of broadband calls to the number of set top box initialisations made. Here we will assume that for every three broadband calls, one set top box initialisation occurs.

Maximum system capacity (as defined mathematically) is the capacity at which the response times rises to infinity. This is not an achievable operating point. In this study, a capacity figure which is 60% of maximum system capacity is termed the 'normal capacity' figure, and it is this figure which is assumed to be the operational capacity of the system (Table 11.5).

Demand on NIP

In this study, broadband calls are IN calls, and as such require SCF functionality. The assumption is that this will be provided by a Network Intelligence Platform (NIP).

We know from Table 11.5 the maximum capacity of the switch node for each functional to physical mapping, and from Table 11.3 we can deduce that for any call the NIP is visited nine times. This corresponds to an average of 4.5 NIP transactions per call if we note that a transaction normally requires two visits (one message received and one message sent). Thus,

Calling rate at the NIP = Normal calling rate of each switch x Number of switches

Transaction rate at the NIP = 4.5 × Calling rate at the NIP

The NIP transaction rate is shown in Table 11.6.

Table 11.5 Capacity of Switch Node

Scenario	No. of visits to switch node per call	Normal capacity [calls/second]
{L1GW,SM}	88.3	1.36
{L1GW},{PSA},{SM}	80.28	1.49

Table 11.6 The Normal NIP Transaction Rates with 50 Switches

Scenario	Normal calling rate at the NIP [calls/second]	Normal NIP transactions rate [transactions/second]
{L1GW,SM}	67.95	306
{L1GW},{PSA},{SM}	74.74	336

11.3 PERFORMANCE ISSUES

Response time

For queuing systems with a finite service time the response time, that is the time for a system to process a transaction, generally increases with increasing load. If the response time for component i is R_i, then the total response time for a system is generally expressed as

$$\text{Total response time} = \sum_i R_i V_i$$

where V_i is the number of times the ith component is visited. Given that the number of visits to each physical node is known from Table 11.3, and the assumed response times for each non-bottleneck node is known from Table 11.4, we can calculate the load independent response time for each call procedure. If we assume the M/D/4 model with additional fixed delay for the bottleneck node, that is the switch node, and we note that the service time for each of the four servers in this node is 16 ms per visit (i.e. an average service time of 4 ms per visit), we can compute an overall load dependent response time for each call procedure in each scenario. These are shown in Figures 11.7 and 11.8.

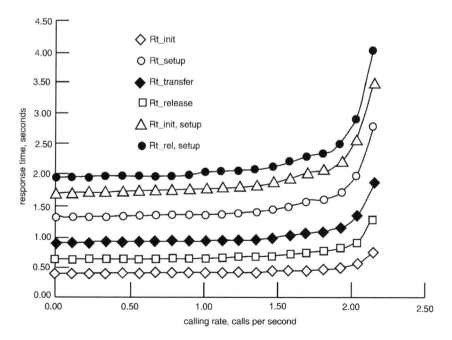

Figure 11.7 Response Time for {L1GW, SM}

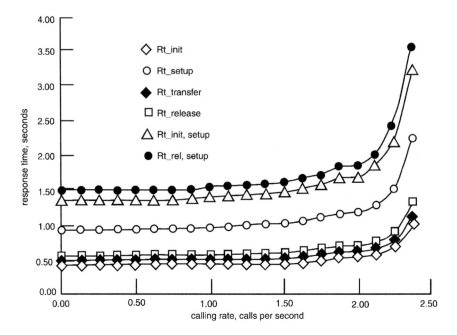

Figure 11.8 Response Time for {L1GW}, {PSA}, {SM}

The legends in Figures 11.7 and 11.8 are as follows:

Rt_init,setup	Response time for Initialisation call procedure
Rt_setup	Response time for Set-up call procedure
Rt_transfer	Response time for Transfer call procedure
Rt_release	Response time for Release call procedure
Rt_init,setup	Total response time for Initialisation and Set-up call procedures
Rt_rel,setup	Total response time for Release and Set-up call procedures

In all cases, the single call procedures (Initialisation, Set-up, Transfer and Release) completed within 1.5 seconds at normal load. The double call procedures (Initialisation-Set-up and Release-Set-up) completed in under two seconds at normal load.

11.4 DISCUSSION

The aim of this study is not necessarily to provide exact answers on topics like response time and system capacity, but rather to point out some major issues which will need careful consideration as the specification of the broadband access network accrues more detail. At present, the specification of a broadband access network has yet to be completed. This gives the performance engineer the freedom to carry out a comparative study of a number of different scenarios to test which is likely to be the best.

The capacity of each broadband switch was found to be in the range 0.99 to 1.76 calls per second with the assumptions made. This can be compared with a single cluster System X unit, which can handle up to about 25 POTS calls per second. This difference is a consequence of a broadband call requiring more messages to complete. A complete telephony call requires about seven messages, a narrowband call using network intelligence such as a One Number call requires up to 16 messages, whereas a broadband call may require anywhere between 138 and 158 messages (excluding initialisation), depending on the network architecture. What should be noted is that, even at this early stage of specification, it is clear broadband networks will require significantly more signalling and thus processing power. This should not be overlooked when considering the performance of the final system design, or when designing the message flows to support the service.

In this study, the design of the broadband network is assumed to follow IN principles. User and IMS server profiles and set top box information is assumed to reside in a NIP. This means that the NIP is involved in a broadband call. Also, the level 1 gateway is assumed to be an IP which requires direction from the SCF which also resides in the NIP. First considerations suggest that the NIP is required to carry out about 4.5 transactions per broadband call. Given a knowledge of the predicted customer base and possible customer calling rates, the total demand on the NIP for transactions per second can be calculated, and this information then used to dimension a NIP.

An attempt has been made to assess the load dependant response time for a broadband network. Two important points come out of this. First, the delay which affects the user's perception of delay is not necessarily the total response time for all call procedures initiation-set-up, transfer and release, but the delay at each of the call procedures, in particular the initiation-set-up delay and transfer delay. The second point is that delay rises dramatically if system capacity is approached, and so for acceptable service to be offered at normal loads, the system dimensioning needs to take this into account.

11.5 CONCLUSION

A broadband call may require anywhere between 138 and 158 messages (excluding initialisation) compared to seven messages for a standard telephony call and around 16 messages for a narrowband One Number call. The signalling traffic generated is therefore set to increase dramatically, with implications for both customer perceived delays and signalling network dimensioning and financial investment. Signalling flows need to be minimised to offer acceptable quality at a reasonable price.

The technique discussed allows the switching capacity per broadband switch to be identified (in calls per second at normal load, i.e. when the system is running at considerably less than maximum capacity).

First considerations suggest that the NIP is required to carry out about 4.5 transactions per broadband call, and so given the predicted customer base and possible maximum customer calling rates, the demand offered by the broadband network to a NIP, and hence the required NIP capacity, can be established.

The call Set-up and Transfer delay perceived by the customer can be calculated, given the network architecture and normal demand for the broadband service. This information is vital in dimensioning equipment levels to yield system delays which are acceptable to the customer.

Response time rises dramatically if system capacity is approached, and so the system needs to be dimensioned so that the occupancy of the bottlenecks is considerably less than 100%, at normal load.

REFERENCES

ISO/IEC, *Digital Storage Media Command & Control*, ISO/IEC 13818-6, Boston, May 1995.

12

Broadband VPN Signalling

Frank Allard

12.1 INTRODUCTION

Virtual Private Networks (VPNs) allow people to make their corporate communications more flexible and cost-effective. In the narrowband network, a VPN allows the customers to share their PINX (Private Integrated service Network eXchange, also known as PBX) features over geographically distributed sites.

Broadband ATM technology, with even greater bandwidth, will permit integration of the voice and data platforms in addition to providing new services such as videoconferencing and point-to-multipoint calls. Until then, many issues such as signalling need to be fully addressed before the migration from the narrowband to the broadband VPN is complete. The chapter provides an overview of how BQ-SIG and PNNI, the private broadband signalling systems, support the broadband VPN.

12.2 WHAT IS VPN?

There has never been a time when the ability to communicate quickly and efficiently within large companies and organisations has been so important. For most domestic and small business users, the telecommunications services provided by the Public Network Operators (PNOs) are adequate for their needs. However, large and geographically distributed corporations who wish to utilise advanced telecommunication products to improve their business efficiency find that neither the costly mesh of private connections required to link networks, nor the Public ISDN (Integrated Service Digital Network), always satisfy their communication requirements.

VPN (Digital Corporate Networks—Interconnect Communications Ltd.) aims to allow business customers to optimise their communication costs, with respect to dedicated private networks, by providing private networking services on the public network or on a network set-up to offer a set of business services. VPN service providers typically offer a number of private services to customers, including private numbering plan, private service features and advanced billing. An important issue for the service provider is to ensure that these can be delivered across the VPN so that customers receive an 'end-to-end' seamless service.

Figure 12.1 shows a typical VPN configuration. The Customer PINXs are connected to the Public ISDN, which connects them together and performs private to public number translation via the Intelligent Network (IN), for instance.

The advantages of implementing corporate telecommunications networks as a VPN apply regardless of geographical location, and may include

- ability to manage the network

- guaranteed quality of voice and data

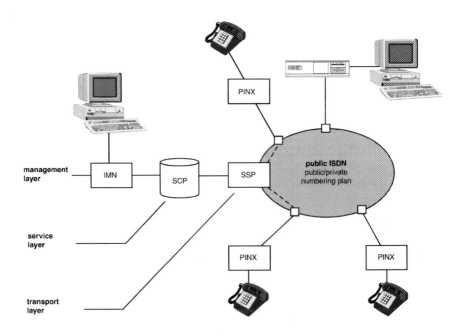

IMN intelligent management network
SCP signalling control part
SSP service switching point

Figure 12.1 An Example of VPN. IMN: Intelligent Management Network; SCP: Signalling Control Part; SSP: Service Switching Point

- improved security
- customised tariff arrangements
- a wider range of facilities that those provided by public networks
- cost effective geographic connectivity
- simplified numbering and addressing
- flexible charging.

12.3 VPN ARCHITECTURE

Public access

User to network signalling for PINX access to the public ISDN is achieved using DSS1 (Q.931) for both Basic Rate (BR) and Primary Rate (PR). The services are provided in a generic manner and not tailored to the customers, as is the case for VPNs. All features are provided by the network and are identically available to all the subscribers.

Private access

Signalling systems have been specified for operation between PINXs (i.e. over the 'Q'-reference point defined by ITU-T). The European version is based on DSS1, and is called Q-SIG (1993). The standards for Q-SIG are being developed by ECMA. These have been adopted internationally by ISO, and this international private network signalling system is called PSS1 (Private Signalling System No. 1). Q-SIG has also been taken from its natural environment to be used in the provision of VPN networks as a VPN access, and not only an inter-PINX signalling system.

Q-SIG is an important signalling system because it supersedes proprietary private signalling and allows migration from them to Q-SIG. Major PBX manufacturers have implemented Q-SIG, and ensure that it can be mapped to their own proprietary signalling solutions.

Although commonality exists between Q-SIG and DSS1, there is a significant difference:

- DSS1 is an asymmetrical access protocol of link significance only. There is a customer side and the services are provided by the network side.
- Q-SIG is a symmetrical network protocol capable of signalling across a network of PINXs/network nodes, which means the core network is only used as a transit platform.

Integrated public/private access

It is generally recognised that most future services provided by the VPN are likely to be implemented using IN, and that further service specific standardisation will be limited. However, currently IN has been standardised around the PSTN and voice services, and it is well recognised that enhancements to the DSS1 standards (as well as other ISDN and IN standards) will be required to fully support services delivered using IN and ISDN.

Standards are currently defining ISDN virtual private networks, which include the convergence of public and private network ISDN. Current options under discussion include the mapping and enveloping of Q-SIG within DSS1, and the combination of Q-SIG signalling and DSS1 within a single signalling system (DSS1 enhanced for VPN).

Network signalling

The main part of the narrowband core network signalling system is referred to as ISUP (ISDN Signalling User Part (Anon 1997)). This system is designed as an international interface for ISDN services. It is used to route the call across the network in conjunction with the IN (when used) to perform number translation for instance between a private and a public number, via INAP (Intelligent Network Application Protocol). Such a scenario including various signalling systems is represented in Figure 12.2.

12.4 ATM BENEFITS TO VPN

Voice platform evolution to ATM

The crucial competitive factor with ATM is quality of service and pricing—ATM itself is just the technology. Voice over ATM does not necessarily mean less quality: considerable research has been done to show how ATM can reach the quality of voice offered by the narrowband switched networks (Alley *et al.* 1995). What is lost with voice over ATM at present is the range of service features associated with private voice networks.

Theoretically, the ATM network could replace the existing dedicated trunk groups interconnecting the narrowband switches, switches to other network components and customer sites, as shown in Figure 12.3. The advantage of the ATM component is that it offers the potential to make more efficient use of the bearer transmission platforms.

12.4 ATM BENEFITS TO VPN

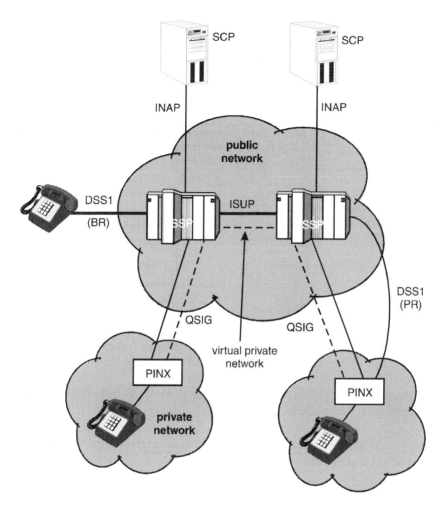

Figure 12.2 Public and Private Network Architecture

In this example, the narrowband PINXs would be connected to the ATM core transport via ATM remote units (or edge switches). The VPN services would be provided by existing narrowband switches interfaced to the ATM transport, or by ATM switches with the narrowband VPN service and signalling.

The narrowband call is set-up in a traditional manner using narrowband signalling. Set-up delays can be introduced with the distance between the two terminals. Constant Bit Rate (CBR) is required as well as high loss priority to guarantee a good quality level.

If a traditional voice call is made, there can be no room for network congestion threatening the conversation. In a voice network, therefore,

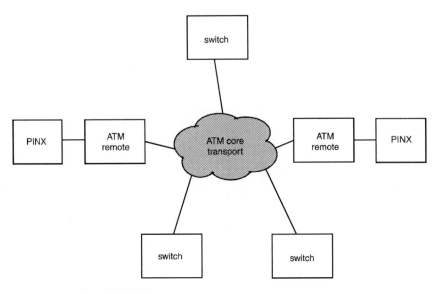

Figure 12.3 The ATM Platform

connections must be delivered with a guarantee; this is the most important aspect which ATM has to prove—its ability to guarantee the Quality of Service (QoS).

It only makes economic sense to change to an ATM infrastructure if it can be demonstrated that

- a wide ATM base could reduce access distances and offer the opportunity for lower access costs
- the solution is cost-effective and the network resilient
- it can offer better voice/data integration.

Voice/data integration

For many large organisations, ATM now represents the only cost-effective method for handling data traffic, while future-proofing their network to enable new developments in communication technology and supporting and developing new routes to their markets. (Anon 1997)

As well as offering high-speed files or image transfer, database access, Local Area Network (LAN) interconnection, video communication and co-operative working, a single ATM network can also support existing real-time data services such as frame relay, Switched Multi-megabit Data Service (SMDS) and voice (Alley *et al.* 1995). That allows people to customise their traffic according to their needs. Until now they have had to constantly redimension their networks to cope with changing traffic needs.

With the introduction of switched virtual circuits, customer equipment will be able to signal for ATM connections. This will unlock the real power of ATM to set up synonymous customer connections on request. It is essential that the Broadband VPN and associated signalling supports these requirements.

12.5 BROADBAND VIRTUAL PRIVATE NETWORKS

The definition of a broadband VPN is that 'a B-VPN is a part of a customers private network supported by the SVC capability of a B-ISDN switched network' (i.e. excludes VP tunnelling, leased lines and equivalents). Similar techniques to those used in narrowband VPN technology techniques (for example, using intelligent network number translation) can be applied to broadband networks. By definition, a network without switching would not be considered or a VPN—and would require the introduction of switching and translation capabilities to be classed as a VPN. A possible BVPN voice network is shown in Figure 12.4.

The traditional telephony-based network model has long embraced this requirement by defining a number of levels. Data networks have tended to

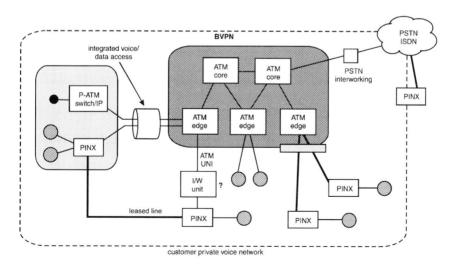

Figure 12.4 Possible BVPN Voice Network

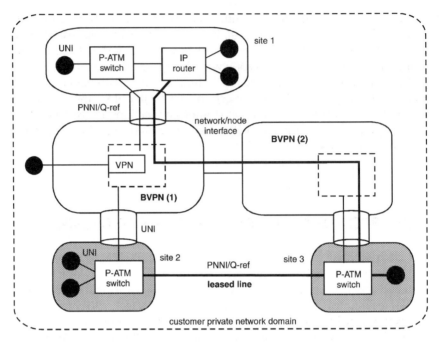

Figure 12.5 Possible BVPN Data Network

be much smaller and have had less need for this approach. As these two worlds coalesce with broadband, the network model may evolve further. A possible BVPN data network is shown in Figure 12.5.

12.6 EVOLUTION TOWARDS BROADBAND VPNS

The role of signalling

In the absence of a signalling interface to the network, the customer will be restricted to using 'Permanent Virtual Circuits' (PVCs). He must contact his network administrator indicating his requirements, such as end points to be connected, bandwidth required and service type. The drawback of this approach is, of course, the time taken and effort involved when the customer wants to change the connection characteristics increase his bandwidth. In a Switched Virtual Circuit (SVC) environment, the situation is quite different (Bale 1995). The end user may now specify the characteristics of his connections, directly, on demand, as shown in Figure 12.6.

To provide any given service to the user, all of the interfaces across the network must support the features required by that service in a compatible manner.

12.6 EVOLUTION TOWARDS BROADBAND VPNS

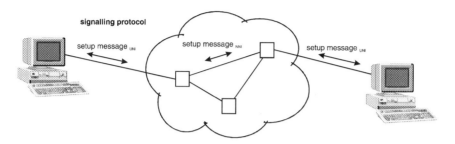

Figure 12.6 Use of Signalling to Set-up a SVC Call

The ITU-T Q reference point is still applicable in a broadband network environment, and is recognised as such by ECMA, and similarly by the ATM-Forum as their Private Network-Network Interface. However, VPNs require additional capabilities to extend private network features beyond their own local domain to the whole of the VPN. The resultant signalling protocol that runs across this interface must be supported by the public network.

The above is a very simplistic view of the network. The larger the network, the greater the need for a hierarchy of switching levels, on economic, functional and connectivity grounds.

Addressing across the public network

In the existing narrowband telephony network, an intelligence 'layer' is used to provide a number translation function which can be applied to offer a number of services, e.g. freephone, premium rate services, flexible charging, etc. It is expected that the number of core functions will be greatly enhanced in the broadband environment, requiring a more complex and powerful 'intelligent network.'

Any network end point must be uniquely identifiable within that network. For the public network, therefore, this demands an addressing scheme which offers worldwide unique numbers—such as the E.164 ISDN scheme. However, many private networks will use AESE within their networks, E.164 is a public numbering plan and, as such, numbers are generally only allocated to operators offering a public service. E.164 numbers are not yet allocated directly to private users for use in private networks, hence the need to translate the customer private number onto an addressing scheme (E.164 or other) that can be used by the network, as shown in Figure 12.7.

One solution could be provided by PNOs who wish to support all address schemes, maybe as a value added service. This sort of service

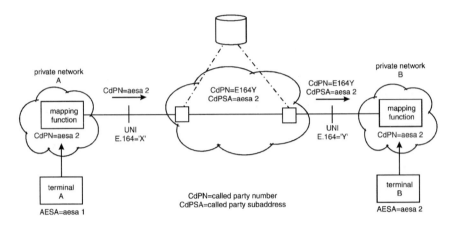

Figure 12.7 Network Addressing

could be provided by an intelligent network in the future, along with other address manipulation services such as number portability. However, at the present time there are no broadband IN standards, and it will be some time before IN technology can be integrated in the broadband area.

Alternatively, operators may choose to use other schemes for VPN addressing, leading to the requirement of global awareness of more than one address scheme if calls are to be successfully routed across multiple public network domains.

Standards for services and signalling in broadband PISNs

Based on experience in developing the Q-SIG series of standards, ECMA has developed a corresponding signalling system for use in Broadband Private Integrated Services Networks (B-PISNs) employing ATM technology. B-Q-SIG (signalling system for use at the Q reference point in broadband networks) operates at the Q reference point between Private Integrated Services Network Exchanges (PINXs) connected together within a B-PISN employing ATM.

B-Q-SIG currently comprises the following ECMA standards:

- B-Q-SIG-GF (Generic Functional Protocol) extends the basic call/connection control protocol by providing generic support for supplementary services and additional network features.
- B-Q-SIG-BC (Basic Call connection/control).
- B-Q-SIG-SAAL (Signalling ATM Adaptation Layer).

12.6 EVOLUTION TOWARDS BROADBAND VPNS

During the development of these standards, ECMA worked in cooperation with the ATM Forum with a view to achieving alignment, where appropriate, between B-Q-SIG and the signalling part of the ATM Forum's Private Network-Network Interface (PNNI) specification. Whereas the PNNI specification provides automatic configuration and dynamic source routing, B-Q-SIG is intended for networks that employ static hop-by-hop routing. B-Q-SIG-BC therefore particularly allows telecommunications networks with their more static structures and their hop-by-hop routing. Consequently, it might become a migration step for such networks towards a fully dynamic and autoconfigured PNNI network.

The future B-Q-SIG activities will basically go in two directions:

- adapatation of existing Q-SIG supplementary services to private broadband networks;

- specification of particular support functions for broadband-specific applications.

Broadband environment

The present capability to provide broadband VPN is to use the broadband public access, Digital Signalling System 2 (DSS2) or ATM-Forum UNI4.0, with the Closed User Group (CUG) Supplementary Service.

SVC Closed User Group

CUG is defined in ITU-T Recommendation Q.2955.1 and in ATM-Forum UNI 4.0 of the supplementary services (B-ISDN). The CUG supplementary services allow users to form groups. A user can be a member of one or more CUGs. Normally, members that belongs to a specific CUG can communicate among themselves, and not outside the group. Some members can have additional privileges, whereby they can be allowed to make outside calls or even receive calls from a member outside the CUG.

In the Signalled CUG options provided, the network needs to check the CUG index with the Calling Party Number and then maps the Interlock Code (IC—used to check the access rights). Moreover, some members of the VPN might require outside, incoming or both way access privileges. A table representing whether a Calling Party Number and IC combination is allowed these privileges has to be recorded in the network.

Address screening is also an option for providing CUG. It also needs a database of matching Calling and Called Party number pairs. The table has to be constantly updated to accommodate new members. On a small scale, Address Screening is a better option. However, it is not possible to make a quick transition from Address Screening to Signalled CUG, as there need to be major changes for this to happen. This could be viewed to be very expensive in the long run. The table sizes in the Addressing Screening option can be considerably cumbersome as the number of members and attributes increases.

In future, the full SVC will be utilised to provide VPN service in conjunction with PNNI and/or BQ-SIG.

Broadband-VPN support for narrowband-VPN

Section 12.4.1 described how a narrowband VPN could be overlaid onto an ATM core transport network. However, if the ATM core network supports broadband VPN, then it would be possible to support narrowband VPN by interworking the narrowband signalling (for example, Q-SIG) with the broadband VPN signalling (for example, BQ-SIG). The drawback with this approach is that broadband VPN does not currently support as many supplementary services and features as the narrowband VPN, nor the ability to transparently carry service information between narrowband PINXs.

12.7 CONCLUSION

Narrowband VPN already permits customers to share their PINX features across sites using the public network as a transit platform to provide services such as private/public number translation.

Using the ATM technology to provide increased bandwidth and service integration is a great benefit to the user. This chapter has described a number of ways in which a VPN can be provided on a broadband ATM network, and the signalling systems used. These are:

- Using an ATM core transport network to provide bearer connection capabilitiy for existing narrowband VPNs, using current narrowband ISDN signalling.

- Using an ATM public broadband ISDN, to provide broadband VPN services using broadband access signalling.

- Using an ATM public broadband ISDN, to provide broadband VPN services using broadband private network signalling.

- Using an ATM public broadband ISDN, to provide narrowband VPN services by interworking current narrowband signalling with either broadband access or broadband private network signalling.

However, many issues need to be resolved before a truly broadband VPN can be deployed. This chapter has highlighted some areas that impact on the signalling systems used for broadband VPN:

- The integration of the IN technology in the broadband network, the convergence between the data and the voice platform.
- The interworking between the narrowband and the broadband signalling systems.
- The addressing schemes to be used to route the calls.
- ECMA's BQ-SIG and ATM-Forum's PNNI will continue to play a leading role in the standardisation of broadband VPN signalling in future, but compatibility between the different broadband signalling systems is essential.

The broadband VPN will offer many benefits to the corporate networks. These include greater bandwidth, flexibility, cost-efficiency and service integration, in addition to innovative broadband services and capabilities. The current signalling systems provide support for simple broadband VPNs, and will continue to progress to support a wider range of VPN services and features.

REFERENCES

Alley D M, Kim I Y and Atkinson A, Audio services for an asynchronous transfer mode network, *BT Technol. J.* Vol. 13, No. 3, July 1995, pp. 80–91.

Bale M C, Signalling in the Intelligent Network, *BT Technol. J.* Vol. 13, No. 2. April 1995, pp. 30–42.

Q-SIG, *The Handbook for Communications Manager*, 1993.

Voice International incorporating World Telemedia Example, *Networking*, 1997.

STANDARDS

ECMA-251 PISN—Inter-Exchange Signalling Protocol—Common Information ANF (Q-SIG-CMN), Second Edition (December 1998)—equivalent to ISO/IEC 15772.

ECMA-252 B-PISN—Inter-Exchange Signalling Protocol—Transit Counter ANF (B-Q-SIG-TC) (December 1996)—equivalent to ISO/IEC 15773.

ECMA-254 B-PISN—Inter-Exchange Signalling Protocol—Generic Functional Protocol (B-Q-SIG-GF) (December 1996).

ECMA-261 B-PISN—Service Description—Broadband Connection Oriented Bearer Services (B-BCSD) (June 1997)—equivalent to ISO/IEC 15899.

ECMA-265 B-PISN—Inter-Exchange Signalling Protocol—Signalling ATM Adaptation Layer (B-Q-SIG-SAAL) (September 1997)—equivalent to ISO/IEC 13246.

ECMA-266 B-PISN—Inter-Exchange Signalling Protocol—Basic Call/Connection Control (B-Q-SIG-BC) (September 1997)—equivalent to ISO/IEC 13247.

E.164 The international public telecommunication numbering plan, ITU-T (05/97).

E.191 B-ISDN numbering and addressing, ITU-T (05/97).

QUESTIONS

1. What is the advantage of using a Virtual Private Network as opposed to a public network or a private network?
2. Why do companies invest in standardised solutions and signalling systems when manufacturers are able to offer proprietary solutions for deployment?
3. Explain the requirement for deploying broadband technology, when compared with obtaining more capacity by deploying additional narrowband infrastructures.

13

UMTS: The Mobile Part of Broadband Communications for the Next Century

Alan Clapton, Nigel Lobley, Steve Dutnall, Mark Dando and Pedro Serna

13.1 INTRODUCTION

This chapter discusses the radio and network requirements and developments for UMTS, and outlines the initial proposals for the network architecture and capabilities of UMTS compared to GSM and B-ISDN.

Within Europe the mobility aspects of UMTS is building upon GSM, with recent architectural developments within GSM acting as the basis for the mobile evolution path for future mobile systems. The transport requirements must also be addressed, and it is here that broadband based upon ATM is the obvious candidate. This will allow support of the dynamic, variable bit rate and asymmetric communications requirements of UMTS to be provided. The objective is to combine the variable, high bit rate capabilities of ATM with the developing intelligence control of B-ISDN and the high mobility and advanced supplementary service support that GSM offers.

One of the major successes of GSM has been its complete system specification in conjunction with the provision of basic standardised service sets and automated inter-operator roaming. While this has enabled operators to rapidly roll out feature rich networks, it has left little room for service innovation and differentiation via operator specific services. The GSM community has addressed this requirement via the development of the GSM CAMEL (Customised Applications for Mobile network Enhanced Logic) feature (GSM CAMEL 1997). CAMEL uses evolved IN capabilities, based on the INCS (Intelligent Network Capability Set) evolution model, to provide operator-specific services in both the home and roamed to (visited) networks. This is achieved by developing architectural SSP and IN signalling capabilities (GSM CAP protocol (1997)) while catering for the inter-operator roaming and mobile communications limitations. The service creation in the CAMEL platform is directly controlled by the home network operator.

Figure 13.1 illustrates the phase 1 CAMEL architecture, and shows the inclusion of the IN capability into the GSM network.

CAMEL provides an evolution step towards the Virtual Home Environment (VHE) feature of UMTS. VHE aims to enable users to access their UMTS services and features in a common configurable manner independent of the environment (e.g. public cellular, business/private, wireless PBX or home). This capability will ease many of the problems of common service support while roaming between the different public and private networks and accesses (termed Access Domains or Environments) shown in Figure 13.2. Access to different environments is already being provided within the GSM community by the use of dual/multi-mode GSM, DCS1800 and DECT terminals. UMTS will build upon these facilities to provide the true seamless access and support of multiple environments.

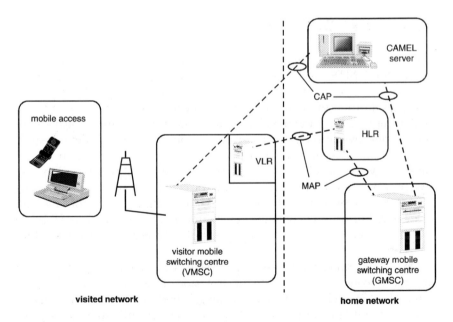

Figure 13.1 The GSM CAMEL Architecture

13.1 INTRODUCTION

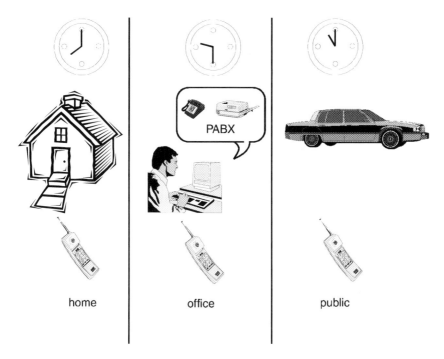

Figure 13.2 UMTS Access Domains

For broadband to support UMTS it must be able to provide seamless service support across all access types. It must therefore be able to communicate across networks at a service level. This will require the visited (serving) network to use information such as location and subscriber profile downloaded from the terminal or home network to handle the service requests of its newly arrived subscriber. Broadband must also provide the capabilities to support other consequences of the 'inbound roamer' and mobile nature of the terminals such as handover, authentication and security.

As well as catering for real-time connection oriented communications such as telephony and interactive video, UMTS will support connectionless/packet type services to cater for less delay critical data applications such as the Internet/intranet. Within the GSM community, support for connectionless communications is being developed via the GPRS (General Packet Radio Service) (GSM GPRS 1998), which uses a modified Internet IP protocol and packet transport mechanism to support direct connections to packet data networks. The handling of mobility management aspects has been achieved by interaction with traditional GSM network elements.

Although GSM supports the mobility requirements and the forerunner to VHE (via CAMEL), the system remains constrained by the (64 kb/s) N-ISDN technology within the core network and the radio interface used.

UMTS is expected to overcome these restrictions by enabling a flexible approach to data communications to enable connection and connectionless (packet) data capabilities, including: the support of asymmetrical data on the uplink and downlink; and variable data rates which allow applications to be added and dropped during communications (GSM HSCSD 1997). For this reason, broadband has been considered to support the variable asymmetric transport requirements. Broadband must therefore progress its developments from the current leased line PVC managed Virtual Circuits to include short duration SVCs. This will allow public access to the variable bit rates predicted for UMTS without paying the costs of 'business leased line' rental. At the same time, there must be intercommunication between the SVC's control and resource allocation functionality, service control; to cater for user's service requirements and preferences (and the subscription which determines if the user can request such a capability), and mobility aspects such as handover and path rerouting. If this cross-interrogation is not available, an illegal request for resources may be allocated, or one which the resources of the environment cannot support; a typical example would be a request for a high bit rate service in a low bit rate, high mobility domain. With the developments of IP in the community, broadband cannot propose to be a UMTS system if it doesn't support common intelligence with IP. At present, IP and broadband are seen as two separate entities, but it is envisaged that as telecommunications become more multimedia based, there will be a greater need for a combined system providing both the time critical and delay tolerant transmission capability. Only a convergence of technologies supporting both broadband and IP will enable UMTS to support the harmonised execution of mixed media communications envisaged.

13.2 THE PHASED INTRODUCTION OF UMTS

Due to the widespread adoption of GSM within Europe and worldwide, ETSI has adopted a phased approach to UMTS to ensure a smooth transition for both operators and end users from second to third generation systems. Table 13.1 provides the typical thinking within the three phases adopted.

The initial usage of UMTS frequencies and the developing radio interface is made within phase 1. Phase 1 introduces the connection of the new UMTS Terrestrial Radio Access (UTRA) to the current 64 kb/s and GPRS infrastructure. This will provide the radio interface requirements to fully support UMTS, including asymmetric transmission, variable bit rate allocation and 'requestable' QoS communications, although in reality it will be limited due to the reliance on the legacy system that it connects

13.2 THE PHASED INTRODUCTION OF UMTS

Table 13.1 The Phased Approach for UMTS

UMTS Phase	Service/customer capability	Network capability and radio spectrum utilised
Phase 0	GSM Phase 1, 2 and some 2+ service set, (Voice and data services, connection and Connectionless, IN service control, world-wide mobility via roaming, data rates up to 64 kb/s) Seamless services (handover and location management) Simple terminal-based (menu) control of supplementary services and features.	GSM radio access at 900, 1800 and 1900 Mhz N-ISDN-based core network, plus an IP-based network as overlay for connectionless data via GPRS. IN integration through CAMEL into connection mode network for both home and visited network-based mobiles.
Phase 1	As above plus data rates up to 144 kb/s via $N \times 64k$ communications. High quality speech (GSM EFR Codec), service management via flexible user control interface. Roaming between UMTS, UTRA network and GSM.	Limited coverage from new UMTS radio interface system, interworking into existing IP and N-ISDN based core networks, mobility management as per GSM. Maintained use of GSM 900, 1800/1900 for mass coverage. Dual mode GSM/UMTS terminal.
Phase 2	Full UMTS requirements. As above plus simplified service control, Virtual Home Environment (VHE) with automated roaming, home business and wide area usage. Support for full variable bit rate multimedia communications, alternate addressing mechanisms, and service adaptation.	First integrated fixed-mobile network. Access network is UMTS radio access system. Core network is connectionless-based transmission supporting connection (real time) and connectionless traffic. Intelligence linkage to manage customer service profile, preferences and limitations, etc., satellite and terrestrial access to network.

to. Despite not actually being part of the UMTS development, the initial phase (termed Phase 0) is considered as 'conditioning the market' towards UMTS via the application of contemporary DECT/GSM/DCS 1800 technology and 'peripheral' intelligent terminal-based solutions as used in IP networks. Phase 2 is the true target scenario, and will offer the full service capability. This will require significant changes within the network to fully exploit not only the UMTS radio access, but also traditional 'fixed' access such as home and private LANs.

13.3 RADIO INTERFACE REQUIREMENTS

The radio requirements for UMTS are critical in ensuring optimum use of the limited radio resource in providing mobile multimedia. The European UMTS forum has estimated that for the type and quantity of services envisaged for UMTS, a total 500 MHz of spectrum is required. Within Europe, UMTS currently only has 2*60 MHz of spectrum allocated for public terrestrial use, hence there is a need to acquire more spectrum to operate in, and/or find more efficient means of utilising the spectrum.

The UTRA solution strives to satisfy the following key objectives:

- To provide a common wireless interface for roaming across all terrestrial environments.

- To provide a flexibly structured radio interface that can adapt to cater for the range of existing and future (as yet unknown) services. This includes a capability to support high bit rates.

- To make efficient use of available radio spectrum.

A range of radio environments have been identified for the UTRA to operate in, including Indoor (residential and business) and Outdoor (urban and rural). Each of these environments has differing radio propagation characteristics that influence the parameters of the UTRA. The constraints that these impose are summarised in Table 13.2.

The radio interface will be able to achieve full dynamic variable bit rate throughout the call (up to the maximum rates specified for the different domains in Table 13.2). To achieve this, the channel Quality of Service (QoS) attributes can be negotiated at call set-up and adjusted throughout the call to suit the specific service required. The attributes that can be varied are:

- Variable bit rate.

- Variable delivery delay.

- Variable duplex symmetry. It is probable that future typical applications will transmit larger volumes of data in one direction than the other (e.g.

Table 13.2 Radio Environment Capabilities

Environment	Maximum user bit rate	Maximum user speed	
Indoor (Low Range)	2 Mbps	10 km/h	(6 miles/h)
Outdoor (Urban)	384 Kbps	120 km/h	(75 miles/h)
Outdoor (Rural)	144 Kbps	500 km/h	(311 miles/h)

Internet browsing—high bit rate downlink but minimal rate uplink). To maximise efficiency in handling such traffic, the radio interface should allow asymmetrical connections to be set up.

- Variable Protection levels to accommodate a range of acceptable bit error rate.

The requests to vary these attributes will be balanced by modification of the link according to quality, network load and radio conditions, in an attempt to optimise the link in the different environments. This highlights the need for network and access communication and QoS negotiation for all resource requests.

13.4 CURRENTLY PROPOSED RADIO TECHNOLOGIES

The standardisation of the UTRA is currently being performed within ETSI's SMG2 group, and is limited to the terrestrial (i.e. excluding satellite) components of the radio interface. This is ongoing, and a time schedule exists to choose the final UTRA technology, which will stem from a combination of the following proposals. The milestones in the decision process were

1. June, 1997: Agreed UTRA requirements and evaluation procedure.
2. December, 1997: Selection of one UTRA concept.
3. June 1998: Definition of the key technical aspects of the selected UTRA.

Several proposals have been presented to SMG2 (UMTS 1996), originating from a range of manufacturers and operators. The key difference between proposals is the multiple access technique chosen, this being the factor that determines most of the other system characteristics. The three fundamental access techniques are FDMA (Frequency Division Multiple Access), TDMA (Time Division Multiple Access) and CDMA (Code Division Multiple Access). All proposals are based on hybrids of these technologies, with the proposed radio access techniques that show similarities being collected into five concept groups to ensure co-operative development.

These concept groups are currently evaluating the technologies against the set of requirements. The final decision as to what UTRA is finally chosen will be based upon their output.

The UTRA highlights the European position for a third generation mobile system. Within other world regions different standards are being developed. It looks increasingly likely that the ITU will not necessarily standardise a single solution, but implement a framework to allow roaming between the different regional solutions (known as the 'family concept'). One suggestion is to standardise a common global broadcast signalling channel which would indicate to a mobile station which air interface technology to operate in that environment. This common signalling channel could also be useful in each region/environment to indicate which portions of the spectrum should be used and which should be avoided. To accommodate regional variations in available spectrum, the radio interface should be capable of operating in any feasible frequency band that becomes available.

13.5 NETWORK REQUIREMENTS

Due to the multiservice and capability nature of UMTS, it is essential to select the correct technologies within the network to provide the required functionality. High level analysis of network intelligence requirements to support mobility and service control has been presented previously (Cullen and Lobley 1996). The underlying issue of the transport technology beneath the network still remains.

When considering the information to be transferred by UMTS, a rough separation can be made in the data types characteristics, summarised as:

- *Message*: Connectionless-'data', time delay tolerant but requiring low BER. This is generally a fixed length single packet, for handshake type operations; as seen for signalling packets.

- *File*: Connectionless-'data', time delay tolerant but requiring low BER. Generally this is variable in length; WWW Internet page download, File Transfer, etc.

- *Stream*: Connection oriented—this can be further separated into two sub-groups:
  — *Stream: delay intolerant* such as 'speech' and 'videophone' like applications, normally characterised as being two-way (duplex) interactive. This is time and delay intolerant, but is more flexible in its BER demand.
  — *Stream: delay 'some' tolerance*, such as in video broadcast, where the terminal may be storing the data but there is no catastrophe if small time delays occur.

To support carriage of the different user information (while still maintaining communications in the hostile mobile environment), a combination of bit transport, call and connection control, charging, service and mobility control is required. The application of a number of contemporary technologies has been considered to satisfy the UMTS requirements:

Circuit switched (64 kb/s):

— PSTN/ISDN/GSM

— 64 kb/s technology

— Fixed bit rates (N*64 kb/s)

— Linkage to 'IN' and 'mobility'

— Potential B-ISDN evolution.

Packet and variable rate switched:

— Continuous and discontinuous

— ATM, GPRS, IP and Mobile IP

— Mainly bit transport

— Flexible, variable rate and QoS control possible

— No external 'control'

— Potential mobility via Mobile IP.

Of the technologies considered, the former 64 kb path is not a realistic solution due to its rigid bit rate and transport capability, however it does support interaction with external intelligence. For the latter technologies, the packet and variable bit rate mechanisms provide a good transport capability, but are currently inflexible with limited interactions to external intelligence.

Current thinking to satisfy the UMTS network transport and control requirements involves adopting ATM/IP transport and including links into intelligence for service control and mobility management. ATM has been selected due to its switching speed and potential QoS parameters, and IP for its current widespread use and simple interworking with legacy systems. This will satisfy the flexible user bandwidth (high and low bit rates), asymmetric data services, multimedia, variable QoS and connectionless/connection oriented requirements, while also enabling the adoption of B-ISDN (Q298X) signalling technology and the developing DAVIC concepts (Donnelly and Smythe 1997) for call/media control.

13.6 B-ISDN AND IP DEVELOPMENTS

In the development of broadband the following stages are seen as major steps to fully support the UMTS requirements. Many of the following proposals are already under consideration or development for other purposes, but to support UMTS it is essential that the mobile specific deltas are included.

The introduction of ATM into the network is via leased line Permanent Virtual Circuits (PVCs). These circuits are permanently nailed up, set up by external management procedures and on a leased line basis. This is a permanent connection providing high bandwidth capabilities with no possibility for service, call control or service interaction. The next step is the introduction of soft PVCs, which support signalling to dynamically re-route PVCs should failure to links or switches occur. Not only does this make the network more resilient, it also introduces a signalling mechanism leading to the next obvious technology step of Switched Virtual Circuits (SVCs), providing dynamic allocation of virtual circuits on demand. SVCs support call control capabilities (in a similar way to N-ISDN's 64 kb bearers) by allowing variable bit rate bearer allocation. The separation of bearer and call control allowing several bearers to be allocated to a single session is the next fundamental step. This not only supports dynamic variable bit rate allocation throughout the call, but also provides a potential solution for mobile handover by allowing legs to be added and dropped within a 'communications session' when required.

With developments in service control to provide mobility procedures, the ability for the intelligence to interact with the ATM transport to add and drop bearers will be essential to physically support handover and re-routing procedures. Handover and variable bit rate allocation will only be efficiently supported in ATM if there is a separation from bearer control and call control. If there is no separation between bearer and call control, complete end-to-end 'calls' will have to be built up, to add or drop calls, which includes the handing of the terminal over to another access point. This obviously has serious impacts on the resources of the system.

IP's development for UMTS builds upon the introduction of some intelligence within the network to support the mobility requirements of Mobile IP. To be fully compliant with the UMTS goal, this mobile information needs to be accessible from other networks, and to be modifiable to update information present in its database. Thus, any other network can have access to the location information of a terminal without the need of sending the packet to the terminal's network only to be forwarded out again.

Another technological hurdle at present is the convergence of ATM and IP. Considering that IP is a connectionless mechanism and ATM a connection-oriented one, there are immediately problems seen in the transport of information and how these two methods can be converged. What UMTS implies is that a common method of 'control' or linkage of information will be available to both mechanisms. This means there will be control information such as mobile information, subscriber preferences and limitations, and security information which relate to both transport choices for UMTS.

It also implies that a single common system will control mobile procedures, call control and resource allocation and modify parameters throughout the session.

For ATM to support a connectionless non-resource allocating mechanism such as IP, a route relationship needs to be established (without the allocation of any set routes or resources until the packet was being transmitted) and then closed down once transmission of the packet was completed. On the other hand, if IP wished to provide a circuit switched type service it would have to introduce call control and resource allocation procedures throughout, clearly going against IP original thinking. It therefore seems unlikely that solely one or the other technology will exist in isolation; however, there are more optimistic views on a common databases and intelligence servers influencing/controlling the lower transport mechanisms.

What has been presented previously are the initial high level thoughts on the ability of broadband and IP (with developments) to support UMTS. The adoption of this evolved ATM and IP with links to intelligence still requires development and more detailed study to consider the interworking functionality required for contemporary systems, plus the more detailed mobility, security and radio related issues.

13.7 THE NETWORK ARCHITECTURE FOR UMTS

The vision of UMTS is of convergence between fixed and mobile networks. The core network will support both wired and wireless access. Figure 13.3 illustrates the conceptual architecture required for UMTS.

The three main components for UMTS (or any telecommunications system) are access (radio or wired), network transport/switching and overlying service and mobility control.

Based upon the overall architecture requirements and the analysis of transport and intelligence options, a conceptual functional network architecture has been developed, as shown in Figure 13.4.

The main areas depicted in Figure 13.4 can be separated out into the Home Server, the UMTS Server, Transport, Access and the Terminal. These are explained as:

- *Home Server*: this is likely to be a single entity containing the subscriber's profile. The subscriber's profile consists of the complete list of services subscribed to, and their constraints, as well as the particular preferences of the user. It also contains the most recent location information and authentication/security information, and hence is interrogated for terminal terminating calls and initial registration when the terminal is requesting attachment to the network.

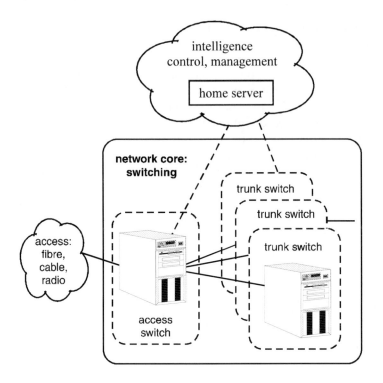

Figure 13.3 Conceptual Architecture for UMTS

- *UMTS Server*: this is seen as a distributed intelligent layer, although depicted for ease as a single entity. The UMTS server controls the resource allocation, call control and bearer allocation. For mobile terminals the UMTS server will control the mobility procedures and associated resource allocation. To enable the VHE for UMTS requires constant interaction with the service profile, and so a copy of the relevant parts of the subscriber's profile is present, downloaded from the home server. Information from the access environment is also essential to identify what resources are available and under what restrictions.
- *Transport*: the transport system encompasses two mechanisms, both packet switched and cell switched (connection oriented). It is assumed that all mobility, call and service control intelligence will be taken out of the switch/router, and decisions on routing and resource allocation are carried out in the distributed UMTS server.
- *Access*: the access connects the network to the user. The transfer of information from the access type to the UMTS server is essential to inform it of the current resource status of the access and the access' capabilities/limitations.

13.8 KEY NETWORK OPERATIONS FOR UMTS

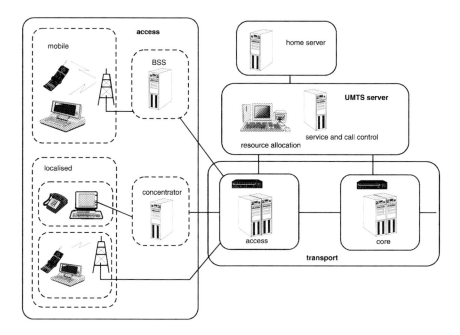

Figure 13.4 Network Architecture for UMTS

- *Terminal*: the terminal will have much greater contact with both the network and access systems; communication will probably be via an Internet front end.

Developments for this architecture include the adoption of the IP switching/router and ATM-based transport in parallel with links to the distributed intelligence processing (home/serving division).

13.8 KEY NETWORK OPERATIONS FOR UMTS

To illustrate the operation of the network, the following introduces a number of typical network operations. Figure 13.5 illustrates initial registration with UMTS, and involves interaction between the terminal and the network intelligence (both home and serving) to check and transfer the users relevant service information (including information such as VHE capability) into the serving network's intelligence. As shown, one potential technology which can be applied to such activity is the developing Mobile IP. This would remove the need for a dedicated channel on registration, thus increasing signalling efficiency over the air interface.

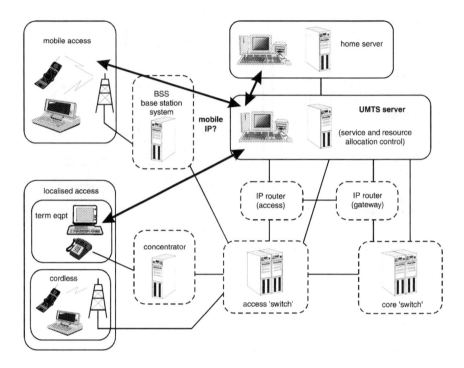

Figure 13.5 UMTS Registration

Figure 13.6 illustrates the procedures for setting up mobile originated connectionless (packet) communications.

The operation involves the terminal interacting with the network intelligence to set up a 'session'. The session defines the requirements for the communications. The intelligence verifies whether the request is legal in relation to the subscriber's profile, and communicates with the packet mechanism of the network and access nodes. The network intelligence will configure the packet network's routing tables to enable the correct packet route. The 'session' will be maintained as the mobile user moves, with the network path being reconfigured accordingly, with data 'cached' and rerouted as required for 'handover'.

A similar operation is performed for mobile terminated connectionless (packet) set-up, as shown in Figure 13.7.

Note that for the mobile terminated connectionless (packet) set-up, the Home Server is involved to configure the packet routing to the serving network (to invoke a terminal initiated 'session set-up'); this also involves a checking of the called mobile's subscribed services and capability. Once the (mobile originated or terminated) path has been set up (via the session being in place), the actual transmission path is open to convey the packets of data. The path adopted is illustrated via the dashed lines in Figure 13.8.

13.8 KEY NETWORK OPERATIONS FOR UMTS

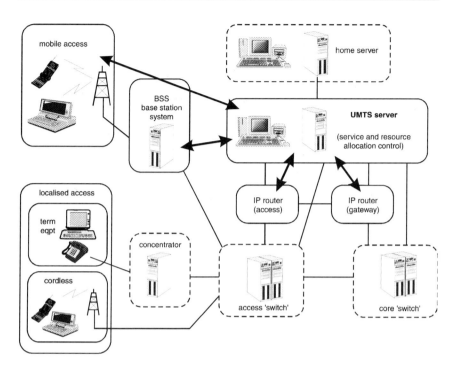

Figure 13.6 Mobile Originated Connectionless (Packet) Set-up

If the path needs modification, the mobile/server will interrupt communications and modify the routing configuration as illustrated in Figure 13.6. A similar operation is applied to connection oriented communications, the only difference being that the 'session' is on a 'per-call' basis. The server will interact with the ATM switching to operate services and support path modification for handover and rerouting, etc., dropping or adding bearers accordingly.

To enable UMTS to cater for variable bit rate multimedia communications, it is essential that the network can modify, add and drop communications components 'in call'. This is achieved by modification of the parameters discussed within the 'session', as shown in Figure 13.9.

Changes in the session can occur due to changes in the user or application requirements and are termed *application initiated*, or are caused by the network and are termed *network causal*. These are typically due to the inability of the network to continue to support the communications in their present form, hence a renegotiation of the session parameters is required. Network causal changes can be further separated into *mobility* aspects due to the mobile environments (e.g. handover, location or user service constraints), and *network resource* aspects to allow more users access to the (limited) network radio resource via the lowering of priority settings for transmission.

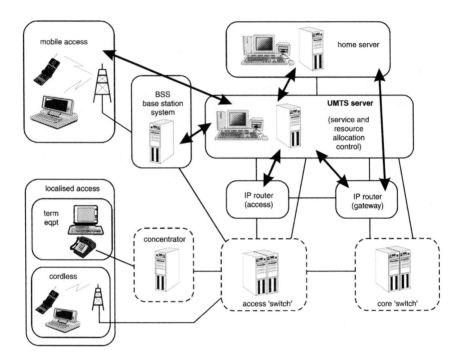

Figure 13.7 Mobile Terminated Connectionless (Packet) Set-up

13.9 CONCLUSION

From a high level perspective, the major development UMTS offers is the provision of multi-environment multimedia mobile communications. The network technology to fulfil UMTS requirements still needs evolution and enhancement; the major development being the linkage of connectionless (packet) and connection oriented (real time) capability, and then enabling this technology to interact with network intelligence.

For broadband and IP to fully support UMTS, developments must encompass the following:

1. Develop broadband interaction with IN.
2. Develop mobility and service control within intelligence on the transport mechanism.
3. Standardise completely the separation of call control and bearer control to allow for the dynamic variable bit rate and handover capabilities required.
4. Develop IP interactions with IN.
5. Develop methods of linking intelligence across IP and broadband systems.

13.10 TIME LINES

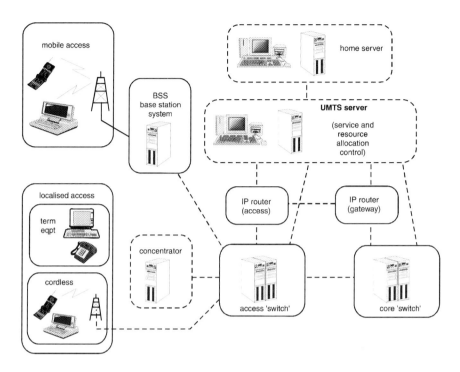

Figure 13.8 Connectionless (Packet) Transmission

6. Develop how both IP and ATM transmission technologies will support common mobility control via the intelligence.
7. Standardise capability to allow for intervendor and internetwork operability on a capability level to support the UMTS requirements such as VHE.

13.10 TIME LINES

The time lines are separated into what has been identified for the radio requirements according to the SMG work plan, and the network requirements that would be necessary to support this added functionality.

Radio requirements

The following is a summary of the predicted time line according to the SMG workplan.

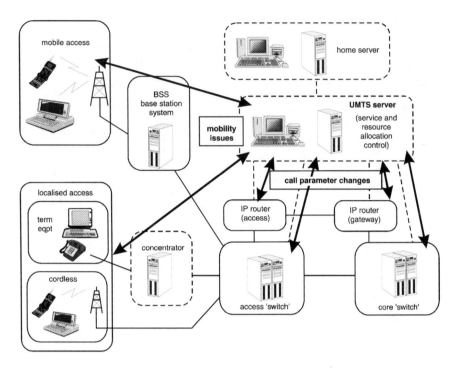

Figure 13.9 Session Modification

1997 Radio access technology chosen for the 2 GHz air interface with up to 2 Mbit support.
1998 Basic standardisation complete on chosen radio access technology.
1999 Test bed trials.
2001 Availability of equipment from vendors.
2002 The UMTS air interface is introduced with interworking between UMTS and the legacy systems of that time GSM 900/1800. This would allow up to 2 Mb/s enabling communications via blocks of N*64 kb/s.

Network requirements

To meet the needs of the requirements identified, the following would need to be achieved within the times specified.

2000 Standardisation complete covering the following areas:
- Switch Virtual Circuit definition.
- Subscriber profile storage.

- Inter-service communication to provide personal mobility and limited VHE.
- ATM separation of call and bearer control providing dynamic variable bit rate allocation.
- Broadband support of mobility procedures for terminal mobility.
- Interaction with IP 'intelligence' to provide common control of mobile and service interaction in the network.

2002 Test bed trials.
2003 Availability of equipment from vendors.
2005 Implementation of UMTS within the network. Evolution of a ubiquitous transport mechanism, to provide efficient carriage of both connectionless and connection oriented call types within the network. Development of full intelligence interaction to provide service and mobility management control to allow handover and operation in all access domains, including the support of service adaptation and full dynamic broadband rate allocation.

With these capabilities, UMTS will be the common network platform to support fixed and mobile telecommunications into the next millennium.

REFERENCES

Cullen J M and Lobley N C, The Universal Mobile Telecommunications System—A mobile network for the 21st century, *BT Technol. J.* Vol. 14, No. 3, July 1996, pp. 123–131.

Donnelly A and Smythe C, A tutorial on the Digital Audio Visual Council (DAVIC) standardisation activities, *E&CEJ*, IEE, February 1997.

GSM CAMEL, GSM Recommendation 03.78, *Digital cellular communications system (Phase 2 +); Customised Applications for Mobile network Enhanced Logic (CAMEL)*, 1997.

GSM CAP, GSM Recommendation 09.78, *Digital cellular communications system (Phase 2 +); CAMEL Application Part (CAP) specification*, 1997.

GSM GPRS, GSM Recommendation 03.60, *Digital cellular communications system (Phase 2 +); General Packet Radio Service (GPRS) specification*, 1998.

GSM HSCSD, GSM Recommendation 02.34, *Digital cellular communications system (Phase 2 +); High Speed Circuit Switched Data (HSCSD)—Stage 1*, 1997.

UMTS radio interface proposals submitted to SMG2. ETSI SMG2#20, Sophia Antipolis, December 1996.

QUESTIONS

1. What functionality provides the major requirement for broadband to support UMTS?
2. What advantages will UMTS have over a combined GSM and GPRS service?
3. Why does the separation of call (or session) control from bearer control have major advantages for UMTS?

14
Signalling with Objects

Paul McDonald

14.1 INTRODUCTION

Conventional Public Switched Telephone Network (PSTN) and Narrowband ISDN (N-ISDN) signalling systems have evolved from the need to support the two-party telephony service. However, the development of Broadband ISDN (B-ISDN), with its associated advanced switching and control capabilities, results in a network that can support services that are far more advanced than telephony. In particular, using Asynchronous Transfer Mode (ATM) permits multiple connections to be established over the same transmission pipe, advanced switching can support more than two parties in a call, whilst in future there will be many more communications services on offer than just telephony.

Thus, there is the requirement for B-ISDN to signal multiparty and multiconnection calls across the network. This is a very complex task, especially once one considers that some terminals may not be compatible with others, may lack facilities for some media types in use, require special resources such as conference bridges within the network, and that parties may join/leave the call.

The network technology already exists to perform all of the above. The difficulty lies in communicating the requirements between terminals and the network. Thus, it can be seen that signalling is key to enabling this. Inadequate signalling can prevent users from fully utilising the capabilities of the network.

At the start of the 1990s the existing signalling protocols, such as the initial releases of Q.2931 and B-ISUP (Broadband ISDN User Part), were inadequate for such advanced services, and it would be necessary to develop enhanced signalling protocols to meet these needs. The European Commission established the RACE 2044 (MAGIC) project to undertake pre-competitive research into exactly these issues (Popple and Glen 1993).

A number of network operators, equipment providers and universities, with BT acting as prime contractor, collaborated between 1992–94 to develop advanced User-Network Interface (UNI) and Network Node Interface (NNI) signalling protocols for the B-ISDN. Their remit was to look to the future, and to provide a longer term view to the international standards bodies for signalling.

The following companies contributed to the MAGIC project: BT (UK), ATEA (B), AT&T (NL), University of Twente (NL), Alcatel Bell (B), Alcatel CIT (F), CSELT (I), Ericsson Telecommunicazione (I), NTUA (GR), Telekom FTZ (D), Technische Universität Ilmenau (D), PTT Research (NL), Bellcore (USA), Italtel (I), RACE Industrial Consortium (RIC).

This chapter discusses the signalling system developed within this project, and describes how it can be used to support complex services. The next chapter describes how some of the techniques developed within the MAGIC project have been incorporated by the ITU-T into the B-ISDN standards.

14.2 LIMITATIONS OF THE CURRENT APPROACH

The first release of the access signalling protocol for broadband ISDN was Q.2931 (DSS2), and early releases were essentially a development of the narrowband ISDN's signalling protocol Q.931 (DSS1), with alterations to accommodate the ATM transport layers. Unfortunately, this evolution means that Q.2931 inherits many of the limitations of Q.931 and its ancestors. When applied to multiparty, multi-connection calls, extending Q.2931 becomes unwieldy and complicated, essentially requiring the signalling state machines to be modified for each new type of service supported. This is due in part to the architecture behind Q.2931 deriving directly from early telecommunications signalling protocols designed purely for telephony (Figure 14.1).

Q.2931 is designed for two-party communication with a single bearer connection between the parties. When the call is established, the calling party controls all of the attributes of the call according to its model of how the call should exist through the network. However, as services become

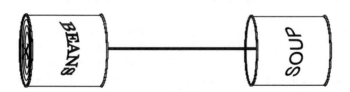

Figure 14.1 Underlying Architecture of Q.2931

14.2 LIMITATIONS OF THE CURRENT APPROACH

more advanced, such an overall view of the call becomes more difficult, in fact in some services it may even be difficult to determine who is the caller and who is the called party. As an example, consider the hypothetical travel agency service that was described in Chapter 4.

A customer uses their videophone to enquire about a holiday. During the conversation the travel agent contacts a multimedia video library to select a movie clip for the holiday in question. The travel agent then instructs the video library to play the movie clip for the holiday directly to the customer. At this point, the call model has the appearance of the arrangement in Figure 14.2.

In the above service we have a mix of parties creating connections, not just the original caller. Some of these connections are bidirectional and some unidirectional. Also, it can be seen that not all parties participate in all connections. For instance, the customer need never be aware of the data link between the travel agent and the video library. Thus, in this case the original caller need not have a view of everything that exists during the lifetime of the call. It can also be seen from Figure 14.2 that the travel agent is setting up the video connection to the customer without using the video itself. Thus, we have an example of the control flows for the call being separate from the signalling used to establish the bearer connections. This requires separation of call and bearer control.

Another issue that has to be dealt with is interworking between the different codecs used to encode information. It is pointless for the video library to establish an MPEG video connection to a customer that only supports H.261 video. What is really required is a mechanism to invoke network- or terminal-based interworking to deal with such mismatches without failing the call.

The travel agency service is a fairly complex service. It is impossible for the network to directly build in support for every such service that may come about. Instead, what is required is a signalling protocol that can flexibly accommodate any such service requirements without needing to know the overall architecture of the service. This suggests that it is necessary

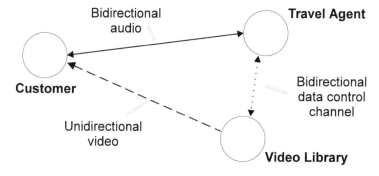

Figure 14.2 Travel Agency Service

for the signalling protocol to permit applications to manipulate connections, support parties undertaking different service roles and using terminals of differing capabilities, effectively treating them as a set of resources, or objects, to be mixed and matched as necessary. Also, with the high degree of complexity of such services, it is necessary for the signalling protocol to permit these manipulations to take place in a controlled way such that the network never ends up in an indeterminate state.

14.3 MAGIC SIGNALLING MODEL

Existing signalling protocols have a number of weaknesses when applied to the complex multiparty multimedia world of broadband ISDN. Therefore, the MAGIC project developed a number of new signalling concepts to counteract these weaknesses, and to build the basis for a flexible signalling system for broadband. This section describes the main concepts introduced by MAGIC (Knight 1993).

Separation of protocols

It was a basic assumption of MAGIC that the network would support separation between call and bearer control, at least within the internodal (NNI) protocols. It was determined within the MAGIC project that call/bearer protocol separation was not necessary at the User Network Interface (UNI), and thus a monolithic protocol was devised for this.

Effectively, Call Control (CC) is an end-to-end negotiation protocol that permits users to determine the call configuration they desire. Once determined, Bearer Control (BC) is used to establish the required bearers throughout the network to support this call. With this separation, not all nodes need implement the Call Control signalling stack, for example Call Control could be included in the local exchanges (LEX) and omitted from the transit exchanges (TEX) (see Figure 14.3). To provide an evolutionary path from existing networks, Broadband ISUP (B-ISUP) was employed as the bearer control protocol.

When the MAGIC project investigated the complexity of multimedia calls, and in particular, the problems involved in allocating special resources such as interworking units, conference bridges, etc., it was found necessary to develop an additional protocol, Resource Control (RC), to manage these special devices within the network. Call Control could not be used for this purpose as resource management would violate the principle of Call Control being an end-to-end signalling protocol.

At the UNI a monolithic protocol was devised to perform the call

14.3 MAGIC SIGNALLING MODEL

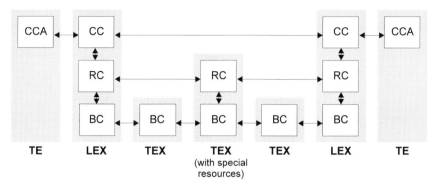

Figure 14.3 CC, BC and RC Signalling Stacks

control and bearer control functions. This entity is termed the Call Control Agent (CCA).

Local views

Traditionally, the terminal creating a call signals all of the information related to that call as applied to all participants in the call. Thus, it could be said that there exists a 'global view' for this call, that is, all of the relative attributes for the call can be described as a single entity.

With the complexity of broadband ISDN, where different parties in a multiparty call are using different communications flows, such a 'global view' can become unwieldy. Thus, in MAGIC the concept of 'local views' was adopted. A 'local view' is the view a particular terminal has of the call. Basically, it only describes the parts of the call with which that terminal has an association, for example, the remote parties that the terminal is communicating with, or the connections that terminal has established. If objects are not of relevance to a terminal, e.g. a connection between two remote parties only, then it need not be represented in that terminal's 'local view'.

Each terminal has a 'local view' that describes that terminal's view of the service. This will be different to the 'local views' of other terminals in the call. With this approach there is no single 'global view' of the call, it is only when all of the individual 'local views' are brought together that the entire call can be described (see Figure 14.4).

Abstract services

With the increased communications options supported by broadband ISDN, interworking between terminals using different encoding standards

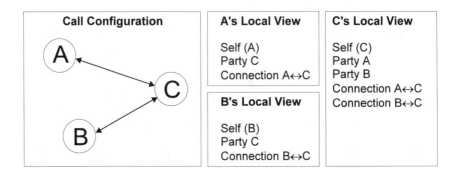

Figure 14.4 'Local View' Concept

for media sources will become increasingly difficult. For example, two terminals may wish to establish a videotelephony call, but one terminal uses H.261 video standards and the other uses MPEG. If the network can offer an H.261/MPEG converter, then in theory this call could be connected. The problem lies in signalling this between the terminals and network.

MAGIC introduced the concept of *abstract service*. This describes encoding independent values for the communications media, i.e. video in the example above. For a videotelephony call, both terminals must accept the same video abstract service, but may signal different encoding parameters for use within their own 'local view' of the call. The fact that they are using different encoding types to the remote terminal is irrelevant as the network deals with the required translation, thus permitting incompatible terminals to interact with the service. Thus, the abstract service enables the global attributes of the communication for that media to be described, whilst the encoding types remain local to each terminal.

MAGIC Call Control Objects

The MAGIC signalling protocols essentially establish calls by manipulating a set of signalling objects. This object-orientated approach permits complex services to be built from reuseable set of resources. This provides a powerful signalling capability based on a set of simple objects and operations on these objects. Figure 14.5 shows the signalling object model, and Table 14.1 lists the object definitions.

Signalling for broadband calls is essentially the process of creating, modifying and deleting these objects in a terminal's local view. For example, to create a two-party telephone call, the initiating terminal would have to create the following objects:

- Party Set (specifying telephony service).

14.3 MAGIC SIGNALLING MODEL

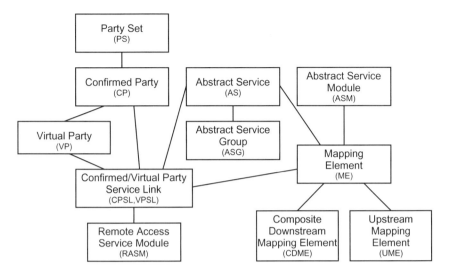

Figure 14.5 Signalling Object Model

- Confirmed Party (called party).
- Abstract Service (for audio).
- Abstract Service Module (for the audio codec used by the caller).
- Upstream/Downstream mapping elements (suitable for audio codec).
- Confirmed Party Service Link linking the remote party to the audio service.

The remote party will have a similar set of objects created except the Abstract Service Module and Mapping Elements may be different to the caller (if interworking is taking place within the network). Once these objects are created, the telephony call will exist in the network.

Atomic Action Protocol

Complex broadband services will require complex sets of information to be signalled. For example, multiple parties may be added to calls, each using different sets of connections, whilst the connections to other parties are being modified. This results in complex interactions that need a mechanism to ensure the system ends up in a stable and correct state.

MAGIC employed an Atomic Action Protocol (AAP) to signal these changes. Atomic Actions are operations which act on entities that must succeed or fail as a unit. This protocol is based upon the ITU X.851 Commitment, Concurrency and Recovery (CCR) protocol.

Table 14.1 Object Definitions

Object	Description
Party Set (PS)	This defines an association between a set of parties, and thus essentially defines a call and the service being supported by this call. All parties in a call must share the same Party Set object.
Confirmed Party (CP)	A Confirmed Party is a party that is directly involved with the negotiation of bringing a party into the Party Set. For example, you add another party to a call by creating a Confirmed Party object for it. This party will similarly have a Confirmed Party object created in its local view representing the initiator.
Virtual Party (VP)	A Virtual Party is a party who is not being directly added to the Party Set for a particular local view. Thus, if a call exists between parties A & B, and A adds a new party C to the call, then C will appear as a Virtual Party in B's local view.
Abstract Service (AS)	This is an abstract 'global' representation of a media communication type, or Service Module as it is termed in MAGIC. For example, this could be video or audio.
Confirmed/Virtual Party Service Links (CPSL/VPSL)	A Party Service Link associates a particular abstract service with the relevant party. Thus, if a remote Confirmed Party is to have audio and video connections, then two Confirmed Party Service Links need to be created to link this party with the audio and video Abstract Service objects.
Abstract Service Group (ASG)	The Abstract Service Group permits a number of Abstract Services to be synchronised in some way, perhaps to ensure the audio and video streams are in lip-synchronisation with each other.
Abstract Service Module (ASM)	An Abstract Service Module encapsulates the details of a user endpoint of a bearer connection. It represents the attributes for that connection such as ATM channel attributes and the encoding standards used.
Remote Access Service Module (RASM)	This is a virtual object that provides information of the Access Service Module used by a remote party. If a Party Service Link does not have a RASM then it indicates that that party has no bearer for this Abstract Service setup. This object also permits third party call setup (where a party sets up a connection between remote parties only).
Mapping Element (ME) Composite Downstream Mapping Element (CDME) Upstream Mapping Element (UME)	A Mapping Element describes a unidirectional flow of information between an Abstract Service and an Abstract Service Module. It can effectively be thought of as a representation of an ATM connection. Two mapping elements are normally needed for a connection, one for the downstream link, and one for the upstream link.

14.3 MAGIC SIGNALLING MODEL

When an entity wishes to change the state of the call (e.g. to add parties to a call), it starts an atomic action that requests the other entities to accept or reject this change. If they all accept the change, then the initiating entity will tell them to commit to the change, and all entities will update themselves to the new values. If any one of the entities fails to accept the change, then the atomic action will be 'rolled back' and all objects will return to their initial state. Thus, in the previous example, all parties being added to the call would have to accept the request, otherwise none would be added (Figure 14.6).

When deciding whether an atomic action request can be accepted, an entity may have to cascade this request to one or more other entities. These other entries may further cascade the entries, building up an atomic action tree. In this tree, unless all of the terminating entities accept the atomic action, then all of the atomic actions will fail. An example of this is adding parties to a call. A terminal will signal the creation of new party objects in an atomic action to its exchange. This may then cascade two further atomic actions, one to create each party object to the exchanges these parties are connected to, and thus to the remote parties themselves. Both parties must accept this creation for the originator to receive the READY indication (see Figure 14.7).

In the example in Figure 14.7, A wishes to create a call to B and C. A initiates an atomic action to add B and C to the call. These cascade two further atomic actions for B and C through the local exchanges. B accepts this with a READY, whilst C refuses it with a ROLLBACK. Since at LEX1 one of the child atomic actions has failed, the entire atomic action tree fails. Therefore, LEX1 sends a ROLLBACK to all remaining atomic actions in the tree.

A further refinement of the atomic action protocol is the ability to nest additional atomic actions within a parent action. The child operations can be labelled as *mandatory* or *optional*, mandatory meaning that the child must succeed for the parent to succeed, optional meaning that the child

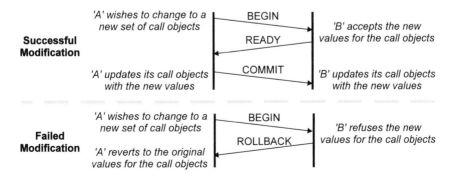

Figure 14.6 Atomic Actions

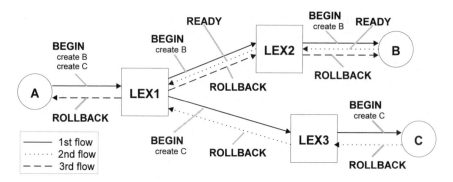

Figure 14.7 Atomic Action Tree Example

may fail without affecting the success of the parent. However if the parent operation fails, then all child operations fail.

This nesting of atomic operations provides the modularity of the atomic action protocol, while permitting optional operations to be indicated. For example in a videotelephony call, one nested atomic action could be used to establish the audio connection, and another optional atomic action to establish the video. A terminal may accept the overall atomic action with the audio while refusing the video (perhaps it does not support video).

14.4 MAGIC SIGNALLING EXAMPLE (VIDEOCONFERENCE CALL)

This section describes an example of setting up a videoconference call using the MAGIC signalling protocol.

In Figure 14.8, terminal TE1 will establish a videoconference call to two other terminals; TE2 and TE3. However, only TE2 will accept both the video and audio connections; TE3 will accept the audio only.

Figure 14.9 shows the signalling flows that occur across the network. These signalling flows are described in Table 14.2. Since, in this example, the bridging of the audio and video streams is performed wholly within LEX1, no Resource Control signalling flows will exist.

By the end of this call set-up, each party will be participating in the service and have a set of established call objects appropriate to their own local view. For example, Figure 14.10 shows the set of call objects that TE1 will have in its local view:

14.5 DISCUSSION AND FUTURE WORK

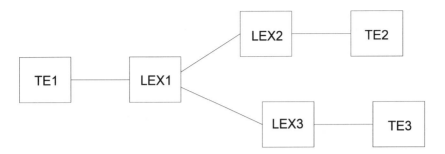

Figure 14.8 Network Configuration for the Videoconference Example

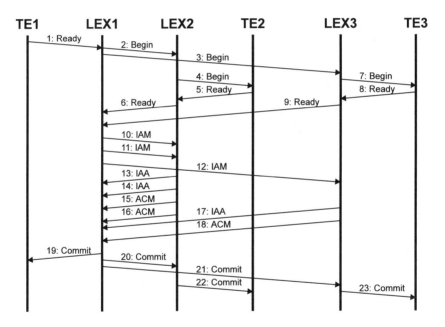

Figure 14.9 Flows for Setup of Videoconference Call

14.5 DISCUSSION AND FUTURE WORK

Part of the work performed by the MAGIC protocol was to build a software simulator to test the protocols being developed. As part of this work, experiments were performed in supporting services with both an extended version of Q.2931 and the UNI Atomic Action Protocol. The latter was found to be much easier to implement, simply requiring basic signalling actions to be performed in the correct order. Using Q.2931

Table 14.2 Set-up Flow Message Description

Message	Description	Example Message Contents

1:Ready	TE1 signals over the access interface that it wishes to establish a videoconference call, with audio and video, to TE2 and TE3.
	This signalling message starts an atomic action to create the objects in TE1's local view of the call, that is:
	
	• The service
	• Two confirmed parties
	• Audio and video abstract services
	• The coding and bearers TE1 will use for the audio and video.
		`1:READY`
		`create(PS:Videoconference)`
		`create(CP:TE2)`
		`create(CP:TE3)`
		`create(AS:Audio)`
		`create(CPSL:TE2/Audio)`
		`create(CPSL:TE3/Audio)`
		`create(ASM:A-law)`
		`create(UM:64kbit/s)`
		`create(CSDM:64kbit/s)`
		`create(AS:Video)`
		`create(CPSL:TE2/Video)`
		`create(CPSL:TE3/Video)`
		`create(ASM:H.261)`
		`create(UME:128kbit/s)`
		`create(CDME:128kbit/s)`
		`create(ASG:audio,video,sync)`

2:Begin 3:Begin	The originating LEX assigns video and audio bridge resources internally. It then continues the call setup process by contacting the destination LEXs setting up similar sets of call objects to those in message 1.	

4:Begin
7:Begin
	The LEXs signal the call attempt to the called TEs.
	(Note how the objects relate to the local view of the called TE, e.g. the other called party is seen as a Virtual Party in this view).
		`4:BEGIN`
		`create(PS:Videoconference)`
		`create(CP:TE1)`
		`create(VP:TE3)`
		`create(AS:Audio)`
		`create(CPSL:TE1/Audio)`
		`create(VPSL:TE3/Audio)`
		`create(AS:Video)`
		`create(CPSL:TE1/Video)`
		`create(VPSL:TE3/Video)`
		`create(RASM:TE1,A-law)`
		`create(RASM:TE1,H.261)`
		`create(ASG:audio,video,sync)`

5:Ready
6:Ready
	TE2 accepts the call. This TE wishes to use both audio and video, and thus selects the encoding schemes that this terminal wishes to use and the mapping elements necessary for the bearer connections. The acceptance is passed back to the originating LEX.
	(Note, most of the call objects created by message 4 are implicitly accepted by this message and are not signalled explicitly. Only the newly created objects need be signalled).
		`5:READY`
		`create(ASM:A-law)`
		`create(UM:64kbit/s)`
		`create(CSDM:64kbit/s)`
		`create(ASM:H.261)`
		`create(UME:128kbit/s)`
		`create(CDME:128kbit/s)`
		`6:READY`
		`create(RASM:TE2,A-law)`
		`create(RASM:TE2,H.261)`

14.5 DISCUSSION AND FUTURE WORK

Table 14.2 (cont.)

Message	Description	Example Message Contents

8:Ready
9:Ready
	Similarly TE3 accepts the call, but only wishes to use audio. Thus, it only creates an ASM and mapping elements for the audio stream (note it also selects a different audio encoding type than that of the other terminals). *Once the originating LEX has received both confirmations from the destination LEXs, and has ascertained that it has the necessary A-law ↔ μ-law interworking unit, call control processing stops and bearer control takes place to setup the network bearers between LEXs.*	`8:READY`
		`create(ASM:μ-law)`
		`create(UME:64kbit/s)`
		`create(CDME:64kbit/s)`
		`9:READY`
		`create(RASM:TE3,μ-law)`

10:IAM 11:IAM 12:IAM	Normal B-ISUP Initial Address Messages are sent to setup the bearer connections. 10: Audio bearer to TE2 11: Video bearer to TE2 12: Audio bearer to TE3	
13:IAA 14:IAA 15:ACM 16:ACM 17:IAA 18:ICM	Standard B-ISUP flows. *When LEX 1 has received all three Address Complete Messages (ACM), the bearers have been established. LEX 1 can now complete call control and finish the setup phase of the call.*	

19:Commit
20:Commit
21:Commit
22:Commit
	Call control will now complete by confirming all of the pending atomic actions. The created call objects are now also confirmed. On the accesses, the Virtual Path and Virtual Channel Identifiers (VPI/VCI) assigned to the different ATM connections are signalled to the terminals.	`19:COMMIT`
		`create(RASM:TE2,A-law)`
		`create(RASM:TE2,H.261)`
		`create(RASM:TE3,μ-law)`
		`create(ASM:A-law,VPI,VCI)`
		`create(ASM:H.261,VPI,VCI)`

essentially required the signalling subsystems to be specifically tailored for each individual service, in order to communicate all of the complex requirements of the service. Thus, it was concluded that the object approach to signalling, as used by MAGIC, was far more flexible and simpler to use. This is, however, a radical departure from conventional signalling protocols, and the object based approach has taken a long time to be adopted by the standards bodies.

For the services examined by the MAGIC project, the object-based signalling model and the atomic action protocol was found to meet all of the requirements of these services. In particular, it was able to signal the

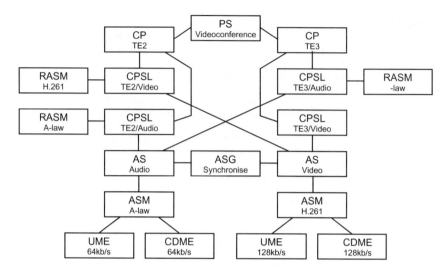

Figure 14.10 TE1's Local View of the Videoconference Call

following requirements, which would be very difficult to do with conventional protocols:

- Support multiparty calls.
- Adding dropped parties during the call.
- Permit parties to undertake different roles (i.e. called parties to add/remove bearers and other parties).
- Provided a mechanism to facilitate the interworking of codecs.
- Supported advanced services like remote control (party A sets up a bearer between parties B & C) and a user-defined service (a party connects to other parties using connections as required, and not to any formal service definition).

The MAGIC work does not necessarily remove the complexity of application control for such services (such as working out the necessary billing mechanisms). What it did do was remove much of the complexity from the signalling protocol. Signalling is there to communicate requirements across the network, not to restrict them.

The major weaknesses of the MAGIC work were its failure to address the requirements of data networks, such as TCP/IP, utilise the capabilities of the Intelligent Network (IN), and address fixed/mobile network convergence. Data networks were not as important when the MAGIC project existed as it is today, and needs further investigation as to how public networks should support such an important service. As for the IN, there appears to be much in common between the services provided by the

MAGIC Resource Control protocol and the IN. It can therefore be envisaged that the Resource Control functions will instead be provided by IN capabilities, as proposed by Knight (1993).

14.6 CONCLUSIONS

Initial signalling protocols for broadband ISDN are essentially an evolution of the simple protocols developed to support the two-party, single connection service of telephony. The evolution of these approaches did not adapt to the multiparty multiconnection services envisaged for broadband. Therefore, MAGIC developed a new protocol for such advanced services. It was designed around the concept of communicating simple objects using an atomic action protocol to build associations between functional entities. By using such a signalling system, it was possible to set up calls for advanced broadband services in a relatively simple way, demonstrating that this approach was indeed a good way of flexibly supporting the advanced services of tomorrow.

The signalling protocols designed by MAGIC were not simply accepted wholesale by the standardisation bodies—they just don't work that way—but ETSI, and after some persuasion ITU-T as well, standardised a call control protocol that had the capability to exchange information by the communication an object model. Whilst the first stage only considered two-party calls, so the begin, ready and commit procedures were not needed, there is enough detail in the ITU-T's call control protocol to determine that its roots are very firmly in the results of the MAGIC work.

REFERENCES

Knight R R, Stage 2 Specification, *6th Deliverable*, RACE2044 MAGIC, R2044/BTL/DP/DS/P/006/b1, September 1993.

Knight R R, Broadband Signalling and the Evolution to an Intelligent Network, *IEE Colloquium on 'Developments in Signalling'*, IEE, November 1993.

Popple G W and Glen P J, Specification of the broadband user/network interface, *BT Technol. J.* Vol. 11, No. 1, January 1993, pp. 86–92.

STANDARDS

ITU-T Recommendation X.851—*Information technology—Open Systems Interconnection—Service definition for the commitment, concurrency and recovery service element*, November 1993.

ITU-T Recommendation H.261—*Video codec for audiovisual services at $p \times 64\,\text{kb/s}$*, March 1993.
ISO CD 11172: *MPEG (Motion Pictures Experts Group)*, (1993).

FURTHER READING

The RACE MAGIC deliverables, starting with
Knight R R, Final Report, *12th Deliverable*, RACE 2044 MAGIC, R2044/BTL/DP/DS/P/012/b1 April 1995.

QUESTIONS

1. Why were the B-ISDN signalling protocols, such as Q.2931, not designed to meet all of the multiparty, multiservice requirements of broadband from the outset?
2. Why have a single complex multiparty, multiservice call instead of many simpler, separate calls?
3. Why use separate Bearer Control and Call Control protocols at the NNI and not at the UNI?
4. What information does the Abstract Service convey?
5. Why separate upstream and downstream Mapping Elements?
6. In what way is the MAGIC Atomic Action Protocol simpler than traditional techniques?

15

The Call Control Protocol in a Separated Call and Bearer Environment

Dick Knight and Bryan Law

15.1 INTRODUCTION

Broadband networks are the future platforms for telecommunications companies, and those networks will only be economically viable if they support a diverse range of teleservices providing scope for the developers of applications. The provision of generic communications systems has long been the aim of Public Network Operators (PNOs), and diverse communications options complicate the problem of controlling a network independently of the application. Any complex problem is simplified by breaking it down into constituent parts and, in the case of a diverse broadband network, the International Telecommunications Union—Telecommunications standardisation sector (ITU-T) have decomposed the problem into the control of the connection by a bearer control and the co-ordination of a series of related connections into the rather more abstract mechanism known as call control. The ability of a separated call control to communicate the requirements for a call has been recognised as sufficiently complicated to warrant a new protocol. This protocol has been developed to provide a generic communication structure that is independent of the network and protocol transport system. ITU-T have separated the features that they require in the signalling systems into capability sets. This chapter concentrates on the features required of the call control protocol to enable broadband signalling capability set 2 (CS2) to be implemented. The capability set approach by ITU-T, the services and the signalling requirements were explained in Chapter 4.

15.2 THE SEPARATION OF CALL AND BEARER

The narrowband telephony network can be regarded, at least as far as the signalling is concerned, as supporting two very similar teleservices. The voice teleservice provides a bidirectional symmetrical connection between two terminals guaranteed to transfer commercial speech band (300 Hz to 3.1 kHz) analogue frequency signals, whilst the data teleservice guarantees to make digital, synchronous bidirectional, symmetrical, connections at 64 kb/s. Neither service, however, is application specific—and, for example, a Group 3 fax machine is an example of an application that takes advantage of the facilities offered by the voice teleservice, or analogue connection, to provide an entirely different service to human voice communication. The commonality between the two underlying teleservices lies in the fact that both services provide a single bidirectional symmetrical connection between two end points through one or more networks. This description of the communications could be referred to as a 'call' and is set up implicitly during the request for connectivity.

Broadband networks, and Asynchronous Transfer Mode (ATM) in particular, offer many more diverse types of connections. The support for constant bit rate, variable bit rate and the ability to separately specify the peak cell rate, sustainable cell rate and many other additional characteristics of an ATM connection (see Q.2961) enable diverse applications to use the same communications network. If those applications are combined together, such as a videotelephony call with additional data exchange (e.g. a shared whiteboard for 'scribbling space') then the connections need to be co-ordinated into a single relationship, the 'call'. This ensures that additional connections are not offered as a new call but set up within the restrictions of the existing call.

Functional model

A full separation of protocols was first identified in ITU-T signalling requirements in an unpublished draft recommendation Q.73, which described the ITU-T's functional model of processing entities involved in a call, and a simplified model is reproduced in Figure 15.1. Separating call control and bearer control protocols provides support for the future convergence of the fixed and mobile networks, the Universal Mobile Telecommunications System (UMTS), as well as negotiating network and bearer resources to prevent needless reservations in calls failing due to terminal incompatibilities. This is increasingly likely in broadband networks, due to the diverse terminals, services and information encoding that may be supported. Additional advantages include reducing the redundant processing of call control information at intermediate bearer control nodes

15.2 THE SEPARATION OF CALL AND BEARER

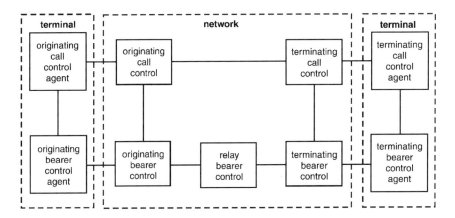

Figure 15.1 Simplified Functional Model of Processing Entities

(Relay Bearer Control (RBC) nodes). By including information directing the protocol messages towards a call control entity, the RBC's are spared from decoding and 'understanding' the whole of the message and its parameters.

The functional model uses the concept of a terminal, or other user equipment, as taking on the role of either originating or destination signalling end point, and the signalling entities within the terminal are referred to as control agents. Splitting the functionality in call and bearer control results in an Originating Call Control Agent (OCCA) and Originating Bearer Control Agent (OBCA) in the call originator's terminal, and similarly, a Terminating Call Control Agent and Terminating Bearer Control Agent in the destination user equipment. The functional entities in the network nodes show Call Control entities only at the edges of the network. The ITU-T model is more general and does allow Call Control entities at other points, such as at gateway locations or other network boundaries. In the centre of Figure 15.1, the Relay Bearer Control (RBC) node provides the procedures to support switching connections, but has no concept of the call and so cannot support service functionality. This means that, for example, it can route connections only to the next Bearer Control node.

Controlling connections

Bearer control is defined as providing the procedures to enable a connection to be set up, maintained, modified and released. The bearer control is responsible for overall control of individual connections, control of the network resources and the collection of traffic count information. Bearer control is also sometimes known as *link control* or *connection control*, and is based upon the signalling protocols mainly developed in the ITU-T and published as recommendations Q.2931, Q.2971 and Q.2761-4 (see Chapters 5 and 8). The overall control of a connection is performed by the protocol undertaking the set up and release procedures associated with a connection and the maintenance of the state of the connection. Connection control is performed across a network and between networks, selecting the connection identifier and protecting networks (e.g. by maintaining timers). Controlling network resources is performed at each node by the reservation and release of network resources, node selection and negotiation, resource interworking (e.g. with existing networks) and the through connection and disconnection of bearers to commence and cease communications. These functions are common to both broadband and narrowband networks, but for broadband networks there are two new resources to control: the parameters concerned with the bandwidth and the overall Quality of Service of a connection; and the functions specific to the handling of connection branches for multipoint connections. The ability to support connections of differing characteristics introduces two concepts new to broadband networks: the procedures concerned with both the negotiation and modification of a connection's characteristics.

Controlling calls

In separating the call from its connections, the ITU-T have replaced the old definition used to describe the monolithic control of the connection and its embedded call with a new definition of call control.

Call control is a signalling association between one or more Terminals (User Applications) and the Network to control the set up, release, modification and maintenance of sets of communication connections to provide related information. Call control is used to maintain the association between parties, and a 'Call' may embody any number of underlying connections, including zero, at any instance in time.

Call control can be realised in a number of ways:

- separating the call information into parameters carried by a single call/bearer protocol;
- separating the state machines for a call control and a bearer control whilst signalling information in a single call/bearer protocol;
- performing a full separation of information and state machines by providing separate signalling protocols.

The full separation of call and bearer protocols enables a number of functions to be realised that applications can use to make communications more efficient. The call can be set up to enable the characteristics of the communications session (the 'service') to be established first, and the connection characteristics to be determined before the connection is established. This is referred to as *call negotiation* and *connection pre-negotiation*.

The full separation of the call control and the bearer control protocols facilitates a simpler upgrading strategy, and allows the separation of the call control software from the telecommunications switches. This provides a step towards making the service provisioning independent from the switching technology and the suppliers by enabling the separation of the call control protocol states from the information it is trying to convey.

15.3 SEPARATION OF PROTOCOL AND INFORMATION

Traditionally, signalling protocols have reflected the states of the call and its embedded connection, reusing names in the protocol states and messages. To take one example, communication with another network endpoint (a 'party') is requested by sending a SETUP message; the destination party may inform the originator that the user has been contacted by sending an ALERTING message, and the confirmation that the communication channel is available is indicated by a CONNECT message. The states that the call is placed in reflect the status of the destination party, although to a lesser extent. The called party's protocol states associated with the three messages above are Call Present, Call Received and Active.

Broadband calls in Capability Set 2 (CS2) have the ability to support multiple end points (multiparty calls), and the states and messages become less clear. In CS1 protocols it is an error to put multiple destination party numbers in a SETUP message. A multiparty call must allow this situation, and the consequential state explosion of attempting to reflect the states of all the parties becomes overly complex. A new approach was needed for any type of call supporting more than one connection or more than one party, or both. The approach adopted was to define a protocol that carries the information concerning the changes of states of parties and connections as parameters to a small set of protocol messages. Storing and manipulating the state changes that the protocol carries is more efficiently performed by applying object-oriented techniques, leaving the protocol states concerned only with the progression of any related protocol actions.

The separation of the protocol from the information it carries can be represented by the general schematic for an object model driven protocol shown in Figure 15.2. The user application manipulates the object model via an Application Programmers Interface (API), and the object model signals any changes through the protocol (in this case the Call Control protocol) to the network. The network has a similar arrangement, with the

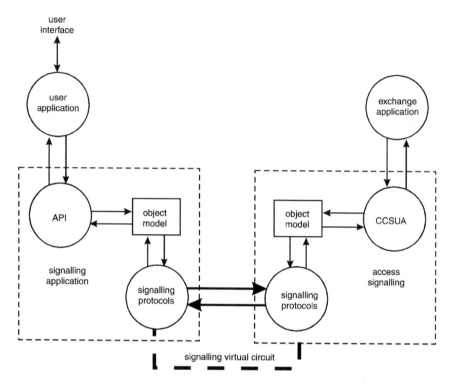

Figure 15.2 Signalling State Changes at the User-Network Interface through an Object Model

changes being recorded within an object model which then updates the exchange application process (in this case, the Call Control Service User Application—CCSUA). More details on the requirements for using an object-oriented model as the basis of a call control protocol to support multiparty scenarios is given in Chapter 14.

15.4 MODELLING A CALL WITH OBJECTS

The object model is concerned with the abstraction of a call into its fundamental parts, and the status of those parts. A complex engineering problem, as this is, can always be simplified by breaking the problem up into constituent parts. A call has a number of constituent parts, each requiring its own particular data. This type of situation readily lends itself to a description based upon object modelling, and the objects that are required to describe the call as an information model are shown in Figure 15.3.

15.4 MODELLING A CALL WITH OBJECTS

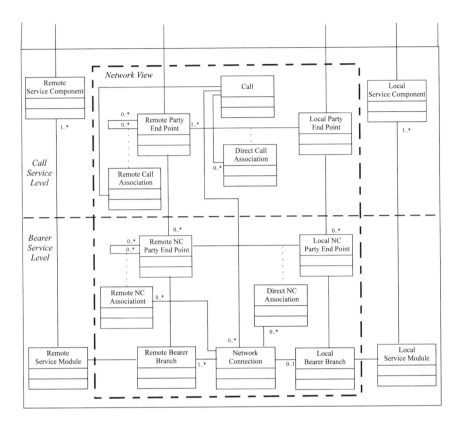

Figure 15.3 Information Object Model

Call objects

The model divides the signalling objects into three types: call control objects, service control objects and connection control objects. Service control objects, which are carried by the signalling but are outside the network view, are concerned with abstraction into attributes of a service. An example of the use of separation of service from the call control is the DAVIC specified Session Control, described in Chapter 10. Call control associated objects represent the aspect of a call that involves end-to-end associations among parties and the network that are independent of the other characteristics of the service. Call control objects refer to particular connections but do not store the characteristics of the connections (for example, Connection Object Identities need to be 'known', but the bandwidth and the other Quality of Service characteristics of the connection are not defined in call control). The model includes a simplified view of the connection control (as it is a view of a connection from the call control perspective). The connection control components shown here do not include any access network or control, as described in Chapter 9. Bearer control objects shown in this view are more directly associated with the communication attributes of the connections (and do not need to show aspects such as physical routing).

The service objects that are placed outside the 'network view' are carried by the signalling, but are not maintained by the network. This reflects the possibility of adding or removing service components without any supporting signalling flow. For example, a telephone may have a 'silence' button, enabling the microphone (a service component) to be removed from the call, without the network being informed of the removal.

Call control is primarily concerned with providing generic communications between conceptual end points (parties) and a network to implement a service. The objects in the call control model must therefore identify who is participating in the call, through the various party objects, and associate data (attributes) with those objects. The passing fashion for proxies, to provide a common signalling element on behalf of a number of applications, demonstrates the need to model the possibility of a call's end point being a different signalling entity to the bearer end point.

Object modelling is primarily concerned with limiting the scope of data within a generic object that can then be replicated without needing to share the data. This makes the object model primarily applicable to a multiparty call, where parties (which consists of a number of end point objects, one at each level) hold the same type of data but must store different values. Table 15.1 lists the attributes of the objects. The following sections provide some examples of how the objects and attributes could be manipulated to dynamically modify the call.

Call object attributes

The call is a view of the communications configuration at a particular point in the network (or an end point). Each call control must view objects as related to the local or remote parties, which requires the call control protocol to provide a translation of the local to remote, and *vice versa*, during any signalling interchange. The call object must store the object references for the end points and also their relationship within the call, defined by the call association objects.

The call segment identifier provides a unique reference for a segment of a call between two call control entities. An associated service reference may be used to identify the session, or other service level view, to which this call belongs.

A call must have permissions, limiting the behaviour of the parties in the call. The call permissions logically define such activities as the right to

15.4 MODELLING A CALL WITH OBJECTS

Table 15.1 Object Attributes for the Information Object Model

Class Name	Class Attributes
Call	Call/Call Segment ID Party End Point Identifier (PEP ID) of local party List of PEP IDs of Remote Parties Associated service reference List of Direct Call Association IDs List of Remote Call Association IDs Call Permissions List of Bearer References
Local/Remote Service Component	Service Component ID List of Call Party End Point Object ID's Service Component Characteristics (inc. High Layer Information) Communication Configuration (Source, Sink, Bi-directional) Service Component Traffic descriptor requirements Service Component QoS descriptor requirements Associated Resource Component ID Associated Service Module ID
Direct Call Association	Direct Call Association ID Call associated with this Association Local PEP associated with this Association Remote PEP associated with this Association
Remote Call Association	Remote Call Association ID Call associated with this Association First Remote PEP associated with this Association Second Remote PEP associated with this Association
Local Party End Point	Party End Point ID Party Address PEP ID of Party owner Party Type List of associated Local Resource Party End Point IDs List of associated Remote Party End Point IDs List of associated Local Network Connection Party End Point IDs
Remote Party End Point	Party End Point ID Party Address PEP ID of Party owner Party Type List of associated Remote Resource Party End Point IDs List of associated Local and/or Remote Party End Point IDs List of associated Remote Network Connection Party End Point IDs

involve new parties in the call, to set up new connections, and so on. Finally, a call must have references to any connections that have been set up.

Party end point object

A party end point can be identified in a number of different ways. This attribute enables a network to provide a party's presentation number (e.g. directory number), the default number, and a network may also wish to provide a network address, for routing purposes.

Service component and service module objects

Bearer control associated objects are directly associated with the communication capabilities of the connections, and are modelled as two object types. The network connection object provides the containment of the connection's attributes—such as peak cell rate, sustainable cell rate, etc. The service module represents the hardware that a terminal may use to interpret the bit stream provided by the ATM adaptation layer. To illustrate this with an example, a bit stream may consist of video pictures (in some format) and sound (in some format). These two formats are combined together and compressed to produce a Motion Picture Experts Group (MPEG) stream, which may be carried in a single ATM virtual channel as shown in Figure 15.4. The type of multiplexing and the compression

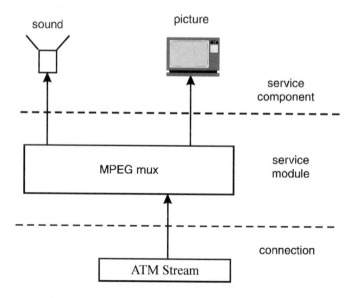

Figure 15.4 Service Components

algorithm applied need to be signalled to ensure that both terminals can communicate, but the information does not need to be interpreted or checked by the network. This demonstrates the difference between the service module and the connection object—the attributes of the connection object must be understood by terminals and the network, while the service module is information that needs to be shared by the terminals during the connection set up. The bearer control protocol is heavily based upon the CS1 Q.2931 protocol, which does not signal information using an object model, but uses more traditional information elements. In the example shown in Figure 15.4, sound and pictures are identified, but the terminals will also need to agree to use the same coding algorithm—such as µLaw, A-law, etc. for sound. This does not have the same sort of effect on the connection characteristics as the service module used. Terminals may wish to determine that the basic service is compatible from one end to the other, before setting up connections, by signalling the details using the service component object.

Network connection object

The objects associated with bearer control are concerned with the control of the connection, and included in the model for reference. Bearer control is not implemented as an object model, and the separation of connection, bearer branches and service module is therefore theoretical, but completes the model from the call control perspective.

15.5 PROTOCOL ARCHITECTURE

Signalling using an object model requires a number of functions to interact in a co-ordinated manner. Bearer control interactions cannot take place until the call has been established, and as has already been mentioned, the connection pre-negotiation procedures in a call. These are just three of the operations that need to be supported in the signalling protocols; others include supplementary services and a means of transporting the call control operations. The mechanisms that bind these functions together to enable them to co-operate in various procedures are represented in a protocol architecture model. This type of signalling communication readily lends itself to an application layer structure which enables functions to be used as necessary, and ignored if not needed. For example, pre-negotiation is only an option to be used as necessary.

Figure 15.5 shows how the signalling protocols interact in a local broadband exchange (an ATM switch). Any modern telephone exchange

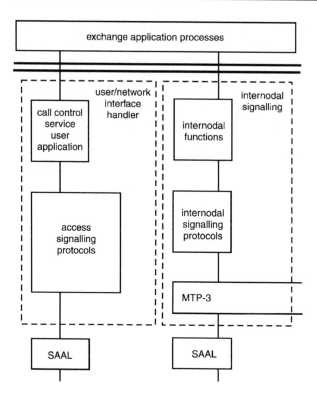

Figure 15.5 General Architecture for Broadband Signalling

is dependent on the software that implements the various functions the exchange undertakes. Many of these functions are either specific to the particular switch manufacturer, or are specific to the particular network operator. These functions, such as call records for charging mechanisms, network management, testing, etc., are exchange application processes, which are not a subject for standardisation and are not covered in this book. Access Signalling, and more generally, the User-Network Interface handler, whose applications are present on the left-hand side of the diagram, have been split into two parts.

The Call Control Service User Application (CCSUA) provides that part of the access signalling concerned with the current state of the parties and the connections in the call. The CCSUA is not concerned with the details of the signalling communication, but the current status of the terminal (connected, answering, etc.) and their current attributes (for example, the address or number), the service and any supplementary services, and similar details for the connections.

The Access Signalling Protocols support the CCSUA by providing the means for the states and attributes (data) held in the CCSUA to change. The Access Signalling Protocols free the CCSUA from the more mundane

15.5 PROTOCOL ARCHITECTURE

actions concerning the procedures for establishing, maintaining and releasing a communications association.

This general architecture enables the information concerning the call to be separated from the protocol, allowing for appropriate methods of specification to be applied. Traditionally, the CCSUA has not been a subject for standardisation. The transport of the information that enables the Object Model to be shared amongst CCSUA entities involved in the call must, however, be standardised to enable communication between equipment realised by different suppliers. The information ultimately takes the form of parameters to the Access Signalling Protocols, providing the object references and its attributes.

Access signalling protocols

The access signalling protocols provide the means to exchange information between terminals and networks. Since broadband calls are not always simple point-to-point mono-connection calls, the signalling will also need to support the call consisting of a single point-to-multipoint connection. This type of call utilises the ATM capability of multicast (often incorrectly termed a 'broadcast call'), in which cells are copied to more than one destination at a time by the switch. To co-ordinate actions between the parties in such a call, a function for party control is specified. There are also functions to support restarting the protocol under error conditions (such as link failures detected by the SAAL or corruption of data within an exchange application process). The full protocol architecture for the user-network interface handler is shown in Figure 15.6.

The structure of the architecture for the access signalling elements is based upon the B-ISUP approach, which predated the access signalling discussions in ITU-T, and took into account the Architecture Framework using Open Services Interconnection (OSI) concepts. The actions that the various elements undertake are linked together to form a set of co-operating procedures by a Control Function (CF). This type of architecture, consisting of a control function and any number of Application Service Elements (ASE), follows the concepts outlined in ITU-T and the International Standards Organisation (ISO) of an Application Layer Structure (ALS). This structure specifies that an Application Service Object (ASO) consists of a CF + ASEs + ASOs, which is a recursive definition. This allows for extensions, if necessary, in CS3 for multiparty support. For certain CS3 calls, especially complex multiparty, multiconnection calls, it is not clear whether a single control function is adequate. However, it is anticipated that the most drastic change will be to add a further ASO, although the current work on the protocol requirements for CS3 have not identified the need for another ASO to be added.

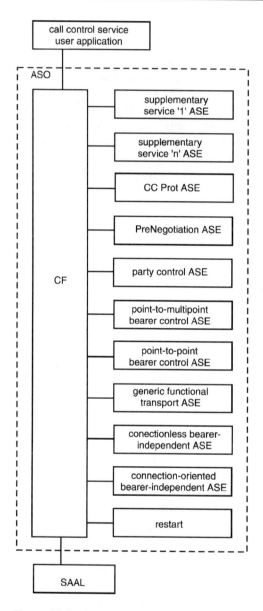

Figure 15.6 Access Signalling Application Service Elements

Interfaces between the individual ASEs and CF are Primitive interfaces. The interface between the CF and Signalling ATM Adaptation Layer (SAAL) is the SAAL Service Interface, described in Chapter 2. All functions also have an interface to a 'Management application', although this has not been defined as a formal primitive interface.

Supplementary Service ASEs can be expected to be added as necessary.

15.5 PROTOCOL ARCHITECTURE

The implementation of Supplementary Services is generally dependant upon the network operator and the capabilities in the terminal. For example, a service such as 'Transfer on no reply' may be considered to be appropriate for the emulation of a narrowband type private network, but is less likely to be found in public networks.

ITU-T have identified that pre-negotiaton could be applied both to call and connection attributes. While this is true in principle, at present no call pre-negotiation has been identified, since the negotiation of the services required within a call may be adequately negotiated during call establishment. For multiconnection calls, a Prenegotiation ASE has been defined to enable this operation to be undertaken either at call establishment, or subsequently—prior to the establishment of a connection. Although it is possible that pre-negotiation might be considered as a sub-operation of call control, ITU-T were of the opinion that the operation is sufficiently complex to warrant a separate protocol, and a 'stand-alone ASE' approach has been adopted.

Party control was first defined as a mechanism within point-to-multipoint connection control, but that was an interim solution to enable simple multicast to be defined very early on. ITU-T accepted restructuring for a harmonised architecture even though Party Control could also be considered as a sub-operation of call control. As with pre-negotiation, a stand-alone Party Control ASE is a better approach, providing a simpler and more generic solution.

The bearer control consists of a number of components: point-to-point, point-to-multipoint, modification and negotiation of traffic characteristics (Q.2961), etc., and while it is possible to make each of these separate ASEs for definition within the protocol architecture, a generic mechanism is preferable, enabling a single BC ASE to be both comprehensive and simple. Only in the case of the combined call and bearer control (Q.2931) and point-to-multipoint (Q.2971) protocols are stand-alone ASEs appropriate.

Internodal signalling

The signalling protocol between the nodes in a network (nominally switches—or exchanges) is known as *internodal signalling*. Since service functions are invoked at the edges of the network (access and gateway points), there is no requirement for call control functionality at these points. ITU-Ts B-ISUP protocol controls the connections within a call, but does not need to co-ordinate the connections. This principle can also be applied to the ATM Forum protocol, PNNI.

15.6 CALL CONTROL PROTOCOL ACTIONS

Two of the major requirements for the call control protocol are co-ordination of connections (multiconnection calls), and the co-ordination of parties (multiparty calls). Since call control is 'a signalling association between co-ordinating entities', a single call control protocol has been derived that is generic enough to satisfy both these cases. The simpler case is the co-ordination of multiple connections within a call. This has the advantage, from the network provider's point of view, of reducing the number of times that a routing address needs to be derived from the end point address supplied by the originating terminal. To co-ordinate multiple connections, the following actions need to take place within the network:

- all connections must be routed to pass through every co-ordinating (call control) entity;
- call control associations must be completed before bearers are established; and
- a call may exist without any bearers (facilitating complex connection rearrangements).

The second requirement is for call control to co-ordinate the actions of multiple parties. This is achieved by using the concept of atomic actions, coupled with a protocol that allows the principle of 'undoing' a request if one party refuses a request. An atomic action is a means of breaking a large number of interactions into a set of smaller, inter-related actions. Multiparty principles and atomic actions are described in the previous chapter.

Call establishment

The Call Control Service User Application provides the means for a serving node co-ordinating a call to hold an object model representation of the call. In practice, the node can only 'see' those parties and connections that the local party is associated with. This is known as the 'local view'. The global, or complete, view of a call may not exist at all—although it helps us to understand how the object modelling works—but it is not necessarily required by any node in the call control. The Call Control Protocol is responsible for communicating the details of the basic information model between co-ordinating entities (call control network nodes). In a simple two-party call, this can take place by a simple request and response interchange to communicate the local view of the call at establishment, as shown in Figure 15.7.

15.6 CALL CONTROL PROTOCOL ACTIONS

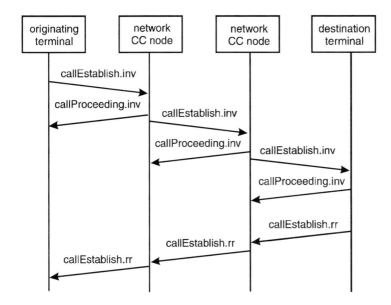

Figure 15.7 Call Establishment in Call Control Protocol

The flows do not show messages between Call Control co-ordinating entities, since messages must also pass through the intervening Relay Bearer Control (RBC) nodes as shown in the functional model of Figure 15.1. Instead, the flows represent the call control protocol operations, which are parameters to the underlying messages. Information on the call control operations must be contained (enveloped) within an underlying protocol transport mechanism to enable the call control protocol flows to pass between nodes in different parts of the network. This is covered later in this chapter, and also explains the '.inv' and '.rr' extensions in the flows.

Call control protocol signalling associations

It has already been shown that the CC entities are not necessarily adjacent, and so the mechanism for making the signalling association is slightly more complicated than with the current DSS2 or B-ISUP. DSS2 uses the connection reference, unfortunately named the Call Reference Information Element in Q.2931, to provide the association for bearer control by indicating which end of the link allocated the unique number used to direct messages to the ASO.

In the CC protocol there may be more than one Call Control node beyond the end of a bearer link, as in Figure 15.8, so two call establishment requests received over a common signalling link may both have the same referencing number. The mechanism adopted is for the originating CC to

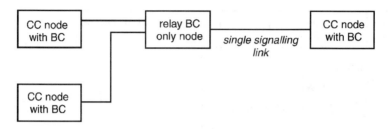

Figure 15.8 Call Control Co-ordinating Nodes

allocate half the reference number (a unique number so far as that CC is concerned) and the destination CC to allocate the other half. The Call Proceeding flow is used by the destination CC to indicate to the originator its half of the reference. This reference number becomes unique between a pair of CC entities, and is therefore named the Call Segment Identifier. The Call Proceeding flow is also used to confirm the routing address of the CC entity, enabling a call to start setting up connections prior to the addressed terminal being contacted.

Bearer establishment

Figure 15.7 does not show the establishment of any connections. The exact point at which connections are established must be service dependant—and can even be application dependant. It is the implementer's choice to decide whether a particular service will be best implemented by setting up the connections as soon as possible, or whether some connection pre-negotiation is more appropriate. To emulate the narrowband ISDN (and POTS) service, ITU-T at first placed a requirement that the first connection can be set up simultaneously with the call. This has caused some heated discussion, but it has been shown that this is not a real requirement. The real requirement is for the connection to be available at the time that the user is offered the call, otherwise 'clipping' may occur. Clipping is a phenomenon due to our habit of picking up a ringing telephones handset and immediately speaking to the caller. If the connection is not immediately available, then some loss of speech can result, dependant on the connection set up delay. A simple solution is to wait until the connection is available before the presence of the call is indicated to the user. This may be added into the diagram as shown in Figure 15.9.

The flows show both Call Control protocol flows and Bearer Control protocol messages. Bearer Control messages are indicated by dotted lines, and the BC initials precede the message name. For simplicity, the flows represent a two-party call between terminals serviced by two adjacent broadband exchanges.

15.6 CALL CONTROL PROTOCOL ACTIONS

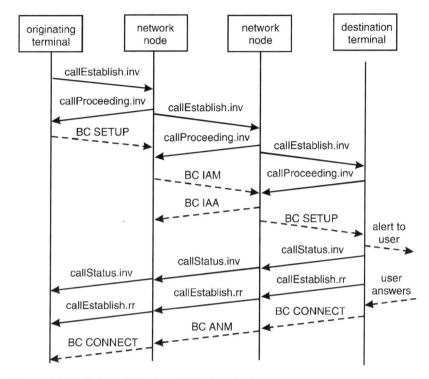

Figure 15.9 Delay of Alerting Indication to User

The originating terminal chooses to implement the fast set up of the first connection by performing a BC set up as soon as the callProceeding.inv is received. The network has to delay the Initial Address Message (IAM) until the callProceeding.inv is received from the destination local exchange. The connection is accepted across the network using the Initial Address Acknowledge (IAA), and once the set up is sent to the terminal, an Address Complete Message (ACM) is sent by the destination exchange to confirm the end user has been reached. The destination terminal chooses not to offer the call to the user until the first connection is available. The user is made aware of the presence of the call by some terminal dependant means, and the Call Control co-ordinating nodes are informed of the change of status of the party by means of an update to the information model in the callStatus.inv flow. This unconfirmed flow updates the status attribute of the party object to indicate that it is now in the 'Alerting' state. In the case shown, the call is accepted and the connection completed in the last two messages.

Flows to enable further connections to be set up do not require any interactions at the call control level, since the call is now established and the protocol interactions take place only at the Bearer Control level.

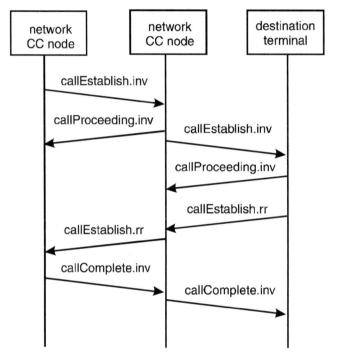

Figure 15.10 Co-ordinated Call Control Establishment

Support for multiparty calls

To enable a smooth transition from CS2 to CS3, the call control protocol incorporates a single flag that enables it to indicate whether the signalling flow needs to be co-ordinated. These actions support the requirements of a signalling protocol that can 'undo' actions in some parties of a multiparty call when another party is unable to accept a request. The requirements for multiparty calls came out of the RACE MAGIC project, described in the previous chapter. The CS2 call control protocol may, therefore, be used by a terminal to act as a terminating party in a multiparty call, provided that that party does not communicate directly with more than one of the other parties in the call. Under these circumstances, a co-ordinated approach needs to be taken, similar to BEGIN, READY and COMMIT in RACE MAGIC. While the message names differ, with the new call control protocol using callEstablish.inv, callEstablish.rr and callComplete.inv for BEGIN, READY and COMMIT, respectively, the actions and underlying procedures are the same. Figure 15.10 shows the flows for two segments of a multiparty call, allowing a participating CS2 destination party. The multiparty co-ordination is not shown in this flow.

15.6 CALL CONTROL PROTOCOL ACTIONS

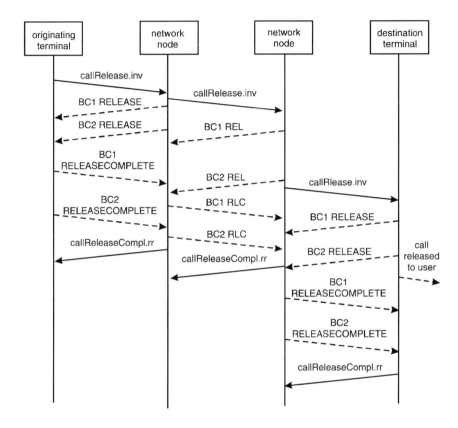

Figure 15.11 Release of a Call with two Connections

Releasing a call

The release mechanism is shown using the same mixture of call control and bearer control flows as in Figure 15.9. The example chosen in Figure 15.11 is the release of a call with two connections being initiated by call control.

The requirements for multiconnection calls stated that a call may exist without any connections, but that a connection may not exist if it is not contained within a call. The release procedures are always local and cannot fail (which means that any signalling entity may not refuse a release request), and so the call does not complete until an entity's co-ordinated connections have been released. It is not necessary, however, to receive a confirmation of the call's release from a succeeding entity before confirming release to a preceding entity (in either call or bearer control). These requirements are a logical extension of the narrowband

and CS1 signalling protocols. It is envisaged that in CS3, the only change will be for a party who requests a call release in a multiparty call merely releasing themselves, and not the entire call.

15.7 PROTOCOL TRANSPORT MECHANISM

The protocol transport mechanism for call control operations expected to be used by ITU-T is the Remote Operations Service Element (ROSE) in X.219, and is enabled by the Generic Functional Transport mechanism (Q.2932.1). This mechanism takes advantage of the existing call/bearer control protocol (Q.2931) to send a Connection-Oriented Bearer-Independent (COBI) SETUP message. The message contains all the required fields to be acted upon by the receiving bearer control signalling entity, except that it does not request a connection. Contained within the fields is a FACILITY information element, which is passed on unchanged until a Call Control co-ordinating entity is reached. The Control Function of the ASO (see Figure 15.6) splits the contents of the message to the appropriate ASE, as shown in Figure 15.12. The advantage of the GFT protocol is that it can be used as part of a private network system, for example using the private network signalling protocol, B-QSIG. The ROSE functions can also make use of the Transaction Capabilities Application Part (TCAP) facilities of Signalling System No. 7 (SS7) (Manterfield 1991) to support the transfer of information between non-adjacent signalling entities within a public network.

The ROSE functions support a generic message interchange of an INVOKE procedure and a confirming RETURN-RESULT, indicated in Figures 15.7 and 15.9 by the 'inv' and 'rr' extensions. A RETURN-RESULT is not always required, and so ROSE provides the capability for the invoking entity to indicate whether the operation is confirmed or unconfirmed. The callProceeding is an unconfirmed flow, since a confirmation is not required in this case.

The GFT protocol in broadband access signalling and the ROSE mechanisms were all imported, largely unchanged, from narrowband signalling. The techniques were first developed to enable supplementary services to be transported, and more details of these generic mechanisms will be found in the narrowband signalling recommendations and articles.

15.8 CONCLUSION

The support of complex services, made possible by ATM, will require communications interactions over more than one channel. The support of

15.8 CONCLUSION

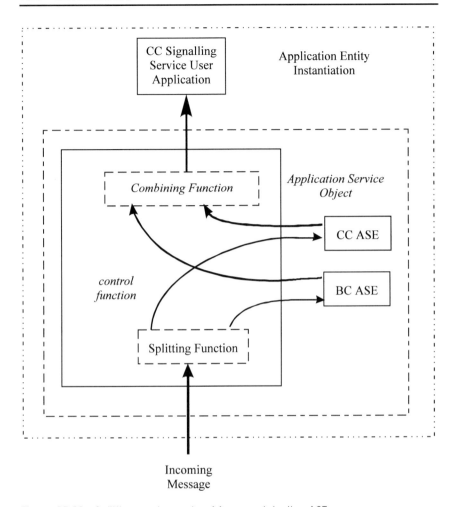

Figure 15.12 Splitting an Incoming Message into the ASEs

such requirements has been shown to be simplified by separating the functions required for the control of the call from the functions of the control of the connection. By extending this separation into the protocols, it is simpler to signal the state changes of a particular connection within a call. A further separation of the information that the call control protocol signals from the states that the protocol may be placed in will enable a smooth transition in future for the support of multiparty calls. In adopting such a strategy, the standards bodies have accepted the strategic results of the RACE MAGIC project (Knight 1995).

Further work has enabled the call control protocol to remain independent of the protocol transport mechanism, and allows call control entities to be situated at the point where co-ordination is required, without regard for the underlying signalling protocol stack or the network type.

REFERENCES

Knight R R, RACE MAGIC deliverable No. 12, 'Final Report', CEC Deliverable No. R2044/BTL/DP/DS/P/012/b1, April 1995.

Manterfield R J, *Common Channel Signalling*, Peter Peregrinus, London, 1991.

STANDARDS

ATM Forum 94-0471R15 PNNI Specification.
Q.771 *Functional Description of Transaction Capabilities*, ITU-T (06/97).
Q.2932.1 *Generic Functional Protocol: Core Functions*, ITU-T (07/96).
Q.2981 *Rose based Call Control*, ITU-T (to be published).
Q.2983 *Separated Bearer Control*, ITU-T (to be published).
X.219 *Remote Operations: Model, Notation and Service Definition*, ITU-T (11/88).
ISO standard IS9545 *Information Technology—Open Systems Interconnection—Application Layer Structure*.
Q.1400 *Architecture Framework for the Development of Signalling and OAM Protocols using OSI Concepts*, ITU-T (03/93).
X.207 *Information Technology—Open Systems Interconnection—Application Layer Structure*, ITU-T (11/93).

QUESTIONS

1. What are the advantages and disadvantages of separating signalling control into call control and bearer control?
2. What is the advantage of using an object model to signal information?
3. Why are the actions undertaken by the Exchange Application Processes not standardised?
4. Why is the 'alerting' information buried in an object attribute, rather than sent as a discrete message?
5. Why is the call control protocol carried using the ROSE and GFP functionality, rather than being defined as a stand alone protocol?

16

Supporting Applications with Network Intelligence and B-ISDN

Gary Bruce and Jon Clark

16.1 INTRODUCTION

There are many institutions working to develop what is essentially the global public network that we all presently use to transfer speech, data and images to one another. This global public network, the telephony network that we have today has, in a large number of cases, reached its limitations, and has been replaced, albeit on a non-global scale, by newer technologies capable of delivering information at faster speeds or in a wireless environment.

While these non-global networks are being successfully delivered, the utopia for many people working on these technologies must be to provide a global public network worthy of the 21st Century. This network must accommodate fixed and mobile terminals in a consistent manner across a network, it must transport a multitude of media types, and it must provide the necessary logic to support and proliferate advanced services. The mandate for the people delivering this utopia is to piece together the innovation from a number of isolated network-specific solutions, and to establish a totally integrated telecommunications network.

To understand the future global requirements, this chapter starts with the history surrounding the development of B-ISDN and IN recommendations produced by the Telecommunications Sector of the International Telecommunications Union (ITU-T) (formally known as the CCITT). It continues with the service requirements from experts looking into areas such as broadband communications, mobile communications, intelligent networks and the Internet. It describes where we are now, and what still needs to be done.

The introduction to Section 16.4 describes some of the possible proposals being considered for the future global telecommunications network, and delivers the rationale for proposing that the integration of B-ISDN with Network Intelligence will provide the best way forward. The rest of the section then describes two parallel routes that development processes are taking towards an integrated telecommunications network. Both of these routes are based on B-ISDN and IN technologies, however while one route is based upon a nearer term adaptation of the equipment and technology we have at present, the other route is based on a longer term evolution of today's equipment and technology. Perhaps not surprisingly, the nearer term route is somewhat driven by the equipment manufacturers, while the longer term solution is largely being promoted by telecommunication network operators within the ITU-T.

The chapter is concluded by speculating on how close telecommunication network operators will be to meeting user requirements for the 21st Century. It predicts that integrated telecommunications networks will not arrive until five years into the next century, and suggests what the likely stop-gap will be.

16.2 THE DEVELOPMENT OF B-ISDN AND IN

The Intelligent Network (IN)

The IN can be considered as the precursor to multi-service information networks, as it was the first time the network was used for services other than connections between two locations. It provides advanced services which cannot be easily implemented in the switch, or are best offered from a central point in the network. To achieve this, the digital switch is enhanced so that it becomes an IN Service Switching Point (SSP), enabling it to notify the Service Control Point (SCP) when processing of the call/connection requires an IN interaction.

The IN Functional Architecture is illustrated in Figure 16.1. The existing software which controls the switch and communicates with the terminal is the Call Control Function (CCF). The Service Switching Function (SSF) is added to provide the interface to enable call processing to interact with the IN control platform. The flexibility of the IN arises from the SCP, which comprises a Service Control Function (SCF) and a Service Data Function (SDF). The SCF runs services via Service Logic Programs (SLPs). The SDF provides service information like translation tables and user profile data. The Specialised Resource Function (SRF) can be temporarily connected to the caller to play announcements and collect digits from them.

16.2 THE DEVELOPMENT OF B-ISDN AND IN

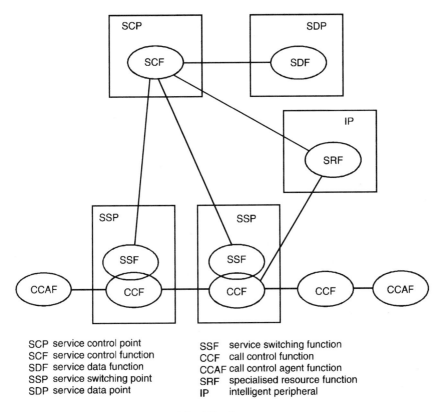

SCP service control point
SCF service control function
SDF service data function
SSP service switching point
SDP service data point
SSF service switching function
CCF call control function
CCAF call control agent function
SRF specialised resource function
IP   intelligent peripheral

Figure 16.1 IN CS-1 Functional Architecture

The ITU-T are developing and releasing IN specifications in stages known as Capability Sets.

IN CS-1 was first released in 1992, and contains the capabilities to re-route calls, modify charging or billing records and obtain extra information from the caller via the SRF. Service Independent Building Blocks (SIBs) are used to construct IN services in SLPs. Ideally, a service built from SIBs would be compiled and run on all SCP platforms. In practice, the different SCP manufacturers have produced proprietary Service Creation Systems (SCSs) which only adopt SIB principles. The SLPs communicate with the underlying SSPs via INAP (Intelligent Network Application Part) operations. As it is easier to compare INAP operations with the requirements for B-ISDN, this chapter will concentrate on limitations in this area as opposed to more generic SIBs.

A list of IN CS-1 INAP operations is given in Table 16.1. This should enable the capabilities of IN CS-1 and the services that can be built up to be envisaged.

Table 16.1 IN CS-1 SSP, SCP and SRF Operations

INAP Operation	Action
Activate Service Filtering/Response	Sets up event monitoring, e.g. Televotes
Activity Test/Response	Test for continued SSF/SCF relationship
Apply Charging/Report	Requests charging data from SSF
Call Gap	Restricts requests to SCF for overload control
Call Information Request/Cancel/Report	Generates call report at end of call
Collect Information	Collect more digits
Connect	Requests the SSP to connect the call up to the specified destination
Connect to Resource	Connects call to SRF, SSF relays all commands to and from SRF
Disconnect Forward Connection	Disconnects call from SRF
Establish Temporary Connection	Connect call to assisting SSF, Intelligent Peripheral (IP) or separate SRF
EventReportBCSM (Event Report Basic Call State Model)	Report of a Dynamic Event
Furnish Charging Information	Provide charging information for SSF Call Records
Initial Detection Point (Initial DP)	Initial Service Request for SCP
Initiate Call Attempt	New call initiated by IN
PlayAnnouncement	Plays an announcement on SRF
PromptAndCollect	Plays an announcement and collects digits on SRF
Release Call	Clears down call
Request Notification Charging Event/Event Notification Charging	SCF requests notification of charging events
RequestReportBCSMEvent	Set up monitoring for dynamic event
Send Charging Information	Allows SCF to set charging rate (e.g. Free)

IN CS-2 is a superset of CS-1. To illustrate the new capabilities a list of new IN CS-2 INAP operations are given in Table 16.2. The most significant change is the addition of leg control, known in IN CS-2 as Call Party Handling (CPH). It enables individual parties to be joined into existing calls to create multiparty calls. Individual or multiple parties (Call Segments) can be split up and reconnected with other calls. The correct terminology and a full definition can be found in the ITU Q1224 specification.

> It should be noted that these CPH enhancements go some way to addressing the multiparty/multiconnection requirements in the B-ISDN. Making them fully consistent with B-ISDN and backwards compatible will be the key task for an Integrated IN/B-ISDN network. See Table 16.4.

16.2 THE DEVELOPMENT OF B-ISDN AND IN

Table 16.2 New IN CS-2 SSP, SCP and SRF Operations

INAP Operation	Action
Activate/Deactivate Trigger Data/Confirmation	Allows SCF to set up Static Events (triggers)
Create Call Segment Association (CSA)	Create a new CSA, does not contain any Call Segment (CS) initially
Disconnect Leg	Disconnects a specific leg, other legs remain in call
Event Report Facility	Reports that an appropriate DSS1 Facility IE has been received
Hold Call In Network	Used for IN call queuing systems
Join	Joins one call into an existing call
Merge Call Segment	Connect Call Segments together
Move Call Segment	A split operation followed by a join on a Call Segment
Move Leg	Moves Legs between Call Segments
Prompt and Receive Message	Request SRF to play announcement and store a message
Reconnect	Re-establishes communication between a controlling leg and held party
Request/Report UTSI	Request/Report User To Service Information (UTSI)
Request/ReportBCUSMEvent (Request/Report Basic Call Unrelated State Model Event)	Requests SSF to monitor for certain call unrelated events from the user
Script Run/Event/Close	Commands to control user scripts on SRF
Send STUI	Send Service To User Information (STUI)
Split Leg	Separates one party from its Call Segment and places it in its own associated CS

Other significant additions to IN CS-2 include call unrelated signalling, which can be used to enable terminal to IN communication without having to set up a connection to an SRF. Also, there is a new SCF to SCF interface to enable Service Control information to be exchanged between network operators. This avoids all calls/connections being tromboned back into the caller's home network. The SRF has also been improved to enable scripts to be run, text information to be sent and received from ISDN phones, and the ability to store messages will enable network-based answering services to be built in the IN.

In relation to broadband, these new features provide very useful capabilities which, with a few enhancements, could be exploited to enable a service provider to set up multiple connections per call as required for videoconferencing, broadcast TV or many other services which are not just single end-to-end connections.

The broadband network (B-ISDN)

During the 1980s, telecommunication network operators were investigating how to develop their networks beyond the plain old telephone service we know today (Chen and Liu 1995). In 1988, Asynchronous Transfer Mode (ATM) heralded a revolution that made the leap from narrowband services to broadband services a possibility. At around the same time, the ITU-T drafted a framework to develop B-ISDN. This framework consisted of three phases, called Releases. As B-ISDN is conceptually providing no more than higher speed N-ISDN services, Release 1 B-ISDN was targeted simply to adapt the N-ISDN services, such as telephony and constant bit rate services, to operate upon ATM technology. Table 16.3 gives the high-level requirements for B-ISDN Release 1. The immediately obvious similarity of B-ISDN Release 1 with N-ISDN is that the call and network connection are tied together when a 'Call/Connection' is established or cleared down. The other comparison with N-ISDN is that only single connection, point-to-point 'Call/Connections' can be configured. As mentioned earlier, B-ISDN Release 1 does offer larger bandwidth 'Call/Connections', however it also supports the additional provision for variable bit rate 'Call/Connections' as an alternative to the Narrowband-*like* constant bit rate 'Call/Connections'.

The signalling recommendations for B-ISDN Release 1 were published early 1995, and network equipment supporting these recommendations are now available. Support for ITU-T B-ISDN terminal equipment, however, is somewhat lagging, the majority of terminal vendors preferring to build their equipment on ATM Forum standards.

Releases 2 and beyond of B-ISDN were then identified to cover new and enhanced services, such as broadcast services, Video on Demand (VoD) and video telephony. To support some of these extra services, the main innovation for Release 2 B-ISDN is the functional separation of Call Control and Bearer Control, to support the new connection configurations required by these new services. Table 16.4 depicts the high-level signalling

Table 16.3 B-ISDN Release 1 Signalling Requirements

Signalling requirements	Connection type
Simultaneous Call and Network Connection establishment Release of a Call and associated Network Connection	Point-to-point, single-connection

16.2 THE DEVELOPMENT OF B-ISDN AND IN

Table 16.4 B-ISDN Release 2 Signalling Requirements

Signalling requirements	Connection type
Simultaneous Call and Network Connection establishment Call establishment without Network Connections Addition of one or more new Network Connections to an existing Call Addition of one or more new Parties to an existing Call Attach one or more existing Parties to an existing Network Connection Detach one party from an existing Network Connection Release a Network Connection from an existing Call Release a Party from an existing Call Release of a Call and associated Network Connections Renegotiation and modification of an active Network Connection Look-ahead at the destination party's availability and terminal characteristics	Point-to-point, multi-connection or Point-to-multipoint, multi-connection

requirements for B-ISDN Release 2. It has been greatly simplified to give a taste of the possible richness of B-ISDN Release 2 capabilities. New capabilities include point-to-multipoint network connections, the ability to support multiple network connections within a call, plus the ability to manipulate the calls independently of the network connection(s), and *vice versa*. Also worthy of note is that the capabilities listed in Table 16.4 can be invoked by any party within the call even if they are not associated with a network connection.

> **Protocol pitfalls** Although the point-to-multipoint protocol Q.2971 was developed in the B-ISDN Release 2 time frame, it is the opinion of the authors that in the haste to provide the point-to-multipoint capability, it failed to address the Call and Bearer Control separation issue. As a result, additional network connections cannot be added after a Q.2971 Call/Connection is established, because it does not support both Call Reference and Network Connection values. The 'Call Ref.' tag used in Q.2971 is a carry over from the time when a call always contained a single connection and, as such, the 'Call Ref.' could be construed as referring to a connection with an implied reference to the call. In a separated Call and Bearer environment, it is likely that the 'Call Ref.' will be used as a network connection reference while a new tag will be generated for a call reference.
>
> Q.2931 is an even more restricted version of Q.2971, and there are current moves to revise this protocol to accommodate Call and Bearer Control separation concepts.

While the signalling requirements for B-ISDN Release 2 were completed in December 1993, the sheer complexity of providing these additional capabilities led the ITU-T into delivering the signalling recommendations for this release in two main steps. The majority of the signalling recommendations for the first step of B-ISDN Release 2 have been approved, and approval for the remaining recommendations will follow shortly. Network or terminal equipment supporting some of these recommendations has yet to materialise beyond trial versions. The signalling recommendations for the second step of B-ISDN Release 2 is still being developed, and it is this step that contains the activity to support the separation of Call Control and Bearer Control. It is expected that approved versions of these recommendations will start to become available in May 1998.

The primary objective for B-ISDN Release 3 was labelled as the integration of B-ISDN with IN, mobile communications and control systems for operating and managing the network. The first phase of the signalling requirements for B-ISDN Release 3 are expected to become stable by October 1998, with signalling recommendations for this phase starting to become available in February 2000. The second, and possibly final, phase of the signalling requirements for B-ISDN Release 3 are expected to become stable by September 1999, with signalling recommendations for this phase starting to become available in June 2001. The high-level B-ISDN Release 3 signalling requirements are given in the Table 16.5.

Table 16.5 B-ISDN Release 3, Phase 1 Capabilities

Signalling requirements	Connection type
Join a Party to an existing Call (Leaf-Initiated Join to a Call)	Point-to-point,
Join a Party to an existing Call and existing Network Connection(s)	Point-to-multipoint,
(Leaf-Initiated Join to a Call and Network Connection(s))	Multipoint-to-point
Simultaneous Call and Network Connection(s) transfer	
Call transfer	Multipoint,
Network Connection(s) transfer	Bidirectional
Call ownership transfer	point-to-multipoint
Party ownership transfer	
Network Connection ownership transfer	or
Branch ownership transfer	Server supported multipoint

16.3 APPLICATION REQUIREMENTS FOR FUTURE TELECOMMUNICATIONS NETWORKS

There are many changes taking place in the type and volume of applications being run by networked terminals. These changes have partly been driven by the high speed, low cost, wide area networks that the terminals are connected to, and partly by the advances in terminal technology. Requirements for new applications have mainly originated from the IT world, however, the main telecommunications players are not far behind.

The next few sub-sections describe some of the main application, or service, requirements coming from experts working in the field of intelligent networks, broadband ISDN, the Internet, IT and mobile networks. Each set of requirements presents its own obstacles that must be crossed. What will be discussed is which of the requirements have already been achieved, and which are currently being addressed. Section 16.4 draws all of these requirements together, and examines what needs to be done to fulfil these requirements.

IN services

Services that can be constructed on an IN platform include routing to non-geographic numbers such as freephone and premium rate numbers, etc., virtual private networks, call screening, closed user groups, call distribution according to load, time of day, etc., authentication for charge cards or account calls and generation of advanced billing records. Most of these services will still be applicable in a broadband network, and there are abundant opportunities for many more services.

The new broadband services are likely to provide new requirements from service providers. These will include the ability to set up, release and move connections to their customer. For example, offering an API to the Service Control for service providers could enable VoD servers to be distributed around the network while still remaining under control of the service providers.

The evolution to broadband will also require a number of enhancements to the SRF. It may now be acting as a service gateway to which the user connects initially in order to choose the desired service. It will therefore need to have the ability to play MPEG data streams, retrieve responses and download software to users' set-top boxes or PCs.

B-ISDN services

The main requirements coming into the B-ISDN domain have come from the Digital Audio-VIsual Council (DAVIC). Requirements from DAVIC have been focusing on the emergence of new multimedia services such as VoD, broadcast and staggercast TV with return channels, videoconferencing and video telephony. While VoD, broadcast and staggercast have been specified within DAVIC, videoconferencing and video telephony have yet to be specified.

Within DAVIC, services such as VoD have led to the development of a Session Control plane above the Call/Connection Control plane as a means to co-ordinate disparate Call/Connections among the parties involved in long lasting sessions.

Within the ITU-T, the VoD type services have led to the development of Session Control and Resource Control to provide a means to co-ordinate not only calls and connections within a particular network implementation, but across many disparate network implementations. Within the ITU-T, Session Control is network implementation independent. This network independence is provided by Resource Control. It will provide functions similar to those provided by Winsock™ and H.245 in that it will provide a single API to Session Control while adapting to different network implementations (i.e. ITU-T B-ISDN, ATM-F B-ISDN, POTS, Internet) and capabilities (i.e. Q.2931, Q.2971) it is running over. Extra flexibility being built into Session Control will also allow end applications involved within a session to negotiate for access network and software resources when these resources are not readily available.

IP/Internet applications

The massive growth of the Internet has brought about a wealth of new requirements placed on the 'net' from more sophisticated applications. Gone are the days of simple UDP (User Datagram Protocol) and TCP (Transmission Control Protocol) support for applications such as file transfer programs and browsers. Newer applications such as NetMeeting™, which supports videoconferencing on a non-guaranteed Quality of Service (QoS) using the ITU-T recommendation H.323, would provide a better service if transported using a guaranteed QoS. Today, services such as the Internet Phone are starting to demand a guaranteed QoS, and it is for this reason that the newer capabilities such as RsvP (Resource Reservation Protocol) are being developed for the Internet.

As with most telecommunications systems, the Internet is both a threat and an opportunity for the existing Intelligent Network. There are no formal IP interfaces specified as part of the IN architecture, although some types of SCP do offer this capability. It could be used to provide customers with access to their IN data via the Internet, or for dial-in PC connections. Other possibilities include CTI type applications, where extra information about the caller is presented on the screen of the person answering the call.

Another area is the IETF PSTN/Internet Interfaces (PINT) Working Group, which is addressing connection arrangements through which Internet applications can enhance PSTN services. An example of such a service is a Web-based Yellow Pages service with the ability to initiate PSTN calls between customers and suppliers. This will require new APIs between the PSTN and Internet.

Mobile services

A mobile terminal's requirements should be the same as those pertaining to a fixed terminal such as basic connectivity and IN support for enhanced services. However, there are a few notable exceptions, the main one being that the mobile terminal should be allowed to roam between base stations or even exchanges. Although roaming is currently achievable using GSM, a more resource efficient method is being specified within the International Mobile Telecommunications (IMT) 2000 initiative. The other main difference is that the total possible bandwidth delivered to a mobile terminal is going to be less than that delivered to a fixed terminal. While this apparently reduced bandwidth to mobile termins is likely to imply a subset of services than those provided to the fixed counterparts, it is this 'bandwidth challenge' that is inspiring services such as SMS, and no doublt others that are unique to the mobile domain will follow.

16.4 ALTERNATIVE APPROACHES TO BUILDING AN INTEGRATED TELECOMMUNICATIONS NETWORK

Proposed solutions for future telecommunications networks

Section 16.3 examined the pressure placed upon any future public network by a number of dissimilar but potentially overlapping user requirements. How to realise a future telecommunications network that would fulfil all of these requirements would be impossible to predict. At one extreme the perfect solution might be to build a distributed environment that all terminals and applications can exist in, while at the other extreme the most pragmatic solution might be to enhance the present terminals and applications that run on the Internet. Either one these proposed solutions could, with some extra effort, quite easily be the dominant future global telecommunications network. The immaturity of object busses in a global telecommunications environment and the issues of quality of service, security and charging surrounding the Internet possibly makes B-ISDN-based solutions the most likely of contenders for future public telecommunications networks. B-ISDN is a mature and proven technology that, amongst other things, provides quality of service, security and charging. However, B-ISDN is not the complete panacea for all user requirements, as there is still much work to be done.

This section poses two alternative approaches to building an intelligent broadband network for the future. The first approach looks at a near term realisation by adapting and enhancing the equipment that is presently available to solve user requirements as and when they emerge. The second approach is a longer term realisation of the unified telecommunications network that has been based upon the initial requirements for B-ISDN Release 3 and enhanced along the way.

Nearer term realisation to building an integrated telecommunications network

Although a lot of the concepts presented here are a long way from being realised in hardware and software, there are a number of services which can be built from systems available now or in the near future. This section covers some of these solutions to illustrate how the fast pace of telecommunications development is stimulating new services to be constructed from current technology.

DAVIC IP browsing user interface

This is a good example of using existing technology to solve the current limitations of user interfaces to the network. The terminal now contains a WEB Browser (i.e. Netscape or Internet Explorer), and this talks to a Call Control function which is basically a WWW Server. The advantage this gives are that the user can be presented with an easy to use graphical interface that can be customised for the particular services a user subscribes to. Figure 16.2 illustrates this solution.

The broadband access network connects the Set Top Unit (STU) to the Internet Service Access Point (ISAP) via variety of technological possibilities not covered here. The HTTP request is for the reserved address of the

16.4 BUILDING AN INTEGRATED TELECOMMUNICATIONS NETWORK

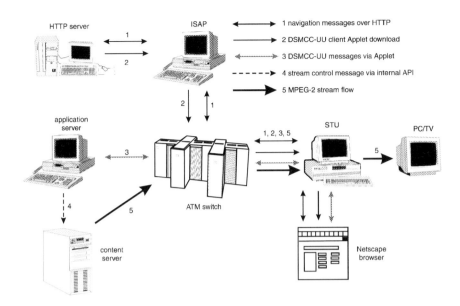

Figure 16.2 IP Browsing User Interface

HTTP Server, which is providing the equivalent of the network intelligence layer. The user will be presented with a range of service options and selects a video, which is provided by the Application Server on this diagram.

The HTTP Server downloads a DSM-CC User to User (see Chapter 10 on DAVIC) Client JAVA Applet which is needed for accessing the controls on the application server. This applet will convert commands from the PC/TV remote control (play, stop, FF, etc.) into DSM-CC UU messages. The connection for the MPEG-2 stream flow is made via DSM-CC User to Network commands from the STU initiated by the JAVA applet. Alternatives are reasible such as setting up this connection from the HTTP Server, which would significantly simplify protocols supported by the STU at the expense of complexity of the switch to server interface.

Alternatives to INAP

Instead of waiting for the INAP specifications to be enhanced to support all the capabilities required and for ATM switches to be upgraded to support INAP, some alternatives are being explored. These include using interfaces to the switch management system or using a proxy signalling interface to set up the connections across the switch.

Proxy signalling was originally developed to allow Set Top Boxes with minimal signalling capability to set up calls on the network. They would communicate with a Proxy Signalling Agent in the local network, which would support the full requirements of a Broadband Edge Switch and set up the call on its behalf.

At the moment, providing a UNI on ATM switches to enable users to set up their own ATM connections (Switched Virtual Circuits) is a higher priority than adding IN interfaces. As proxy signalling is only an enhancement to UNI signalling, this interface is currently more realisable than an ATM IN Switch.

The limitations of this approach relate to the User to IN interface, which is not clear because of the lack of IN Basic Call State Models to represent the Call set up states. This may be overcome by terminating the user signalling in the IN SCP instead of the CCF, although the Browsing interface is particularly appropriate for this purpose.

Longer term realisation to building an integrated telecommunications network

Within the ITU-T the goal is for a truly unified network that is rich in network capabilities for all end users. This vision of a 'Unified Telecommunications Network' which, although based on B-ISDN Release 3 requirements, has also been influenced by the Internet, DAVIC, TINA-C and multimedia groups.

To progress the Unified Telecommunications Network the ITU-T is currently using a method to develop B-ISDN and IN signalling requirements, and hence capabilities. It is expected that future mobility and multimedia signalling requirements will be developed using the same method. The method used is derived from I.130 and object-oriented techniques, especially the Open Distributed Process Reference Model (ODP-RM), to provide a co-ordinated framework to develop the standards. Figure 16.3 shows an overview of the framework.

Having acquired service requirements and generated the service definitions, a Business Model is generated to provide a common platform of understanding for the experts from different disciplines involved in pulling together the unified network. Precise definition of the data and behaviour of the system at a fine granularity object level is carried out. This shows

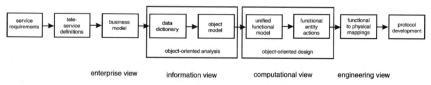

Figure 16.3 Framework to Develop Unified Network Standards

16.4 BUILDING AN INTEGRATED TELECOMMUNICATIONS NETWORK

information that must be passed around the system when each of the capabilities of the system are invoked. Using the object model, the objects in the telecommunications 'system' are partitioned into courser granular objects which make up the functional model. The necessary information exchanges around this system are expressed as information flows and functional entity actions which are produced to verify the analysis and the functional model. These show how entities co-operate and what information is passed between co-operating entities to realise a capability. At this point, this information could be passed on to generate the necessary protocols, however, as protocols only need to be standardised at physical interfaces, functional entities are mapped onto physical entities to identify physical interfaces and reduce the number of protocols needing to be standardised.

The following sections show in detail some of the more interesting processes and results of this framework, namely the Business Model, the Functional Model, Functional Entity Actions and the new capabilities.

The business model

The Business Model captures business requirements to structure and orient the system of interest. The model is derived from the Open Distributed Processing (ODP) enterprise viewpoint, to partition the total business enterprise into business domains. These business domains represent, for example, the users, owners and providers of information processed by the system.

The proposed Business Model is shown in Figure 16.4, and is composed of six types of business domains, connected by interdomain logical reference points. These logical reference points identify the communication channels available to exchange information between these business domains.

The model illustrates the following business domains:

- **Consumer:** enrols for and consumes information. A consumer may be an individual, a household with a small number of end user, or a business with multiple end users.

- **Broker:** provides location information that the business domains use to locate other entities for the purpose of establishing communications.

- **Retailer:** provides the optional initial contact point for the Consumer to arrange for, and the contact point for the Service Provider to offer, services which employ communication services as the delivery mechanism.

- **Service Provider:** provides any of a variety of services which employ communication services as the delivery mechanism.

- **Network Connectivity Provider:** provides the communication services that transports information that may either be control plane information or user plane information.

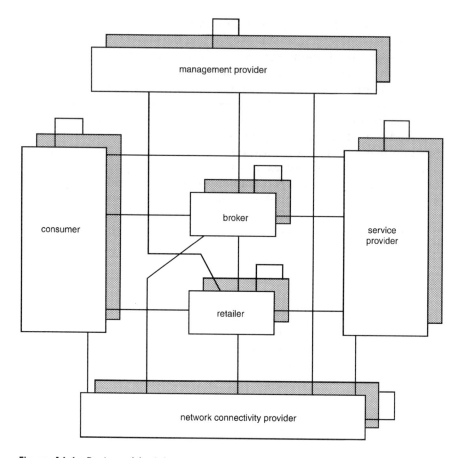

Figure 16.4 Business Model

- **Management Provider:** manages, in whole or part, a number of business domains and provides the necessary administrative capabilities.

Note that nothing in the model precludes any combination of these six domains within a single physical entity. For example, an IN SCP which may typically be associated within the broker domain may contain functionality specific to the retailer domain.

The unified functional model

The Unified Functional Model (UFM) partitions the system into Functional Entities (FEs) performing individual functions that will be conducive to flexibility in support of new services and in the deployment of service processing in the network. The model combines and integrates the aspects

16.4 BUILDING AN INTEGRATED TELECOMMUNICATIONS NETWORK

of the IN Distributed Functional Plane (DFP) architecture, the IMT 2000 Functional Network Architecture (FNA) and the B-ISDN architecture.

The proposed model is shown in Figure 16.5, and is expressed as FEs which communicate via FE relationships. The FEs are independent of physical deployment and implementation in any of the business domains, providing the flexibility for multiple physical network configurations.

The model has been arranged to show one possilbility of how the functional entities map onto the Business Model. Although complex in its functionality, the salient components of the Unified Functional Model, however, are described below.

The UFM is best described by dividing it into horizontal functional groupings given below:

- **Network/Terminal Element (X):** although not strictly an FE, the network/terminal element represents an aggregation of network or terminal elements under control of a single functional entity.

- **Fabric Management (FM):** manages the resources contained within a particular network or terminal element.

- **Access Control (AC):** provides access channel control to establish, for example, a signalling channel on a mobile air interface.

- **Bearer Control (BC):** supports the control of network resources on a network link by network link scale to provide end-to-end carriage of information.

- **Call Control (CC):** provides the control of functions on a network-wide scale. CC is not technology dependant and can be used over a number of different concatenated transport technologies to co-ordinate a call.

- **Resource Control (RC):** uses the appropriate network infrastructure to provide the capabilities required by Session Control. This network infrastructure may be an ATM network, Internet, Wireless, etc. (note that this RC is not the same as the Resource Control in Chapter 14, defined by RACE MAGIC).

- **Session Control (SeC):** supports the control of communications sessions between user applications, including end-user (customer), network (broker) and third-party (retailer) based applications. Through an Application Programming Interface (API), a communications session supports a single instance of a telecommunications application.

- **Service Control (SC):** provides end-to-end brokerage type functions between terminals. An example could be for advertisement of the identifier for a broadcast or conferencing service required by third parties prior to initiating a leaf-initiated join to the call. The need for this functionality is currently undergoing investigation.

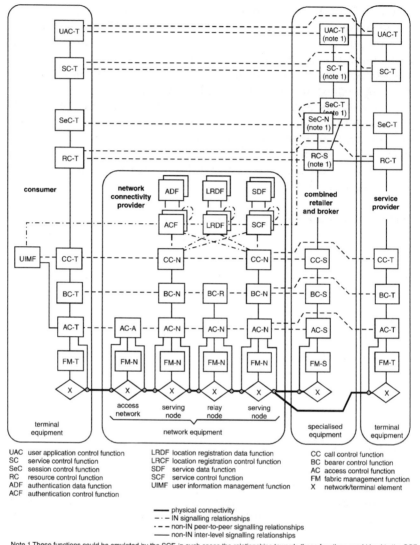

Figure 16.5 The Unified Functional Model Overlaid upon the Business Model

- **User Application Control (UAC):** provides the control of the application itself. Examples would be pause, rewind, book marking and content selection for a VoD application.

Additional support functions contained in the UFM are:

- **User Information Management Function (UIMF):** provides the end-user's service profile to allow for service mobility.

16.4 BUILDING AN INTEGRATED TELECOMMUNICATIONS NETWORK 319

- **Access Control/Data Functions (ACF/ADF) and Location Registration Control/Data Functions (LRCF/LRDF):** performs processing and provides access to data that is specialised for particular non-call-related (call unrelated) mobility applications. It is likely that the ACF and LRCF will be merged into a courser-grained function, possibly to be called a Call-Unrelated Service Control Function (CUSCF). The same applies to merging the ADF and LRDF. The possibility of these merging of functions is currently undergoing investigation.

- **Service Control/Data Functions (SCF/SDF):** performs processing and provides access to data that is specialised for particular call-related service applications.

In addition to the speculation given above, one further point of contention about the UFM within the ITU-T is the relationship between the SCF and the BC, where the BC exists without the CC (i.e. in transit exchanges). The rationale for this relationship's existence is to provide IN support, such as number translation, for the BC function. Opposition to this view is that IN support such as number translation should be done by the CC function in the originating serving node, and that the translated number should be passed on to the BC function. For a possible way forward on this issue, see interaction of SCF with B-ISDN separated call and bearer state machines.

Functional Entity Actions for B-ISDN and IN Release 3

Functional Entity Actions are used to illustrate high level overviews of each signalling capability contained within B-ISDN and IN Release 3. These high level overviews are expressed as a succession of information flows, showing the type and content of the flows, which describe the signalling control actions associated with each of the signalling capabilities. Essentially, it is these signalling capabilities that collectively make up the set of B-ISDN and IN Release 3 signalling requirements which are input into the protocol development stage to produce the signalling protocols.

New capabilities and features for B-ISDN aspects of Release 3

A number of new call and network connection related capabilities are being described using functional entity actions for phase 1 of B-ISDN Release 3. Table 16.5 illustrates these new capabilities. To simplify the table, only the capabilities that do not exist in B-ISDN Releases 1 and 2 are shown. New connection types for B-ISDN Release 3 are multipoint-to-point, multipoint, bidirectional point-to-multipoint, and multipoint supported by a server or bridge function. The capabilities, already available for B-ISDN Releases 1 and 2, have naturally been extended in Release 3 to support these new connection types. It should also be noted that the capabilities associated with B-ISDN Release 3 have not been extended so that they may be invoked by the IN, as well as by the individual parties.

Additional features not relating to Call and Network Connections are being added to phase 2 of B-ISDN Release 3. These include additional capabilities such as Access Control to handle mobile terminals and access networks such as VB5. Session Control and Resource Control together will provide APIs to allow enhanced multimedia applications to operate in network-rich environments. This means that applications will be able to operate over a multitude of network types such as B-ISDN, Internet or PSTN. Finally, adding standardised vertical interfaces on the B-ISDN Release 3 protocol stacks will provide greater flexibility when provisioning network equipment.

New capabilities and features for IN aspects of Capability Set 3

Work on IN CS-3 is now complete and addresses a number of problems. The IN and B-ISDN systems have developed independently. This leads to interworking problems as the relationship between the two systems is not clearly defined, and there is an overlap of services which are offered by both systems. For example, some ISDN supplementary services can be provided by the IN such as charging and call screening services.

More recent networks have extra capabilities, which the IN may need to procoess. Some of these are relatively simple to add, such as the extra information associated with new connection types. Other more complex areas will require further research. These include how to support multi-point/multi-connection calls, and how broadband will affect the IN call models. As there are many things that need to be added in IN CS-3, this Capability Set was subdivided into CS-3.1 and CS-3.2, although CS-3.2 has now become CS-4.

IN CS-3 concentrated on realistic changes that are feasible in the timescales planned. Therefore, it does not offer any major new features, but it will build on what is already within IN CS-2 and hopefully fix problems identified such as complete IN/ISDN interworking. It provides better support for the new network capabilities, in particular point-to-point broadband connections. Broadband networks will have a semi-integrated IN system in that IN will support point-to-multipoint connections, but only where the IN point of attachment is at the 'root' or point end of the point-to-multipoint connection. IN CS-3 does not support Call and Bearer separation, however, additional IN triggers and Q.2971 information elements will be accommodated for and supported by the SCPs. Table 16.6 shows the Q.2971 parameters that currently can now be passed to the IN.

These new parameters are optional, so the network operator will be able to specify which information to export from the SSP to SCP when triggered,

16.4 BUILDING AN INTEGRATED TELECOMMUNICATIONS NETWORK

Table 16.6 New IN CS-3 B-ISDN Parameter

ATM traffic descriptor
Quality of Service parameter
Broadband bearer capability

Table 16.7 B-ISDN Network Features

INAP feature	Action
Point-To-Point connection	User initiated simultaneous call and connection set-up, bidirectional point-to-point connections are supported; multimedia communication shall be supported at call/connections setup.
Network initiated Point-to-Point connection	Establishment and release of point-to-point, bidirectional connection via invocation of B-ISDN signalling.
ATM traffic capabilities control	For network initiated B-ISDN connections, IN shall be able to control the Quality of Service at call set-up.
ATM traffic capabilities notification	For user initiated B-ISDN connection, the SSF is able to report to the SCF a notification of the traffic parameters for the indicated, modified and negotiated ATM traffic capabilities.
AESA addressing	ATM Service Endpoint Addressing is supported in the operations relevant for IN number translation requests.
SCF triggering from B-ISDN signalling	Trigger processing and parameter population rules appropriate for DSS2 and B-ISUP support are being provided.

if required by the IN, see Table 16.7. An example of why this data is needed would be to select a suitable SRF for the calling terminal which may now be incompatible with certain types of SRF, e.g. playing voice announcements to a Set Top Box should be avoided.

Other improvements in CS-3 include enhancements to the Feature Interaction Manager to limit service interaction problems, and new capabilities for the SCF to SCF interface, particularly in the area of billing. Developments to the Call Unrelated Signalling should make this feature more generic and useful.

A clear recognition of the importance of the Internet is the addition of new services to enable a PSTN all to be initiated via a web site and the IN, as shown in Table 16.8.

An example of an application of this feature would be online shopping: A user is browsing through an online catalogue, and clicks a button thus

Table 16.8 Interworking between IN and IP networks

INAP operation	Action
Request-to-Call-Back	A user is able to initiate a telephone call by clocking a button during a Web session. The call can be first set up in the direction of the requester of the call, or first be set up in the direction of the party the requester wants to be connected to.
Request-to-Call	A user is able to initiate a telephone call by clicking a button during a Web session. The requested call is to be set up between two parties identified by E.164 addresses, which are connected to the switched circuit network. The requester him/herself may or may not take part in the call to be set up.

inviting a call from a sales representative. In IN the request could be handled depending upon the availability of an agent, the time of day, etc.

Requirements for IN CS-4

IN CS-4 is where new proposals that require longer timescales additional to IN CS-3 will be considered. Additional B-ISDN parameters like ATM adaptation layer parameter, connection identifier and end-to-end transit delay may need to be passed to the IN. Making these, and some of the B-ISDN parameters listed in Table 16.6, into triggers will enable the IN to be used only for certain call types. Other possibilities include new basic call state models, one for the Call Control level and one for the Bearer Control level, and enhancing the SRF to support videoconferences, act as service gateways and have the ability to download software to users.

Call and Bearer separation is the biggest issue facing the IN under broadband. Should the call models be split and reflect the individual state of both the bearer and the call control? At the moment, the IN has the equivalent of bearer control, because it needs to establish connections to SRFs and the final destination. This requirement is not likely to disappear with the migration to broadband. Similarly, the equivalent of call control is occurring when the IN interacts with the caller, e.g. processing dialled digits or playing announcements.

Interaction of SCF with B-ISDN separated call and bearer state machines

IN CS-1 INAP and IN CS-2 INAP still refer to a call as a call containing a single connection. To progress INAP into a separated call and bearer

16.4 BUILDING AN INTEGRATED TELECOMMUNICATIONS NETWORK

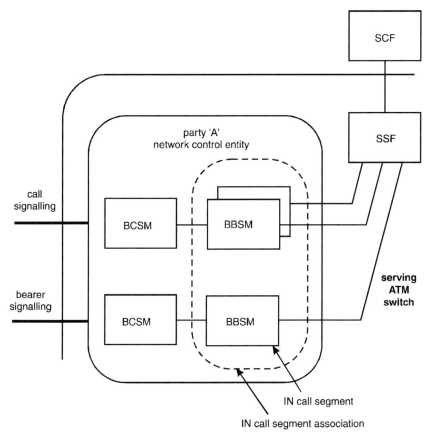

Figure 16.6 Separated Call and Bearer Environment

environment, it is being suggested that the interaction between IN and B-ISDN is restricted to the establishment, modification and release of individual bearers associated with the party end points. The benefit of this approach would be that the IN call model need not change, and that the IN SCF would not be required to be different if the underlying protocols have or do not have the capability of controlling multiple connections per call. This issue can be better described by reference to Figure 16.6.

It is the authors' opinion that the IN CS-2 Call Segment is equivalent to the B-ISDN Release 2/3 Basic Bearer State Machine, and that the IN CS-2 Call Segment Association represents the set of all B-ISDN Release 2/3 Basic Bearer State Machines for a given party. Using this supposition, any enhanced INAP protocol could operate on a BBSM as if it were an IN CS-2

Call Segment, and any IN and B-ISDN interworking may be arranged in such a way so that the IN does not need to know Party 'A's' signalling protocol capabilities.

The other issue is how the Basic Call State Models (BCSM) should be modified to support the new networks. At the moment, the BCSMs are based on the telephony call model, although they should ideally be generic for all call types to enable the same application on an SCP to be used for all call types. Perhaps different call types will require unique BCSMs to allow all these new network capabilities to be utilised by the IN. The default BCSM would remain for simple services or legacy networks.

16.5 CONCLUSION

This chapter has described the requirements that different users will impose on the global telecommunications network for the 21st Century. It has also illustrated how some of the technology detailed in companion chapters in this volume will be utilised to provide these new services. Out of the many proposed alternatives for supporting these requirements, the chapter has examined just two in greater detail.

The current intelligent network provides a good basis to move towards the Unified Telecommunications Network. A wide variety of services can be built from what is already specified, and the ability for the network operator to develop their own services via a service creation system will enable services currently not foreseen to be constructed relatively easily.

While a solution building on existing equipment will definitely be ready by the turn of the century, it is doubtful if it will initially achieve all of the requirements placed upon it. The Unified Telecommunications Network, however, is designed to handle all existing requirements, and to be flexible enough to handle many future requirements. The drawback with this solution is that it is unlikely that there will be any equipment available until 2004.

So what does this mean for telecommunications users? If it means that telecommunication network operators will not be able to provide public telecommunications networks to satisfy the expectations of many users in the year 2000, then the telecommunication network operators have failed. However, what this most likely means is that, to manage expectations, a balance needs to be struck to provide users with the most pragmatic and innovative solutions based on existing equipment and emerging standards until a truly integrated telecommunications network is delivered by 2005, at earliest. Perhaps the millennium is coming just five years too early for the Unified Telecommunications Networks!

REFERENCES

Chen T and Liu S, *ATM Switching Systems*, Artech House, 1995.

FURTHER READING

Dufour I G (ed), *Network Intelligence*, Chapman & Hall, London, 1997.

STANDARDS

H.322 *Visual telephone systems and terminal equipment for local area networks which provide a guaranteed quality of service*, ITU-T (03/96).
H.245 *Control protocol for multimedia commmunication*, ITU-T (02/98).
X.901 & 10746-1 *Information technology—Open distributed processing—Reference Model: Overview*, common text from ITU-T and ISO (08/97).
Q.1200 *General series Intelligent Network Recommendation structure*, ITU-T (09/97)
Q.1201 *Principles of intelligent network architecture*, ITU-T (10/92).
Q.1202/I.328 *Intelligent network—service plane architecture*, ITU-T (09/97), published with double number (for details see I.328).
Q.1203/I.329 *Intelligent network—Global functional plane architecture*, ITU-T (09/97), published with a double number (for details see I.329).
Q.1204 *Intelligent network distributed functional plane architecture*, ITU-T (03/93).
Q.1205 *Intelligent network physical plane architecture*, ITU-T (03/93).
Q.1208 *General aspects of the Intelligent Network Application Protocol*, ITU-T (09/97).
Q.1210 *Q.1210-series Intelligent network recommendation structure*, ITU-T (10/95).
Q.1211 *Introduction to intelligent network capability set 1*, ITU-T (03/93).
Q.1213 *Global functional plane for intelligent network CS-1*, ITU-T (10/95).
Q.1214 *Distributed functional plane for intelligent network CS-1*, ITU-T (10/95).
Q.1215 *Physical plane for intelligent network CS-1*, ITU-T (10/95).
Q.1218 *Interface recommendation for intelligent network CS-1*, ITU-T (10/95).
Q.1219 *Intelligent network user's guide for capability set 1*, ITU-T (04/94).
Q.1220 *Q.1220-series Intelligent Network Capability Set 2 (CS-2) Recommendation structure*, ITU-T (09/97).
Q.1224 *Distributed Functional Plane for Intelligent Network Capability Set 2*, ITU-T (09/97).
Q.1228 *Interface Recommendation for Intelligent Network Capability Set 2*, ITU-T (09/97).
Q.1290 *Glossary of terms used in the definition of intelligent networks*, ITU-T (05/98).

Supplement 6 to Q Series Recommendations *TRQ.2000—Roadmap for TRQ.2XXX series technical reports*, ITU-T (03/99).

Supplement 7 to Q Series Recommendations *Technical report TRQ.2001—General aspects for the development of unified signalling requirements*, ITU-T (03/99).

Reservation Protocol (RsvP)—IETF RFC (Internet Engineering Task Force).

QUESTIONS

1. Why does the unified functional model have so many layers of control?
2. Is it necessary to use all of the unified functional model's layers for simple telephony-like services?
3. Application Programmers Interfaces (APIs) above the session control layer facilitate the support of telecommunications applications. Does this imply that only first party (Consumer, service provider, broker or retailer) applications can be deployed?

Appendix A: Answers to Questions

CHAPTER 1. ATM SWITCHING

1. Suggest a simple method of allowing multiple terminals to share a single network connection

There are three requirements to satisfy in this scenario. Each terminal needs to be easily identifiable; each terminal needs to operate independently, and the terminals must not interfere with each other.

To identify terminals, each needs a unique address or number. This could be mapped to a permanently allocated (provisioned) virtual path. This would enable the controlling exchange to communicate directly with the addressed terminal without the complication of multiple terminal call offering.

A provisioned virtual path facilitates the independant operation of terminals, allowing each to use multiple virtual channels if it wishes.

To avoid the terminals interfering with each other, generic flow control should be used. This can force off a high usage terminal and allow other terminals to use their alloted bandwidth. The dynamic allocation of bandwidth (using signalling) would greatly enhance the capability of an exchange to manage bandwidth and resources on a link.

There are some restrictions to this type of working which should be recognised. There is a limit of 256 virtual paths at a UNI (and therefore this solution is restricted to 256 terminals), and each terminal is restricted to around 64 000 virtual channels (or communication links).

2. What information is required to enable a service to use connection Type 3?

Section 1.4.2 describes the capabilities offered by AAL Type 3/4, stating that it offers sequencing, framing and source information, and that the source information enables the support of Type 3 connections. To use these capabilities, it is necessary for the sink end point (the receiver) to determine which source end point (the sender) is represented by the AAL Type 3/4 source information. This information should be exchanged by the end points before any AAL Type 3/4 cells are transmitted. This defines a signalling requirement for Type 3 connections.

The scenario has limitations, mainly in the number of source end points that can be supported. This provides a further exercise for those interested.

3. Why is AAL Type 5 more efficient than AAL Type 3/4 for large frames?

AAL Type 3/4 supports error detection at the cell level, enabling the retransmission of cells with detected errors. This implies that all cells received after the faulty cell must be stored until the faulty cell has been retransmitted. Whilst this type of retransmission is an efficient use of bandwidth, it places a higher requirement on the amount of buffer space that must be provided. Since up to 64 000 octets can be supported as a single block transfer, then this is the required size of the buffer to allow for all cells to be transmitted and then faulty cells retransmitted afterwards. The 64 000 limit could also introduce inefficiencies if the amount of information is larger than this amount. Each end of a communications link must provide some additional adaptation to reduce the data to 64 000 octet segments, introducing further inefficiencies.

CHAPTER 2. INTRODUCTION TO SIGNALLING

1. Describe an in-band signalling mechanism in common use today

The commonest method of invoking telecommunications services remains the telephone. Communication of service requirements, including destination telephone number, is provided by push buttons which cause mutli-frequency tones to be transmitted within the speech channel. The frequencies used are audible, and therefore in-band.

2. How does a signalling system differ from a signalling protocol?

Signalling is a means of communicating service information between a user and the network, within networks and between networks to establish, maintain and release communications channels(s). A protocol is a mechanism providing the structured and ordered exchange of information between peer entities. A signalling protocol is therefore a structured and ordered exchange of information, invoking specific procedures to undertake services related to the establishment, maintenance and release of one or more communications channels.

3. Why is it necessary to protect a signalling network?

Protection of a signalling network should only be necessary under failure conditions. Some of the common failure conditions that could affect a signalling network include loss of communication due to transmission failure (regarded by level 3 as a level 2 failure) and overloading the signalling network (usually due to traffic congestion). In the first case, the support of alternative signalling routes (such as using an STP) ensures that a single failure on one link does not necessarily cause a failure on all routes between two exchanges. In the second case, it is important to recognise when congestion is beginning to occur and to take steps to prevent it. By protecting the signalling network against overload, new user plane traffic can be refused using signalling procedures, thus alleviating the traffic congestion.

4. What are the disadvantages of the Signalling Transfer Point (STP)?

A signalling transfer point provides a means of communicating messages between non-adjacent entities. A signalling transfer point does not provide any facilities except communication, and this includes address translation or resolution. The originating end point must translate routing information into a destination point code that represents the next service node that the connection must reach. MTP3 then uses the DPC to decide which signalling link to route the message. Since MTP3 must be able to reach all the signalling links, this implies that the processing must be performed on an exchange-wide basis, which is a potential bottleneck for performance.

The ability to use a single number as an address for routing of information is identical to the mechansim that an ATM switch uses for cell routing. An ATM switch is a hardware implementation of this mechanism, and does not need to be repeated in the exchange software if no other function is provided.

CHAPTER 3. SIGNALLING STANDARDS

1. Why are standards important?

Standards enable choice by customers, network operators and manufacturers (vendors). It is national standards, based upon regional and/or international standards that enables customers to be able to choose the terminal equipment from different vendors, and then to be able to choose which network operator to connect the terminal to. Standards enable network operators to interconnect and provide service across concatenated networks; they also enable network operators to purchase network equipment from more than one vendor. Vendors also gain from standards, being able to reuse existing proven solutions in new equipment, being able to use standard testing methods and tools to prove conformance, and designing solutions that will be ready 'off the shelf' for more than one customer.

Standards provide an environment for competitive solutions to achieve successful communications with economies of scale in manufacture and network operation that reduces the cost to the end customers.

2. What added value does ETSI provide to the standardisation process?

ITU recommendations are a consensus of agreement from countries and organisations worldwide. In achieving that consensus, compromises are made that often consist of providing options. ETSI and ANSI both provide regional standards that are variants of the ITU recommendation which select the option to choose. However, ETSI expends considerable effort on conformance testing that enables the network operators to procure equipment that is truly compliant, and for national bodies to licence tested products for sale to customers.

3. Which sector of the telecommunications industry started the ATM Forum, and why?

Terminal equipment manufacturers and private network suppliers originally set up the ATM Forum to 'sharpen the standards making process'. There was some frustration in the industry that ITU-T, with its longer timescales and slower standards-making process, was not meeting the requirements of this sector of the industry. The terminal and private network industry does not require the same level of investment as a large network operator requires to provide a national network.

CHAPTER 4. REQUIREMENTS FOR SIGNALLING—BROADBAND SERVICES

1. Briefly describe the I.130 method for B-ISDN service categorisation and why such a model was thought to be necessary to aid signalling protocol specification.

The I.130 method provides three stages to the definition of a signalling protocol: stage 1 describes the services; stage 2 describes the network capabilities; and stage 3 describes the signalling and switching. The model was derived as a method for defining signalling for narrowbband ISDN services, and the aim was to produce a coherent set of signalling protocol recommendations that support a managed set of network capabilities which will deliver the services required.

2. Why doesn't the B-ISDN have a single signalling protocol for each service it supports?

To separate each service into a single signalling protocol removes the capability of seemless interworking between similar applications (using different B-ISDN services). Referring to Figure 4.3 of Chapter 4 for an example, it can be deduced that there is a relationship between image communication and image retrieval applications, even though one application is classified as using conversational services and the other retrieval services. To provide separate signalling protocols for each of these services would provide undue complication for the application designers and builders. From the point of view of the network, knowing which application can be addressed, so that the correct signalling protocol is used, is an added complication that merely restricts applications, rather than providing an environment that enables innovation and diversification.

3. Explain the difference between a bearer service and a teleservice.

A bearer service provides an application with connectivity through the network (optionally within a common format—the AAL). A teleservice provides a complete definition of the communications scenario and includes details of the applications as well as network capabilities.

4. What are supplementary services? What are the difficulties of having more than one supplementary service active on the same call at any one time?

Supplementary services are services that cannot be invoked directly. They only exist within the context of a call or under a subscription option, and in conjunction with bearer services or teleservices to offer additional information (or additional services) creating value-added features. Care must be taken in the definition of supplementary services to avoid conflicts between them. An example would be subscribing to both Calling Line Identification Presentation (defining the presentation of a user's address when a service is invoked) and also Calling Line Identification Restriction (which prevents the user's address being revealed when a service is invoked).

5. Compare the merits and drawbacks (from a network operators viewpoint) of offering standardised teleservices.

Standardised teleservices simplifies the task of specifying network equipment (procurement), customising interfaces for specific applications, and reduces the cost of network equipment through manufacturer competition. However, new services have a consequential time penalty for their introduction, due to the delay that the standardisation process introduces, and perceived operator differentiation.

# ACCESS SIGNALLING

CHAPTER 5. ACCESS SIGNALLING

1. Why was a two-stage approach taken when specifying B-ISDN signalling?

It was believed that any attempt to produce an advanced signalling mechanism would take too long to meet requirement dates for the early services.

2. There is a significant difference between the release procedure used in N-ISDN access signalling and B-ISDN access signalling; what is it?

B-ISDN access signalling uses a two message procedure to clear a basic call (RELEASE and RELEASE COMPLETE). The DISCONNECT message, used in the three message clearing of N-ISDN calls, is not used.

3. Can you name three of the procedures that are supported in point-to-multipont signalling?

Any three of these procedures:

- adding a party at the originating interface
- add party establishment at the destination interface
- party dropping
- restart
- handling of error conditions
- notification procedures.

4. What are the two information elements that have been specified to support negotiation in DSS2 capability set 2 signalling?

The two information elements are:

- the alternative ATM traffic descriptor, which can contain a number of alternative traffic parameters;
- the minimum acceptable ATM traffic descriptor, which is the lowest value that user will allow.

5. What limitation of Q.2931 prevents it from providing full signalling support for multiconnection calls?

The monolithic structure of Q.2931 prevents the addition and deletion of connection to a call, since the call and connection signalling of Q.2931 are bound together.

CHAPTER 6. THE ATM FORUM SIGNALLING PROTOCOLS AND THEIR INTERWORKING

1. What are the key features of the ATM Forum's main signalling protocols?

UNI v3.1 is broadly a subset of the ITU-T's Q.2931, with additional support for: point-to-multipoint connections, additional traffic parameters, and private network addressing. UNI signalling v4.0 adds support for a number of further capabilities, which include the Available Bit Rate (ABR) service, parameterised quality of service, and leaf-initiated join.

PNNI offers both a control signalling protocol (based on the UNI) and a routing protocol. The signalling protocol supports both point-to-point and point-to-multipoint connections, and includes mechanisms to support source routing, crankback procedures and alternative routing of call set-up requests in the event of set-up failure.

B-ICI v2.0 supports both point-to-point connections and unidirectional point-to-multipoint connections. The B-ICI v2.1 addendum adds support for Variable Bit Rate (VBR) service, Network Call Correlation Identifier (NCCI), and ATM End-System Addresses (AESAs).

2. What are the protocol options for interconnecting ATM networks, using ATM Forum protocols? How do they differ?

Two networks may be interconnected using BICI, PNNI or AINI. BICI is based on the ITU's BISUP protocol, with which it has many common features. PNNI includes both a signalling and routing protocol, and hence is capable of transferring routing information between networks.

AINI is based on PNNI, but does not include the routing protocol. It may thus be more suitable for interconnecting networks where exposure of routing capability is not desirable. AINI has been designed to facilitate interconnection of PNNI-based networks, both to other PNNI-based networks and also to those using BISUP internally.

3. How do E.164 numbers differ from AESAs?

A native E.164 address may be up to 15 digits in length, and comprises a country code, a national destination code and a subscriber number.

There are currently three types of AESA supported by the ATM Forum in existing specifications: Data Country Code (DCC), International Code Designator (ICD), E.164-type, each being 20 octets in length. The various types each comprise an Initial Domain Part (IDP) and a Domain Specific Part (DSP). The IDP defines which type of address it is by specifying an administration authority responsible for assigning values to the remaining part of the address (the DSP).

4. What was the rationale for the Anchorage Accord? What signalling capabilities did it include?

The ATM Forum created the Anchorage Accord in April 1996. This identified a set of some 60 specifications that it had produced to form a stable foundation on which to build an ATM network, and encourage backward compatibility. The main signalling specifications included in this set are:

- B-ISDN inter-carrier interface (B-ICI) specification v2.0 and addendum
- interim inter-switch signalling protocol (IISP)
- private network/network interface (PNNI) specification v1.0
- user/network interface specification (UNI) v3.1.

CHAPTER 7. THE ATM FORUM'S PRIVATE NETWORK NETWORK INTERFACE

1. What is the role and what are the responsibilities of a Peer Group Leader (PGL)?

The PGL is responsible for collecting, aggregating and storing information (i.e. PTSEs) flooded between ATM nodes within its PG. The PGL then uses this summarised information to describe or represent the Peer Group's (PG's) capabilities as a single node, which is used by its parent node in the next layer up in the hierarchy. The PGL is also responsible for collecting PTSE information from neighbouring peer PGLs and flooding this information down throughout its own PG.

2. What additional functionality is required for a border node?

A border node differs from a logical node, in that it has at least one link to another PG. This means that a border node must find out the existence and identity of an attached PG and maintain communication with it. This is achieved via Hello packets. Although the border node can exchange Hello packets with its opposite border node, it is not allowed to exchange PTSEs between PGs. Once the border nodes discover the neighbouring PG, they advertise uplink information in PTSEs flooded within their PG. This uplink information is used by the PGL to help create the higher level peer group connectivity. Also, a border node is responsible for handling the DTLs and crankback procedures.

3. How does topology aggregation affect routing performance and scaleability?

P-NNI was developed to scale to support a very large number of nodes. To do this effectively with source-based routing, there needs to be some aggregation or summarisation of routing information. Therefore, the more information that is summarised, the greater the network size that can be effectively supported. However, when information is summarised, naturally some information is lost. Therefore, the greater the summarisation, the less detailed this routing information becomes for the nodes to use. Also, the larger the network, the greater is the time taken for PTSEs to be dispersed or flooded throughout the network. These factors can adversly effect the path selection performance within a P-NNI network, resulting in more call failure(s) and/or crankback(s).

4. What are the main steps in P-NNI path selection?

Path selection is the process of applying a mathematical algorithm (i.e. Dijkstra) to a P-NNI topology database to compute a suitable route or path to the destination. The main functions performed by a source node in determining a route to a destination are as follows:

1. First, the source node receives an incoming call request. The parameters of the incoming call are then determined to understand the route or path requirements. These requirements can be QoS, service category (CBR, VBR, ABR, UBR), connection bandwidth, etc.
2. The destination node is selected upon the advertised longest address prefix, which has been flooded around the network, and subsequently stored in the source node's database. It is the source that calculates or builds the route to the destination node.
3. It is likely there will be many different paths or routes that can be used to reach the destination node. However, not all these available paths may be acceptable, depending upon path requirements. The path requirements in P-NNI are identified in terms or topology constraints. Therefore, the route needs to be 'pruned' to find the route or route(s) that are the most appropriate to this incoming call. This pruning process is performed by the GCAC. The parameters used by the GCAC are dependant upon the service category defined in the incoming call. The function of the GCAC is to determine if there is enough available bandwidth on a link to support the call. If the bandwidth isn't available then, the link is not used in the next step.
4. The next stage is to use an algorithm to calculate the optimum or 'least cost' path from the source to the destination node. If the Dijkstra algorithm is used, this only allows a single optimisation criteria (i.e. path metric) to be used. The path metric can be configurable on a per service category basis. AW, CDV and CTD are path metrics used in P-NNI. It should be noted that path metrics are accumulative along the path or route; this is a requirement for the Dijkstra algorithm.
5. The Dijkstra algorithm may record the optimum or a number of 'low cost' paths to the destination. Any of these paths that are returned by the algorithm are valid to be used for the call.
6. The paths are then evaluated in terms of satisfying the other P-NNI path metrics. An acceptable path is then selected to build the DTL, and

to provide load balancing. The DTL is the placed within the DTL IE in the P-NNI SETUP message.
7. If no acceptable path is found, then the call set up is released back towards the user, with an appropriate cause code.

It should be noted that the route selection algorithm is not standardised in P-NNI, and hence vendors can use it to differentiate their products.

5. What are the benefits (and pitfalls) of using a dynamic routing protocol for a public ATM network?

Advantages include:

- Dynamic routing algorithm is better suited to ATM's characteristics.
- More flexible, shortest path or least cost route can be determined using different criteria.
- Not all nodes need to use the same routing algorithm.
- Routing information generated and managed automatically, leading to possible reduction in operational support personnel.
- Theoretically, dynamic routing should perform better under failure or congestion conditions.
- Research studies have indicated that dynamic routing protocols make better use of network resources or capacity.

Disadvantages include:

- Extra complexity associated with the dynamic routing protocol.
- Security implications when interconnecting with another network.
- Dynamic routing protocols need properly planned configuration so that excessive routing updates do not cause instability.

Note: although not a strict disadvantage, dynamic routing is a radical divergence from traditional hop-by-hop routing and there is presently a lack of practical experience regarding the performance and manageability within public networks.

CHAPTER 8. B-ISUP, ITU-T'S BROADBAND SIGNALLING PROTOCOL

1. What is the purpose of the four main ASEs defined for B-ISUP? Why does the B-ISUP architecture keeps these ASEs separate?

- BCC: the Bearer Connection Control ASE specifies the protocol procedures that relate to establishing and releasing connections/bearers between adjacent switches.

- CC: the Call Control ASE specifies the procedures for establishing a call, releasing a call, suspending and resuming a call, and for transferring a call.

- MC: the Maintenance Control ASE specifies the maintenance procedures, to block and unblock VPCs, check the status of a user part, to reset a Virtual Path (VP) or Virtual Connection (VC) or signalling Identifier, etc.

- UI: the Unrecognised Information ASE deals with all unrecognised messages and messages with unrecognised parameters it receives from the SACF.

Note: for all the above ASEs there is an incoming and outgoing part.

The ASEs are kept separate to allow for a more modular approach, thereby allowing one ASE to be upgraded without affecting the others. ASEs can be thought of like sub routines or functions within a software program. For more information on the OSI Application Layer Structure (ALS) and its association with ASEs see the following:

- ITU-T Recommendation X.207: Information technology OSI Application Layer Structure (ALS).

- ITU-T Recommendation Q.1400: Architecture Framework for the development of signalling and OAM protocols using OSI concepts.

2. When a destination switch receives a B-ISUP message, how does it determine which B-ISUP AE the incoming message is intended for?

The destination switch knows that an incoming message is intended for itself, by matching the DPC value within the incoming message with its own value. To determine which instance of a B-ISUP AE the incoming message is intended for, the switch examines the Destination Signalling Identifier (DSID) within the B-ISUP message. This DSID enables the switch to correlate the incoming message with the correct AE.

3. Why does the IAM normally take an ATM switch longer to process than other B-ISUP messages?

The IAM generally takes longer to be processed by the switch as it contains many more parameters than other B-ISUP messages. As an example, an IAM can contain upto 24 different parameters, some of which can be repeated. For more information, see ITU-T Recommendation Q.2764, which contains a listing of valid parameters for each B-ISUP message type.

4. If an unrecognised B-ISUP message is received at an ATM switch, what actions could the switch take, and which parameter in the message might the switch examine?

A switch may receive unrecognised signalling information (i.e. messages, parameter types or parameter values). This can be caused by the upgrading of the signalling system used by other switches in the network. Compatibility procedures are invoked to ensure a predictable network behaviour. This is achieved by:

- every message containing a Message Compatibility Information field, and
- every parameter containing a Parameter Compatibility Information field.

The compatibility information field contains instructions which inform the switch how to handle the error condition(s). These are referred to as *instruction indicators* in B-ISUP.

Unrecognised messages

Depending on the instructions set in the Message Compatibility Information parameter, a switch receiving an unrecognised message will either

- transfer the message transparently
- discard the message
- discard the message and send a Confusion message or
- release the call/connection.

Unrecognised parameters

Unexpected parameters (i.e. a parameter in the 'wrong' message) are treated the same as unrecognised parameters. Depending on the instructions set in the Parameter Compatibility Information field, a switch receiving an unrecognised or unexpected parameter will either

- transfer the parameter transparently
- discard the parameter
- discard the message
- discard the parameter and send a Confusion message
- discard the message and send a Confusion message or
- release the call/connection.

Unrecognised parameter values

If a switch detects a recognised parameter, but its contents are unrecognised, then the procedures as stated for the unrecognised parameters apply. There is no specific compatibility information field for each parameter value.

5. How does an ATM switch correctly know how to decode B-ISUP messages and parameters containing variable lengths?

Within each B-ISUP message there is a message length parameter. This parameter is mandatory in each Signalling Information Field (SIF), and is defined in octets six and seven. The message length value defines the total length of a B-ISUP message, excluding the routing label, message type code, message compatibility and the message length parameter itself. Each individual parameter contained within a B-ISUP message also has a mandatory parameter length indicator. Again, the length indicator only defines the length of the parameter contents, and does not include the mandatory fields present at the beginning of the parameter, namely the parameter name, length indicator and parameter compatibility information.

6. If a parameter needs to be repeated in a B-ISUP message, how is this coded?

How B-ISUP codes repeated parameters is different from that in DSS2. DSS2 uses a dedicated Broadband Repeat Indicator Information Element (IE). However, the B-ISUP signalling protocol has no dedicated parameter for indicating when a parameter is to be repeated. Instead, every parameter contains a sub-field which indicates if the parameter is to be repeated or not. When interworking with DSS2, the B-ISUP sub-field and the DSS2 Repeat Indicator IE need to be mapped to one another, to convey the information correctly on an end-to-end basis.

CHAPTER 9. THE VB5 INTERFACE

1. What is the difference between a SNI and a UNI?

A SNI (Service Node Interface) is located between an access network and one or more of its service nodes, whereas a UNI (User-Network Interface) is located between a user and a network (which may be an access network). A UNI provides service to customers, whereas an SNI is concerned with providing connectivity in the access, based on a service request.

2. Why does the VB5 interface specifications not specify the physical layer of the SNI?

Because there is such a wide range of physical layers possible. These include point-to-point SDH networks, SDH rings, and other, non-synchronous interfaces.

3. What does the abbreviation RTMC stand for and why is it so important?

RTMC stands for Real-Time Management Co-ordination, and is used to guarantee real-time co-ordination between access and service node functions which are not available via any other means.

4. Do the VB5 interfaces support only ITU-T terminals?

No. It was a prime requirement that the VB5 interfaces would support other terminals, such as those conforming to the ATM Forum's UNI 4.0 specification.

5. Is there likely to be a need for another VB5 interface, such as a VB5.3?

In general not; it is anticipated that VB5.2 covers most of the functions required, though there may be a need for further enhancements to the VB5.2 standard to cover functions such as intraaccess network switching (i.e. direct connectivity between UNIs on an access network with routing the user plane to the service node).

CHAPTER 10. THE USE OF SESSION CONTROL IN DAVIC TO PROVIDE INTERACTIVE MULTIMEDIA SERVICES

Why is a service gateway element needed?

1. The Service Gateway Element (SGE) provides a 'front-end' to the service content within the Service Provider System (SPS). The SGE presents the presentation layer to the user with a DSM-CC User-User (S2) interface to the application objects representing services stored upon the SPS.

2. What are the two types of information flows in DAVIC and how do they differ?

There are effectively two types of information flow within DAVIC, user plane and control plane. The user plane is represented by S1 and S2 interfaces which provide the user with a media presentation, browsing and command & control interface to applications. The user plane invokes control over the network resource within the control plane at the session level using the S3 interface (DSM-CC User-Network). DSM-CC U-N then invokes call control of network resource using Q.2931 (S4) to establish a new session, represented by S1 and S2.

3. Why did DAVIC choose ATM rather than the Internet as the base technology for its studies?

The DAVIC release 1.0 specs were published in December 1995. ATM was chosen as one mechanism for reliably guaranteeing the quality of service connections required by on-demand services. A multitude of developments in the field of IP QoS, multicast, and even the World-Wide Web explain why future DAVIC releases embraced Internet technologies.

CHAPTER 11. DESIGN FOR PERFORMANCE OF BROADBAND SIGNALLING AND SERVICES

1. What is the key performance parameter from the user's perspective?

If a system is under-dimensioned, so that the demands on the system components are excessive, the user will experience either a poor grade of service or long response times to service requests. If the grade of service is poor, then the user will have difficulty accessing the service, whereas if the response time is large, the user will have difficulty using the service. Generally, it is better to only allow access to a service if one can assure reasonable response times, since a user's perception of performance is most strongly affected by poor response time behaviour.

2. What techniques can be used to evaluate system performance?

A good starting point for system evaluation is to look at the message sequence charts to determine how services use the functional components. Next a functional to physical component mapping needs to be derived, since performance analysis can only be applied to real physical systems. Then look for the potential bottlenecks in the physical system, as these will most likely determine the load dependant response time behaviour. Finally, analyse the bottleneck component using a suitable queuing model to reveal mean and percentile response time figures.

CHAPTER 12. BROADBAND VPN SIGNALLING

1. What is the advantage of using a virtual private network as opposed to a public network or a private network?

Public networks allow customers to be connected to a set of public services that do not require much investment. However, the services are provided in a generic manner and are not tailored to the customers. Private networks allow users to benefit from a wide range of services provided by their local PBX with virtually no cost once the investment has been made. However, it starts being very costly once one wants to connect customer sites that are geographically separated, and nearly impossible if these sites are located in different countries. VPN therefore aims to allow business customers to optimise their communications costs, with respect to dedicated private networks, by providing private networking services on the public network or on a network set up to offer a set of business services.

2. Why do companies invest in standardised solutions and signalling systems when manufacturers are able to offer proprietary solutions for deployment?

Standardised solutions provide economies of scale, and simpler interconnect possibilities that simplify the incorporation of new geographical sites into an existing VPN. Corporate users of VPN solutions that use standardised solutions and signalling systems can more easily undertake competitive tendering and network maintenance and support.

3. Explain the requirement for deploying broadband technology, when compared with obtaining more capacity by deploying additional narrowband infrastructures.

Broadband not only provides connections of bandwidths that may be larger than 64 kb/s, but also promises a vast range of services integrated onto a single network. Broadband has been designed to enable different types of communications to be transported on a single network. Broadband signalling provides bandwidth on demand, tailored according to the type of service that will be required, from simple telephony to videoconferencing. This means that connections can be customised to take account of the traffic requirements, rather than adapting the traffic to the connection characteristics. A narrowband VPN needs to be constantly redimensioned to cope with changing traffic needs.

CHAPTER 13. UMTS: THE MOBILE PART OF BROADBAND COMMUNICATIONS FOR THE NEXT CENTURY

1. What functionality provides the major requirement for broadband to support UMTS?

Major requirements are for AAL2 and 5 capabilities in the radio access network to support time-critical applications in radio access combining.

2. What advantages will UMTS have over a combined GSM and GPRS service?

GSM and GPRS are intrinsically two networks in the core network and dedicated resources in the access network. UMTS is very much about the move to support multimedia, and with the split in current GSM between voice on GSM and data on GPRS it is very difficult to realise a genuine single call related multimedia service. However, the evolution of GPRS towards a broaderband quality enabled IP solution very quickly does provide support of multimedia. So the question is slightly frivolous in that UMTS will be an evolved GSM/GPRS network, not a replacement of them.

3. Why does the separation of call (or session) control from bearer control have major advantages for UMTS?

When introducing mobility within GSM on a narowband ISDN network model, one of the fundamental issues that has been identified is the capacity. The support of mobility protocols and other related protocols causes severe loading on the local exchange equivalent nodes.

Also, in mobile networks the access architecture is generally one where there is a specific radio bearer being applied which is of local significance between terminal and radio access network, then another bearer between the radio access network and local exchange equivalent. Call control runs directly between the terminal and local exchange equivalent. It is therefore essential that there is a separation between bearer and call on the access to allow this multiple bearer architecture to be realised.

CHAPTER 14. SIGNALLING WITH OBJECTS

1. Why were the B-ISDN signalling protocols, such as Q.2931, not designed to meet all of the multiparty, multiservice requirements of broadband from the outset?

Multiparty, multimedia, multiservice calls are very complex to implement. When the original broadband signalling protocols were developed there was a need to rapidly support the new ATM transmission technologies. Thus, it was decided to implement B-ISDN signalling in a number of separate phases, with increasing complexity, to meet both the short- and long-term requirements in a timely manner.

2. Why have a single complex multiparty, multiservice call instead of many simpler separate calls?

Although the simpler/separate calls approach is possible, and indeed is what is currently used in PSTN/ISDN networks, it does have limitations:

- It is harder to ensure synchronisation between bearers. If the network is

aware of the relationship between media, then it can simply effect this synchronisation.
- Some signalling information may need to be shared between many terminals, which can be difficult with independent calls.
- It is possible to have flexible billing mechanisms with a single call, since separate calls makes such billing more difficult to co-ordinate and integrate.

3. Why use separate bearer control and call control protocols at the NNI and not at the UNI?

Separation of the protocols allows optimisation of the bearer connection routing where the network resources are located, and nodes which have to implement the complexity of call control. As the terminal always needs both bearer control and call control functions, and has a single access connection to the network, it was not necessary to have physical separation of the protocols. However, the object model used by MAGIC does provide logical separation of the bearer and call connection signalling flows.

4. What information does the Abstract Service convey?

Essentially, the Abstract Service is the 'user's view' of a media. As well as the basic media type, it contains details of the QoS that the selected codecs will have to provide to implement the service (e.g. the resolution of an image).

5. Why separate upstream and downstream Mapping Elements?

Symmetrical services, such as telephony, can be thought of as the exception rather than the rule. Many services are asymmetric or unidirectional (e.g. TV broadcasting). Handling upstream and downstream bearers with separate objects provides the flexibility for signalling this asymmetry.

6. In what way is the MAGIC Atomic Action Protocol simpler than traditional techniques?

The key simplifications come from the fact that MAGIC operations in a call break down to basic create/modify/delete operations on simple objects. These operations are implemented by a simple set of transactions that accept/refuse the modification of the objects. Contrast these simple state models with the Q.931 protocols for telephony. Here there are many different signalling messages and state transitions that may occur. Imagine the complexity of extending this to more complex service configurations!

CHAPTER 15. THE CALL CONTROL PROTOCOL IN A SEPARATED CALL AND BEARER ENVIRONMENT

1. What are the advantages and disadvantages of separating signalling control into call control and bearer control?

Intermediate switches in communications system are concerned with connection control, and not service control. The requirement to process information at a node and then make decisions on its relevance to the node can degrade the signalling and processing performance of a node.

The separation of call control and bearer control also provides a simplification of the procedures to support multiconnection and multiparty operations. Taking each case in turn, the advantages of separation for multiconnection include:

- no need to repeat call information in subsequent bearer establishment,
- ease of co-ordinating release of all bearers at the end of the call,
- ability to route service requests (call control) into a network or service provider without having to route a connection or dummy connection as well,
- ability to route different connections through different transit networks without the overhead of call set up in each transit network.

The advantages of separation in multiparty calls include:

- ability to co-ordinate negotiation without changing intermediate connections,

- support mandatory party involvement by simultaneously signalling to multiple parties,
- ability to perform simultaneous multiple party actions.

While it is possible to support both multiconnection and multiparty operations with a monolithic call and connection control protocol, there is a repetition of information and duplication of messages.

The main disadvantage of separated call and bearer control is that there appear to be additional messages exchanged when compared with a monolithic protocol in the two-party, single connection case. In this case, the overhead of co-ordinating separated protocols is thought to outweigh the advantages gained by reduced processing at the intermediate nodes.

2. What is the advantage of using an object model to signal information?

Signalling protocols and systems that need to cater for multiparty communications scenarios, need to distinguish between the information (data) that is required for all parties and that which is pertinent for individual parties. One method of doing this is to direct information using party-specific messages (i.e. only using one destination in the address). Another method is to use data containment methods to identify the required relationship between the data and the party. As object-oriented software is a well known and well tried method of containing data within set boundaries, this method can be used to reduce the number of messages that need to be interchanged across signalling interfaces.

3. Why are the actions undertaken by the Exchange Application Processes not standardised?

The standardisation process enables the interconnection, interworking and interoperability of equipment from different manufacturers and different network operators. In defining a standard for a signalling system, it must be recognised that some actions have a consequence on hardware or software within an exchange at a particular instance in time. For example, the point at which bidirectional communication is established can be defined in terms of the receipt or transmission of a particular signalling message. To satisfy the more formal methods of signalling specifications (such as SDL), something somewhere must be informed of events, but the actions that are taken to complete the event (e.g. cell acceptance enabling) do not need to be defined in standards.

4. Why is the 'alerting' information buried in an object attribute, rather than sent as a discrete message?

Alerting is a state of a party and need not be treated as a separate state in the signalling. The internodal signalling has no need to change anything in the intermediate service or connection control logic, and so does not need to process this message. By providing a generic information message that changes the state of the party, only the nodes serving the party need to analyse and process the information.

5. Why is the call control protocol carried using the ROSE and GFP functionality, rather than being defined as a standalone protocol?

To make the call control protocol a standalone protocol, it would need to interface to layer 2 (SAAL), rather than rely on the services of DSS2 or B-ISUP. This means that a separate VCC is needed for this protocol, and the expected bandwidth required of the protocol is very low; to have to reserve its own bandwidth would be inefficient given the minimum bandwidth of ATM networks (64 kb/s). The addressing mechanisms offered by the GFP enable the protocol to be designed without concern for how the message is transported to a non-adjacent node. The ROSE procedures are a tried and tested solution to remote operations and enable vendors to reuse existing code as far as possible.

CHAPTER 16. SUPPORTING APPLICATIONS WITH NETWORK INTELLIGENCE AND B-ISDN

1. Why does the unified functional model have so many layers of control?

The unified functional model is designed to be modular, therefore each control layer has a specific purpose, for example, the resource layer is concerned with obtaining the most appropriate network resources during a session. By making a modular model, it is possible to select only those modules needed for a specific application, and a single protocol that is overloaded with functionality is avoided.

2. Is it necessary to use all of the unified functional model's layers for simple telephony-like services?

In the answer to question 1, we have established that the model is modular, and therefore we can achieve a service using only the absolute minimum layers of control. For a simple B-ISDN Release 1 service, only CC, BC and FM functionality would be used.

3. Application Programmers Interfaces (APIs) above the session control layer facilitate the support of telecommunications applications. Does this imply that only first party (consumer, service provider, broker or retailer) applications can be deployed?

The unified functional model implies that network connectivity provider applications are deployed by network-specific IN functions (SCF, LRCF, ACF). However, it is possible to provide an API above the network connectivity provider so that third party applications can also be deployed. This type of solution has also been suggested in initiatives by the Parlay consortium.

Appendix B: The Development of Broadband Signalling Platforms

Chris Shephard, Peter Hovell, Steve Boswell and John King

B.1 INTRODUCTION

This appendix describes a range of development projects, all of which have resulted in the construction of a Broadband ISDN platform, and extending as far back as the early 1990s, up until the present day. In (approximately) historical order, these are: the RACE I BUNI Demonstrator project; the ADP1 Signalling Demonstrator; the RACE II TRIBUNE project; and the FINDS platform. Each of these very different developments is described in further detail below; in particular, the important lessons learnt from the development are spelt out in each case.

B.2 THE BUNI DEMONSTRATOR

Introduction

The European Community funded a number of projects in the early 1990s that shared a common aim: the exploration of broadband communications. Under the umbrella of the RACE I (Research and development into Advanced Communication technology in Europe) programme, many companies and institutions were brought together to work on the emerging broadband technologies. The RACE I programme had the further aim of bringing together the various pieces of technology being developed in the various projects to form a comprehensive demonstration of the emerging broadband technology. The Broadband User Network Interface Demonstrator project, commonly known as the BUNI Demonstrator, was one of these projects that integrated various pieces of equipment developed in the various RACE I projects with a signalling and control architecture to form a comprehensive demonstration of broadband technology.

The BUNI Demonstration network included equipment ranging from multiservice terminals, HDTV encoders/decoders and displays, customer premise equipment, customer access systems, ATM switches, and the all important signalling and control.

At the height of the project, upwards of 18 European companies worked on the project; some of these companies also represented other consortia, and hence the total number of companies and contacts associated with the project was large. The composition of companies ranged from the large telecommunication suppliers, through network operators to small start-up companies, so a good mix of cultures, working ethics and styles were represented—it made for an interesting time, as BT was the prime contractor.

Demonstration environment

The BUNI demonstrator was housed in specially designed accommodation at the laboratories of PTT-Research in Leidschendam, The Netherlands. The purpose built demonstration facility was designed to display the terminal equipment in three separate, though linked, areas, each reflecting the architectural style of the three distinct 'markets' being addressed. Thus, the domestic equipment was displayed in a 'mock up' of a home environment, e.g. a lounge, whilst the business equipment was shown in an office environment and the broadcast in a studio environment. There were other separate areas to house the network equipment and a reception/lecture area.

The services that could be demonstrated were: conventional and HDTV distribution; multiservice terminals which had a range of N-ISDN services accessible via the broadband network; a range of domestic service and that included video surveillance, TV, VTRs, telephony, etc. The equipment, both network and terminal, to support this range of services was a combination of equipment developed within the BUNI Demonstrator project and from other RACE projects. The one common element to all the equipment, whether developed as part of the BUNI demonstrator project or from other RACE projects, was that they all had to conform to the BUNI signalling specification or its related control architecture to enable full interoperability. Figure B.1 shows the position of the Broadband User

B.2 THE BUNI DEMONSTRATOR

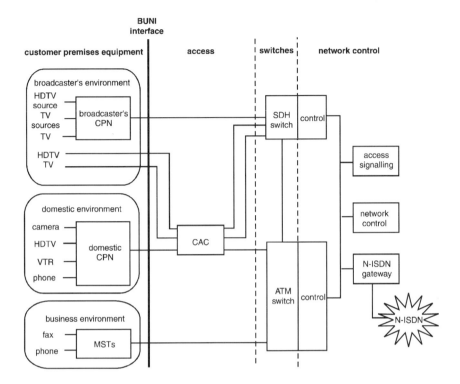

Figure B.1 BUNI Demonstrator Architecture

Network Interface (BUNI) in relationship to the customer premises equipment (CPE, TV encoders and terminals) and the network equipment (CAC, switches, N-ISDN gateway and signalling and control network).

Signalling specification

The major interface in any broadband network is the Broadband User Network Interface (BUNI); this is the point at which the customer premises equipment meets the network operators' equipment. This specification was therefore the major interface within the BUNI demonstrator project and, as such, considerable effort went into the development of the specification for this interface. BT led this activity, and was supported by a number of other partners in the project consortium.

The specification followed the cubic approach, as depicted in Figure B.2. At the time at which the specification was produced, the international standards activity was still ongoing, especially in the access signalling area. The approached adopted when producing the specification was that

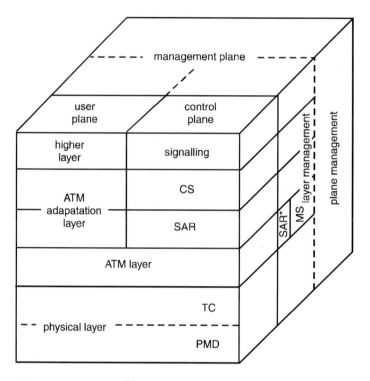

CS convergence sublayer
SAR segmentation and reassembly
SAR* segmentation and reassembly associated with metasignalling
TC transmission convergence
PMD physical medium dependent
MS metasignalling

Figure B.2 Cubic Protocol Reference Model

it should be high in functionality and contain enough detail to be unambiguous, and hence allow interoperability between the various implementors' equipment. The lower layers of the specification (i.e. physical, ATM layer and ATM adaptation layer) were fairly well specified, in draft form, by the likes of ITU, ETSI, etc. The major task of the BUNI demonstrator partners was to pick which options of the various specifications were to be implemented, and to add a certain amount of implementation detail. The area that required the most development was the access signalling specification. At the time at which the BUNI specification was being produced little had been produced in the standards arena. Several other RACE projects had produced an outline 'concept' specification for the BUNI based on the ITU N-ISDN Q.931 specification. The BUNI Demonstrator project used all the available information and, after a considerable amount of effort, produced a specification that fulfilled the following requirements:

B.2 THE BUNI DEMONSTRATOR

- Followed the emerging standards as far as practicable;
- Contained enough detail to be implementable;
- Had a high degree of functionality (service components, etc.);
- Inter-worked with N-ISDN.

The most notable aspects of the BUNI specification are that it included the concepts of metasignalling and service components. The metasignalling protocol allows the assigning, checking and removal of signalling channels. Within the BUNI Demonstrator, this provided little demonstrable functionality; it was concluded that metasignalling had little to offer a broadband network. This has recently, however, changed with the introduction of Passive Optical Network (PON) technology into the local access. Metasignalling can now be used to assign access signalling channels, when needed, and hence save bandwidth on the PON. The other novel aspect of the BUNI Demonstrator access signalling specification was the service component idea. This idea allows a broadband session to be broken down into a number of rudimentary components (e.g. speech, video and fax). A user requiring a speech connection would send a SETUP message to the network with one service component. At a later stage if the user wished to send a fax to the same called party, a second service component (fax) could be added to the original call. The concept of service components was further refined by assigning each component a mandatory, optional or potential flag. The initial call set-up could therefore contain a number of components, some of which are mandatory (the call fails if these components cannot be set up), some of which are optional (the user would like these components, but the call still succeeds even if these components fail) and some of which are potential (the user may wish to set up these components later in the call, and is checking the compatibility of the called terminal). An example of this type of arrangement would be a video call: the speech component would be marked mandatory, the video component optional, and a possible fax component marked potential. This would result in a call if the speech component succeeded; the user would also like a video component, but if network congestion or terminal incompatibility caused it to fail, the call would still go ahead with the speech element; and the user would like to check the called party's fax capability, as a fax component may be added to the call later. This arrangement allowed great flexibility for access signalling and service development but, as yet, has not been progressed to any extent within standards.

Another area that was partially addressed within the BUNI specification was the management plane (layer and plane). What was found was that the individual layers of the specification were fine, but to allow a fully flexible terminal to be developed, a large degree of co-operation between the various layers of the specification was required. The type of actions that management had to co-ordinate were the spawning of new layer 3 access signalling entities for each new call, the binding of these access signalling entities to the higher and lower protocol layers, the creation and binding of user plane ATM adaptation layers to the higher layer application, etc. The BUNI specification included guidance on these issues.

The net result of the BUNI specification work was an informal document that was approximately 200 pages long, weighed 2.1 kg and cost £10 to be reproduced on a single floppy disk.

So far we have only discussed access signalling and its user and network access signalling entities. On the network side, the network access signalling entity was supported via call and bearer control, these two elements not being to any international standards. These three entities formed part of the network control architecture, which also included the N-ISDN gateway and the configuration processors of the two ATM switches. The physical layout of the network control architecture had more to do with the split of work between the various companies involved than any technical requirements. All the parts that constituted the network control architecture were physically interconnected via an Ethernet LAN transporting BUNI Command & Control Message Set (BCCMS) messages; these messages were developed within the project.

Implementation

The implementations of the various signalling components were all hand coded, the user side access signalling being implemented by several companies, some within the BUNI Demonstrator project and some from other RACE projects which brought their terminals to the BUNI Demonstrator for testing purposes. An informal specification, numerous companies being involved from all over Europe (not all English speaking) and hand coding resulted in lengthy integration.

Lessons learnt

The process of developing a specification for the BUNI and the subsequent implementation of that specification by several companies, coupled with the network control architecture, created a number of learning opportunities:

- The development of a specification is a time consuming process—much better to use standards wherever possible and customise them for your specific application. In most cases, for prototype/demonstrations international standards will be too complex, but it is easier to cut things out rather than put them in.

- Once an agreed specification has been produced, a rigorous change control procedure is required. This is very important when numerous companies throughout Europe are implementing the specification.

- If at all possible, go for a formal specification; SDT was chosen for TRIBUNE (see Section B.4), the BUNI Demonstrator's successor. The use of a formal language cuts down the ambiguities introduced in a plain English specification.

- Implementation errors were only found when integration between two different implementations was tried. An implementation always works on the implementors' test harness!

- Performance needs to be designed in, not retrofitted. Within the BUNI Demonstrator, the equipment was integrated and then the call set-up time measured and found to be rather slow. It is better to design in performance than try to improve an existing situation.

- The protocols on the control LAN used in the BUNI Demonstrator were over-complex; keep interfaces simple for performance, reliability and ease of implementation.

B3   THE ADP1 SIGNALLING DEMONSTRATOR

Introduction

The ADP1 signalling demonstrator was an internal BT project that built on the work done within the Advanced Broadband IN Demonstrator (ABIND) project. The ABIND project had explored access signalling, IN and distributed processing, the ADP1 project added a videoconferencing application and a real transport layer that comprised ATM switching and PON local access technology. The final videoconferencing demonstration running over ATM was completed in early 1996, and was exhibited at Innovation 97 and the UK Alliance Symposium.

Demonstration environment

A videoconferencing application was developed for the ADP1 signalling demonstrator. The application included a number of user clients and a centralised service provider element. The dynamics of establishing a videoconference was that a user would send a videoconference request to the service provider; it was then the responsibility of the service provider to invite and establish connections to the other invited conferencees, i.e.

the service provider provided a co-ordination and maintenance function. The service provider also performed the video/audio mixing function. The user terminals were PC-based, and included a special video card that allowed the user presentation control of up to four images (conferencees). The control was similar to the standard Microsoft windowing system, i.e. you could position, size, minimise and maximise any of the four windows. The fully functional PC-based conferencing terminals were augmented by two pseudo-terminals that comprised VTRs.

The network which supported the videoconferencing application included all the futuristic elements that BT foresees being in the network in the next five to ten years, i.e. PON local access technology, ATM switching, network control running on a distributed computing platform, personal numbering, etc. Figure B.3 shows the overall design.

Signalling

The basic concept behind the ADP1 signalling demonstrator was that all connections were controlled (established/released) via the access signalling protocols. These access signalling protocols controlled two distinct types of connections: the low speed control connections; and higher speed video/audio connections. The low speed connections were established between the user and service provider and carried a bespoke message set that requested the establishment of a videoconference session, invited a user to a session, allowed a user to leave a session, etc. The higher speed video/audio connections carried the users' encoded audio/video in the user-to-service provider direction, and in the reverse direction, the connection carried an audio mixed signal and a quadriture encoded image of the videoconference participants. The quadriture image was used by the special video card to display the conferencee images in the separate user controlled windows.

The network control elements included the network access signalling termination, call control, bearer control and a rudimentary IN layer. The IN layer provided number translation, personal numbering and user authentication type services. To allow personal numbering to function, the user has to register their location with the IN. This was achieved by extending the access signalling message set to include a non-call related message based on the N-ISDN FACILITY message. The scenario was that before a user could use any services, they would first send a FACILITY register message to the network. This message would contain user identification and password information, and at which location (line id) they wish to be registered. This message would filter through the network access signalling and call control software to the service control element of the IN function. The user's information would be authenticated and, if

B3 THE ADP1 SIGNALLING DEMONSTRATOR

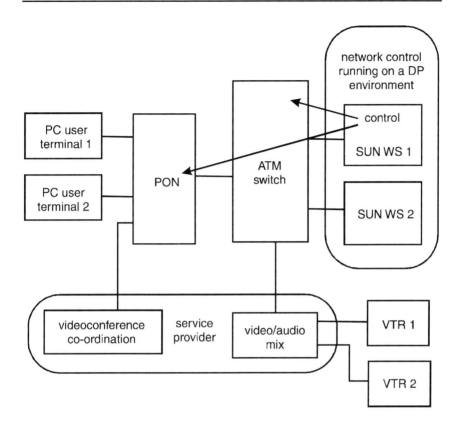

Figure B.3 ADP1 Signalling Demonstrator Architecture

correct, the information held by the IN database updated. Subsequent requests to establish connection via the access signalling protocols would include the personal number of the called and calling parties, and the PIN of the calling party. With this information, call control would request first user authentication from service control and then number translation. Once call control knew the line IDs of the called and calling parties, it would request bearer control to establish the connection. Bearer control was responsible for the allocation of VPIs/VCIs and the requests to the ATM switch and PON to physically set-up the connection.

As can be seen from this environment, the access signalling message set needs to be extended to include the FACILITY type register message, and also to include additional information in the SETUP message for personal number and PIN information. In practice, the implementation of access signalling was rather weak, and only included the message set but not the correct information elements within the message set or the correct dynamics, i.e. no retransmissions or timers implemented (see FINDS, in Section B.5, for a more standards-oriented implementation).

The only part of the jigsaw not discussed so far is the distributed computing element. All the network control software ran on a BT designed distributed computing platform. The DP system was advanced in that it included trading and process creation features, but not in a standardised way. The basic concept was that when a message was received, a request would be made to the trader for the identification of a service resource manager that could create the required process to handle the message. Once the identity of the relevant service resource manager was known, it was requested to create the required process. This was repeated several times as a message filters its way through the system.

Implementation

One of the important lessons learnt from the BUNI Demonstrator project was that integration can be very difficult and time consuming if informal specifications are used. The decision was therefore taken within the ABIND and the ADP1 projects that the SDT toolset from Telelogic should be used. This toolset is based on the Specification & Design Language (SDL), and includes a graphical input notation, simulation and auto-code generation. All of these elements were used on the development of the network code.

Lessons learnt

The development and integration of a large amount of software and hardware into a realistic demonstration that comprised most of the elements of a final system provided a number of important learning opportunities:

- It was found that the Telelogic SDT toolset was good at developing a system in a very short timescale. The basic inputting of the SDLs and subsequent simulations was relatively straightforward. The final stage of auto-code generation for a particular target machine required a fair amount of 'black magic'. However, once the development environment had been established, the specification to code cycle was extremely short. To some extent this type of system requires managers to have 'nerves of steel'; there is no code until very near the end of the project, but when it appears it works nearly first time!
- Integration of the software element that had been developed with the Telelogic SDT tool using the SDL and automatic code generation was fairly easy. Few errors were found because the system had been fully simulated prior to integration. Of the problems experienced, they could be categorised into two types: logical and timing. The logical problems

were easily found, and fixes were relatively straightforward i.e. go though the cycle of updated the graphical SDL's, simulation and re-autocode generate, a cycle that took less than 30 minutes. The timing type problems were more difficult as the SDT toolset did not have were good debug facilities to capture this type of problem.

- Once again the performance was not designed into the system, and hence call set-up times, etc. were rather slow. The requirement to investigate distributed computing necessitated the 'suck it and see' approach.
- The trial of an early distributed processing system was a good learning experience, but as it was not to standards the lessons learnt have to be carefully used. What is obvious it that any extra layer of complexity is going to slow down the system, and plugging extra processors in does not always improve the performance—your software and the task being undertaken has to be appropriate to the technology.

B4 THE TRIBUNE TESTBED

The TRIBUNE project was an element of the RACE II Programme. RACE is an acronym for **R**esearch and technology development in **A**dvanced **C**ommunications technologies in **E**urope. The aim of the programme, part funded by the European Commission, was to set in place a common infrastructure for advanced communications services, in particular the Broadband Integrated Services Digital Network (B-ISDN), within Europe.

TRIBUNE is an acronym for **T**esting, **R**atification and **I**nteroperability of the **B**roadband **U**ser **N**etwork interfac**E**. This expression neatly encapsulated the essential project goals. There are two components to this project title. The first is the **B**roadband **U**ser **N**etwork interfac**E** (which is derived from B-UNI), which identified this key interface as the focus of concern for the project. The second component of the project title states what was to be done to this interface, i.e. to prove that the interface was correct and complete by ratifying its definition, by showing that it could interwork to other services and networks, and finally, by showing how this interface could be thoroughly tested. To verify the user network interface in this way, the project designed and constructed a testbed located in the R&D laboratories of the Dutch PTT (KPN), near the Hague. The essential project goals were as listed below:

- To define and implement a specification of the Broadband User Network Interface.
- To provide feedback to the standards bodies on the base recommendations.
- To construct a thoroughly tested testbed, available to third party users.

- To demonstrate the interworking between TRIBUNE and other network types.

- To develop and utilise the necessary test methods and tools required by such an interface.

The TRIBUNE project was managed by a consortium of collaborating organisations. The TRIBUNE consortium comprised some 12 partners, from eight European countries. The project commenced in 1992, and ran till the end of 1995, by which time a total of 100 man-years and some 15.4 million ECUs (~£10M) had been invested in the work.

Background

The TRIBUNE project came about as the natural successor to the RACE I BUNI Demonstrator project (see Section B.2). TRIBUNE built on the successes of the BUNI Demonstrator project by using the now available text from the ITU-T on the User Network Interface, by designing robustness and performance into the network side to provide a stable and reliable testbed, by the use of sophisticated proprietary tools and the development of a class of purpose built test tools, and by proving a wider range of interconnectivity types.

Testbed environment

In this section the design and use of the TRIBUNE testbed by the project is described. The decision by the project consortium to design, construct and operate a testbed was taken early on. The nature of the testbed was determined essentially by the range of services that were targeted by the project design team. The types of services envisaged early on included:

- TV and HDTV services—to demonstrate the high bandwidths available through B-ISDN.
- Multimedia services—to demonstrate the flexible transport and switching capabilities of ATM.
- PC applications—to demonstrate the use of the broadband Application Programmers' Interface (API).
- Testers—to demonstrate a number of important testing features of the B-UNI.
- ISDN services—to demonstrate the interworking between broadband and narrowband ISDN networks.

- LANs—to demonstrate the interconnection between standard office LAN equipment and B-ISDN.
- External broadband services—to demonstrate the interoperability between separate B-ISDN testbeds.

TRIBUNE was essentially a pragmatic project; having developed the specification of the B-UNI (see the next sub-section), it was considered important to demonstrate that such a paper definition would work in practice. In the opposite direction, practical feedback into the B-UNI specification from the testbed was anticipated. A number of objectives were set for such a testbed, the most important of which are listed below:

- The testbed should be quick to design. At the start of the project the expected duration was three years. To progress from initial project goals to a complete integrated testbed with public demonstrations within such a period, all within the context of an international, collaborative organisation, was a significant challenge

- The testbed was to serve as a reference model. From the beginning it was required to use the TRIBUNE testbed to prove a range of interworking scenarios. In addition, it was hoped to make the testbed available to as many user projects as possible. By widening the audience for the TRIBUNE B-UNI specification, we would achieve a greater degree of feedback from implementers

- This last aspect of the TRIBUNE testbed, its use as a third party testing facility, was a double-edged sword. Together with the advantages gained by such an approach, there were also constraints placed on the testbed. The first of these concerns robustness. In order to support a wide selection of equipment types being connected to the TRIBUNE testbed it must be assured that the network side could run unattended in a continuous and reliable fashion. It was also considered important that the testbed provide a reliable service to the 'users', and one within the expected performance margins agreed within the project.

The design of the complete TRIBUNE testbed is shown in Figure B.4. There were four essential parts to the Testbed:

- *The switching 'fabric'.* Within TRIBUNE there were two separate switches. One of these was an SDH cross-connect, used for the distribution of TV and HDTV signals; the other was the ATM VCI/VPI switch; this represented the core of the network switching and transmission function.

- *The testbed control network.* This collectively represented the central B-ISDN exchange intelligence. At the centre of the control network was the TRIBUNE Control Server (TCS). This element was responsible for

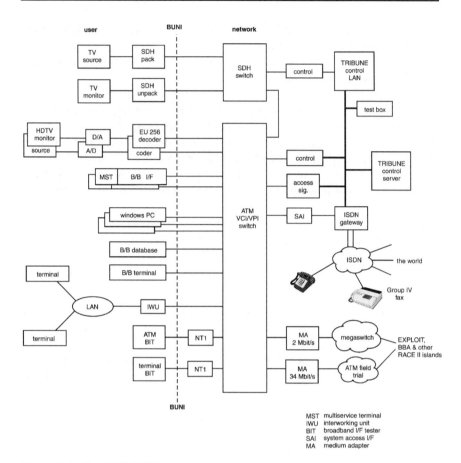

Figure B.4 The TRIBUNE Testbed Design

the control of calls across the TRIBUNE testbed, co-ordinating the actions of the other controllers as required in the establishment, modification and release of all the calls through TRIBUNE.

- *Gateways.* These extended the range of TRIBUNE by providing access to other network types. For example, one of the gateways had connections to the Dutch public N-ISDN; another one was connected to the European-wide ATM Field trial network, which was then undergoing trials.

- *The users.* This term refers to the complete collection of user terminals and CPEs that connect to TRIBUNE over the published B-UNI.

Signalling aspects

The TRIBUNE B-UNI specification was derived primarily from ITU-T recommendations and approved documents, where possible, and draft recommendations where no fully agreed standard was available. At the time when the original specification was being compiled during late 1992 and early 1993, the ITU-T Recommendations were incomplete in a number of areas, particularly the ATM Adaptation Layer and Access Signalling. In these cases, additional information was taken from relevant ETSI material and ATM Forum specifications.

Considerable effort was made in compiling the TRIBUNE specification to make it complete, consistent and implementable. The majority of areas left 'for further study' in the standards documentation were resolved by the project.

A major objective in developing the TRIBUNE specification was that it should be *implementable*. This required that the text be consistent, but also that it be understood as easily as possible. Information was added to several parts of the document to illustrate the dynamic characteristics of its operation. A key aspect of this was the inclusion of overview message flow diagrams which were included to clearly explain the operation of the protocol. These were easily interpreted manually, giving an unambiguous, readable description of the protocol.

During the implementation of the specification, it was found that the text taken from ITU-T Recommendations could in many cases be clarified to aid correct implementation. The recommendations were open to a degree of interpretation that was not suitable for a project such as TRIBUNE. A number of standards contributions were made, and modifications made to the TRIBUNE specification to clearly specify what was required from implementations.

An essential element of the TRIBUNE B-UNI specification was the use of formal methods in the upper layers of the associated protocol stack (see Figure B.2 for the Protocol Reference Model). Two separate SDL tools were used to specify the Access Signalling and ATM Adaptation Layers of the B-UNI (on both sides of the interface). The use of SDL provided a much more rigorous (i.e. less ambiguous) definition of the User Network Interface than was possible in the text-based version used for the BUNI Demonstrator project (see Section B.2). The tools also provided automatic code generation, as well as test case specification. This increased automation of the higher level functions of the UNI resulted in a considerable saving in integration (and in later, further adoption of the UNI to the ATM Forum definition of the interface).

The lessons learnt

The TRIBUNE project followed on, both in terms of its scope and in terms of the development teams involved, from the RACE I BUNI Demonstrator project. In this way, it benefited in a number of ways:

- *Use of SDLs.* The use of SDLs (and the toolsets) was critically important. Although more time was spent 'up front' in the specification phase of the interface, this was more than compensated for by the shorter development time and the much shorter integration time than had been experienced by 'hand coding' methods.
- *Call processing times.* The response time of the BUNI Demonstrator network to new call attempts was poor (> 5 to 6 seconds to establish a point-to-point call). In TRIBUNE there was a requirement to return the equivalent of dial tone to an originating call attempt in less than one second. This target was comfortably met by local code optimisation, by exploiting O/S features where appropriate, and by re-using the Access Signalling message set (defined at the UNI) over the Service Access Point Interface on top of the Access Signalling Layer. In particular, this gave the network Call Control Functions (CCF) full access to whatever information passed over the UNI that it needed. This design choice resulted in a number of benefits later on in the project.
- *Integration of N-ISDN access.* This was never properly achieved in the BUNI Demonstrator project. In TRIBUNE full access to a range of narrowband services was accomplished through a gateway device. The design of this unit was simplified by making the network CCF aware of the Access Signalling message content. Since these messages were themselves derived from their narrowband counterparts, it was a straightforward task to map between the two message sets in the N-ISDN Gateway.
- *ATM Forum Interface.* The project was extended to a fourth year to provide the Testbed with an ATM Forum interface (V3.0). The use of SDLs in both the user and network sides of the UNI, and the extension of the Access Signalling UNI message set up into the network CCF meant that it was possible to enhance the TRIBUNE Testbed in a short space of time so that it could support both the ITU and ATM Forum definitions of the Broadband UNI (and indeed, to demonstrate interworking between the two interfaces).
- *Proxy Signalling.* Providing the UNI Access Signalling messages into the network CCF meant that it was a simple matter to support calls between two terminals, only one of which had an Access Signalling capability. This form of proxy signalling (which was referred to as a $\frac{1}{2}$-call) was effected by offering a simple HMI to the network support personnel. This HMI offered the requisite information that a normal,

signalling, terminal would provide on top of the Access Signalling interface to the CCF. In effect, the CCF was almost unaware of the differences between a conventional call and the $\frac{1}{2}$-call.

B5 THE FINDS PLATFORM

Background

The FINDS (Fast Internet Demonstrator & Studies) Project was a continuation of the ADP1 project described in Section B.3 and, as such, inherited an evolving broadband network demonstrator. Completed in early 1997, the FINDS demonstrator provided the capability of offering a wide range of broadband service demonstrations, including one-way, interactive and two-way services. The demonstration covered aspects of a broadband network including core switching, network/service management, access, IN and signalling and advanced services.

The aim of FINDS was to enhance and refine the ADP1 demonstrator in the following areas:

- Upgrade the IN/signalling network in the areas of performance and demonstrability.
- Provide a standardised access signalling (Q.2931) over a 25 Mb/s B-UNI interface.
- Provide IN services based on the registration and mobility concepts.
- Provide an upgraded form of the ADP1 video conferencing application.
- Demonstration of fast Internet access.

Demonstrator environment

IN services

The FINDS demonstrator was able to offer a number of IN services. The two services that were implemented were a Personal Numbering service and also a Name Calling Service. The former was based on a traditional PN service. The latter gave subscribed users the ability to make a call to another customer by using an alphanumeric name that was customised for the calling user, such as 'office', 'home', 'mum', etc.

Signalling demonstrability

One of the inherent problems of a demonstrator which focuses on signalling is the ability to display the actual signalling that is taking place. Viewers tend to concentrate on the application and take the signalling for granted. This was a particular problem that was found with the ADP1 project. For the FINDS project, a Graphical User Interface (GUI) was developed which was a graphical representation of the signalling modules. This displayed the message flow between the modules, either in real-time or in a deferred message mode.

The interface also provided information on the logical state of each module, the data held within the IN databases, and details of the Virtual Paths and Channels being used on the ATM switch at any given time.

Applications

In addition to connection oriented calls, applications may also make connectionless calls across the FINDS network. The ATM switch is configured for LAN emulation, and thus it is possible to communicate between terminals using IP signalling. This is particularly useful for applications that need to send or receive information regardless of time constraints. An example is a registration command between two terminals.

For initial demonstration purposes, a videoconferencing application was developed for the FINDS network. This application made use of both connection-oriented and connectionless communication. Each terminal had an IP address for connectionless communication and a terminal directory number for connection-oriented calls. The videoconference demonstration consisted of PC terminals at which the service users reside, plus a service bureau terminal from which the videoconference service was driven. At the user's PC terminal, the videoconferencing application was operated using a WWW browser. Each user would log into the browser, the login being sent and confirmed via a connectionless communication with the service bureau. The service bureau retained a database of who was logged into the service so that conferences could be requested between people logged onto the service. When a videoconference session was initiated, the service bureau was informed by the conference requester, again via connectionless communication. The service bureau would then request connection-oriented paths between itself and the participating terminals. It was then the job of the service bureau to receive, mix and return the video via the connection oriented channels. Figure B.5 shows the architecture of the demonstrator.

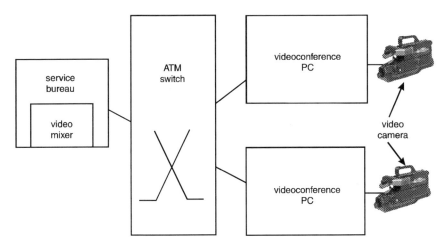

Figure B.5 FINDS Videoconferencing Demonstrator

Signalling

Figure B.6 shows the modules that make up the Network Control Signalling software. Each module runs as a separate UNIX process, and communicates via socket interfaces. The modules have the following functions:

- The *Call Control* module links the two ends of a call, the calling and called parties. It achieves this by co-ordinating the Access Signalling messages to/from each user with the appropriate physical end-to-end connection that is needed to transport the information for the requested service—be it voice, data, images or whatever.
- The *Service Control* module adds intelligence onto Call Control. Call Control can typically establish a basic call to a Directory Number when requested by the calling party, and Service Control is used for the IN type calls such as personal numbering and name calling.
- The *Bearer Control* module is sometimes referred to as *connection control*. It manages the physical connection established between the two ends of a call. Bearer Control receives requests from Call Control during the lifetime of a call, and interfaces to the ATM switch to establish or remove the relevant circuits.
- The *Access Signalling* module manages the Q.2931 protocol that operates between the user and the network. This gives the user (terminal equipment) the ability to access and control the wealth of broadband/multimedia services, and capabilities that are inherent within the B-ISDN network.

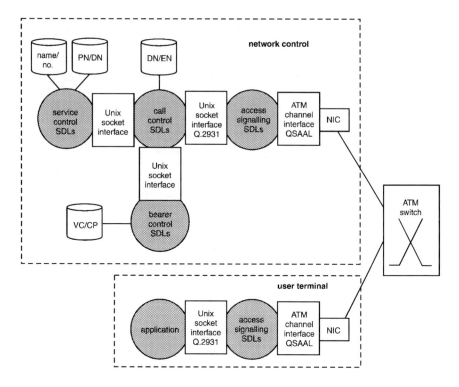

Figure B.6 FINDS Architecture

Access signalling

The access signalling software for both the user and network sides is derived from SDL specifications. The base SDLs used in the FINDS project were taken from the RACE II TRIBUNE project (described in Section B.4). Two packages of SDLs from the TRIBUNE project were used, the Q2931 and QSAAL systems. Editing of the SDLs was required to make them meet the needs of the FINDS project both architecturally and in terms of the messages and information elements supported.

With the SDLs in the required state, and simulations of the SDL behaviour produced to verify their operation, conversion of the SDL to C code was performed; the derived C code being used to produce the final executable access signalling module. In conjunction with the SDL derived code, 'hand produced' C code was written to provide an interface into the SDL produced behaviour. Figures B.7 and B.8 illustrate the architecture of the access signalling module, both for the network and user sides respectively, and show the added interface functionality.

B5 THE FINDS PLATFORM

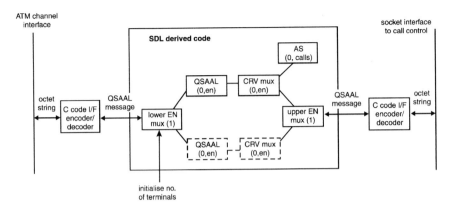

Figure B.7 Network Side Access Signalling Components

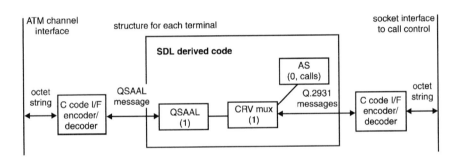

Figure B.8 User Side Access Signalling Components

The structures for user and network side are very much the same. In the network side, a QSAAL/CRV Mux instance is required for each terminal that the network supports. The QSAAL process supports the low level QSAAL connection between user and network. The CRV Mux is used to initiate an Access Signalling (AS) process for each active call between a user and the network. Initially, there are no calls and no AS processes exist. When a call is made either from or to a terminal, the CRV Mux instantiates an AS process, the CRV Mux recording which AS processes are driving which calls via the CRV. The AS process is equivalent to Q2931. In the network the Upper EN Mux and Lower EN Mux are used to send the correct signalling messages to the correct QSAAL/CRV Mux instance. The EN Mux's use the equipment number of the terminal at the user's side to send the signalling messages to the network processes tracking the calls to that terminal.

Outside the SDL derived code area, 'hand written' coder/decoder functions are used to convert the SDL formatted messages into octet strings that can be passed to other software modules. The coding of the messages comply with the Q2931 standard. The coded octet strings are, at the QSAAL end, sent to and received from the ATM Network interface card both at the user's terminal and the network control terminal. At the Q2931 end, the coded octet strings are sent to and received from socket interfaces to other modules of code (call control at the network side, the controlling application at the user side).

For the FINDS project a defined access signalling API exists to aid application development. A defined FINDS API is required since the project only uses a subset of the full Q2931 signalling messages, and within those messages only a subset of the Q2931 information elements. Part of the API package is the FINDS Q2931 coder/decoder mentioned above. As long as the access signalling messages are structured correctly in the resident application, the coder/decoder will ensure that the octet string passed to the access signalling module is correct. In addition to defining the Q2931 messages and information elements, the API describes the process for registering a service type and for requesting a CRV. Since the FINDS network has been designed so that multiple applications may reside on each users' terminal, we needed some mechanism for ensuring that access signalling messages are directed to the correct application. To do this, an application must register the service it provides (e.g. video-conference, 64K Speech) prior to any message exchange. All incoming call set-up messages arriving at a user side access signalling will have a defined service type. The access signalling uses the service type to determine which application should receive the set-up message. Additionally, CRV registration must take place within the access signalling module. This ensures that none of the applications are using the same CRV value. An application wishing to initiate a call must apply to the access signalling module for a CRV prior to sending the call set-up message.

Implementation

As with the ABIND and the ADP1 projects discussed in Section B.3, the SDT toolset from Telelogic was used to build the code for each of the software modules shown in Figure B.6. This was an obvious decision having experienced its advantages from the earlier projects, as well as having the base Access Signalling SDLs.

Figure B.6 shows the basic software modules that were built using the SDT toolset, along with the UNIX Socket Interfaces which were manually coded. The access signalling socket interfaces, also incorporated a message coder and decoder to convert the access signalling message set to/from the SDT text message format to the bit message format as specified by the Q.2931 standard. This gave the platform a standardised access signalling interface.

Lessons learnt

The evolution of this broadband network demonstrator reiterated a number of lessons previously learnt, as well as providing some additional learning opportunities.
- The performance of the demonstrator was significantly improved over that of the ADP1 demonstrator. Call set-up times were reduced to an acceptable two seconds. This was primarily due to the vast reduction in the number of dynamic UNIX processes that were previously being created during a call set-up.
- The performance was also improved by removing a third party distribution layer between the software modules and replacing it with an integrated socket interface. This restates that an extra layer of complexity slows the system down.
- It was reiterated that using the Telelogic SDT toolset was a good approach to developing a system in a short time scale, by simulating and testing a set of SDLs before automatic code generation.
- One of the more interesting developments during the testing of the FINDS network was the performance of remote links to the access signalling, the requirement being to allow the videoconference application to be physically tested remotely from the FINDS network. This was achieved by configuring the access signalling module on one of the terminals to accept access signalling messages across an IP socket. The IP socket connection used the BT Labs' site backbone as the transport medium, this being connectionless with no guaranteed quality of service. This type of testing provided a useful gauge of the differences in performance and quality between terminal resident access signalling and proxy access signalling. What was noted in the process was that the reliability of the network between application and proxy access signalling is an important factor in the success of this method. Over the BT Labs' network the connection was not always reliable, and on some occasions remarkably slow. Over a much shorter distance, such as a single LAN subnet, the proxy signalling connection exhibited similar connection and speed characteristics as a terminal resident signalling connection.

Future plans

As mentioned previously, the FINDS network was initially developed with only the videoconference application to provide a demonstration of the potential services that may exist on the platform. The network was intended to provide a base for operating any broadband application, the only criterion placed on the application developer being adherence to the defined access signalling API, described above, for creating connection-oriented connections across the network. It is hoped that future services will be developed for evaluation using the FINDS network. One current development is the porting of the DAVIC broadcast TV demonstrator onto the FINDS network. The DAVIC demonstrator uses a similar ATM network configuration to FINDS. The DAVIC demonstrator uses a different access signalling method to FINDS, proxy signalling being preferred to terminal resident signalling. To port the application onto the FINDS network requires an alteration to the existing interface to access signalling. In terms of the DAVIC 'S' layers, a FINDS 'S3' layer is required to interface between the current DAVIC 'S1' and 'S2' layers and the FINDS access signalling 'S4'.

Another current activity is a student project to develop a Java terminal to run on top of the FINDS network. At its heart, the Java terminal consists of a harness to enable Java produced access signalling messages to converse with the user terminal access signalling module (see Figure B.9). At its lower, access signalling interface, the Java harness is based upon the FINDS access signalling API. The upper interface to the Java harness is JTAPI (Java Telephony API) compliant. JTAPI is a standardised Java interface for telephony type Java applications. This JTAPI interface to the Java harness allows any Java application, written to use the JTAPI interface, to operate over the FINDS network. Since Java is a fast developing language, and BT is keen to exploit its use, the existence of the JTAPI interface on the FINDS network will enhance its usefulness.

B.6 CONCLUDING REMARKS

This appendix has described four projects which vary considerably in their nature and scope; they were all concerned in an essential way with the design, development and support of broadband signalling interfaces. A small number of common themes have arisen from these projects:

- Whenever possible, resist the temptation to devise your own interfaces. They are almost always more difficult to design and maintain than

B.6 CONCLUDING REMARKS

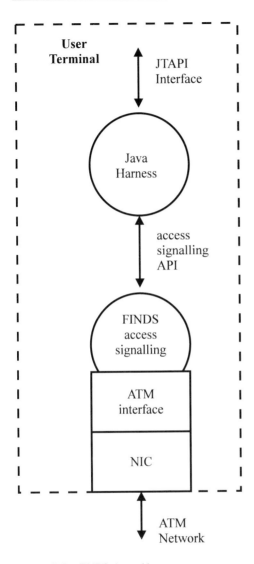

Figure B.9 FINDS Java Harness

would be first imagined. Use standards-based interfaces when available, or at least the best parts of such.
- As a corollary to point 1, make every effort to reuse other interfaces within your project. In TRIBUNE the B-UNI Access Signalling level interface was used in its entirety at the AS-SAPI (i.e. into Call Control within the network). This not only saved the effort of maintaining a separate interface-definition but (paradoxically though it might seem) also contributed to performance optimisation.

- Keep interfaces as simple as possible. The FINDS project has shown the benefits of employing standard definitions such as IP sockets, and is now exploiting the portability features of JAVA interfaces.
- The definition of all but the simplest levels of protocol behaviour should use formal techniques based on SDL specifications and associated tools.
- Non-functional requirements (e.g. performance criteria, reliability, etc.) should be designed in from the outset, not added on afterwards.

Appendix C: The Broadband Call Control Demonstrator—a Demonstrator Platform for ITU-T, DAVIC and TINA-C Implementations

Paul Reece, Richard Macey and Peter Clarke

C.1 INTRODUCTION

To provide future broadband services in a timely and efficient manner, a network is needed which is both powerful and flexible. Key to this requirement is the need to form broadband connections on demand, with specific bandwidth and Quality of Service (QoS) characteristics. To be in a position to better understand this level of 'control', work has been underway investigating the requirements of future networks and their associated signalling systems. A key part of this work has involved the development of a broadband call control demonstrator platform, on which new concepts could be investigated. The platform has proved to be a powerful tool in terms of understanding the operation and interworking of signalling protocols. Its origin, current state and future evolution are covered by this appendix.

Specifically, the approach taken to understand the requirements of 'broadband call control' is presented, covering the evolution of the broadband call control demonstrator, starting with the low-level call/connection control mechanisms, and then examining the higher layer session control. Ultimately, future object-oriented and object request broker techniques are looked at.

The original purpose of the demonstrator platform was as a learning tool to develop expertise, but has since evolved into a comprehensive foundation for other teams to explore broadband networking concepts. Its main focus is still to concentrate on call control issues, and how they would be adopted in a full scale network deployment.

Signalling architecture adopted by the broadband call control demonstrator

Later sections describe in greater detail the aspects of signalling and control explored within the project. This section introduces the basic concepts and architecture of the broadband call control demonstrator.

To simplify the approach to signalling in a large complex network, the end-to-end connectivity can be broken down into simpler elements.

At the lowest level are the elementary point-to-point, link-by-link 'connections'.

Slightly higher, an end-to-end view is given by concatenating the links to create a 'call'. Currently, the call/connection levels are considered as a single layer, controlled using a single signalling protocol, as described later. This supports a single point-to-point connection whereby either party can initiate or clear the connection.

Using the architecture developed within the Digital Audio Visual Council (DAVIC—see Chapter 10), the layers of control can be represented, as shown in Figure C.1. The call/connection layer provided the 'S4' control layer of the demonstrator.

Moving higher still, a set of associated calls can be aggregated into a 'session'. This view allows more complex services to be co-ordinated, where connections may be set up, released or transferred, etc. during an active session. To provide this role, the team adopted the use of DAVIC session control. This used the Digital Storage Media Command and control Set (DSMCC) User to Network (UN) protocol, and allows an abstraction of the complex connection control mechanisms.

The DSMCC UN protocol provided the 'S3' session control layer of the demonstrator (see Figure C.1). End-to-end user information is exchanged over the 'S1' and 'S2' information flows. The 'S2' flow, provided by DSMCC UU (User to user) protocol, supports the bidirectional user-to-user control, used for browsing and controlling service streams. The 'S1' information flow is the service stream itself, e.g. a Motion Pictures Expert Group (MPEG) video stream.

C.1 INTRODUCTION

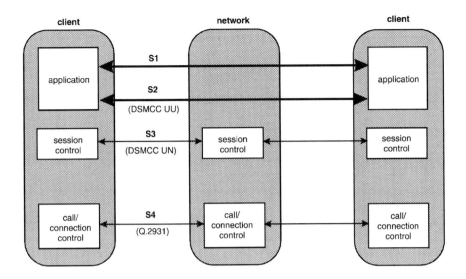

Figure C.1 Broadband Call Control Demonstrator—Control Architecture

Basic demonstrator platform

The requirement was to develop expertise in the emerging signalling protocols being supported within the International Telecommunications Union (ITU-T) and the ATM Forum. At the time work started on the broadband call control demonstrator, the ATM switches available only supported Permanent Virtual Channel (PVC) functionality. To support Switched Virtual Channels (SVCs) there were effectively three options:

- to use a proprietary signalling mechanism, such as SPANS (Simple Protocol for ATM networks);

- to buy third-party software which could be ported onto the relevant switch platform;

- to write our own software to implement a 'standards-based' signalling protocol.

It was decided to implement a sub-set of the ITU-T Q.2931 standard, which was developed on a number of Sun workstations to represent both the user and the network side functionality. Figure C.2 represents the basic architecture adopted.

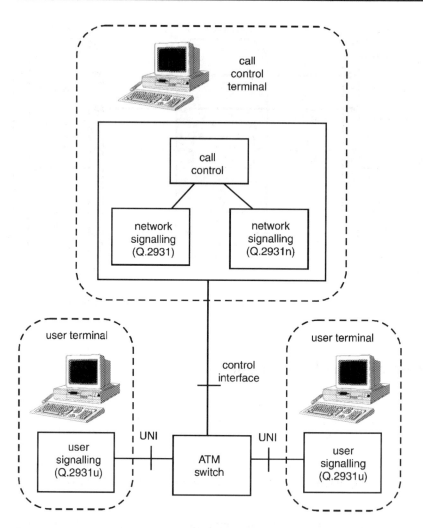

Figure C.2 Physical Demonstrator Architecture

C2 SWITCH CONTROL

The control software developed by the team implemented signalling, call control, address server, Connection Acceptance Control (CAC), connection control and resource manager functionality. The ATM switch was controlled by a Simple Network Management Protocol (SNMP)-based interface that allowed connections to be configured across the ATM switch at the request of the switch control software.

Switch control architecture

Using the switch control architecture as described in the RACE 1044 TRIBUNE project (which provided one of the first implementations of a user/network interface (UNI) signalling) as the basis, a switch control architecture was developed.

The architecture, shown in Figure C.3, was developed in a modular fashion with defined interfaces between the functions. Once completed, the architecture allowed for the flexible exploration of broadband call control issues independent of the ATM switch software and flexible enhancement to support additional functionality.

There are three phases in the configuration of an ATM connection: reservation, allocation and release. Figure C.4 shows the interaction between the switch control functions during the reservation phase. The functions are described in more detail in the sections below.

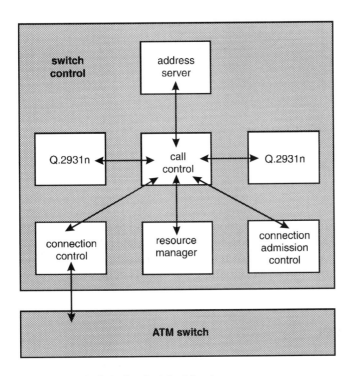

Figure C.3 Switch Control Architecture

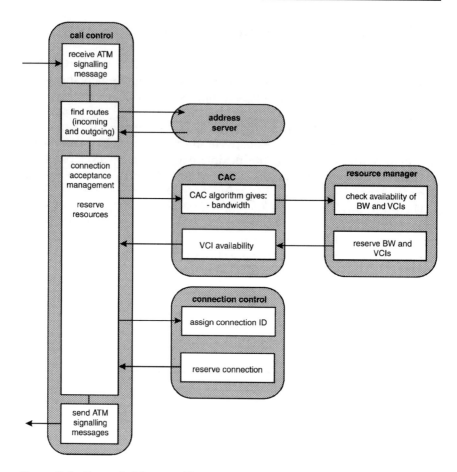

Figure C.4 Example Message Flows—Reservation

Call control

Call control co-ordinates the reservation, allocation and release of resources. On receipt of an ATM signalling message, the appropriate internal messages are generated to reserve, allocate or release the requested resources. Once complete, the appropriate ATM signalling message is generated.

Address server

The ATM signalling messages contain a calling party number and called party number that must be mapped to physical ports and Virtual Paths

(VPs) on the ATM switch. The address server contains a database which is manually configured with these numbers and their associated physical ports and VPs.

Connection acceptance control

The Connection Acceptance Control (CAC) uses the Quality of Service (QoS) parameters from the ATM signalling messages to calculate an effective bandwidth across the ATM switch. This effective bandwidth is then used in requests to the resource manager to see if there is enough free resource to accept the connection. To demonstrate the basic operation of the CAC process, two algorithms were implemented: peak cell rate and linear. The peak cell rate algorithm only makes use of the peak cell rate parameter in the ATM signalling message. The linear algorithm makes use of the peak and sustainable cell rate to calculate an effective bandwidth.

Resource manager

The resource manager keeps track of ATM switch resources in terms of bandwidth, VPs and VCs. Requests for resource are mapped against what is currently allocated and a decision given on whether the request can be accepted. The resource manager also provides an appropriate VPI and free VCI for both incoming and outgoing links.

Connection manager

The connection manager is responsible for mapping the requirement of the switch control software in terms of ports, VPs, VCs and bandwidth into a request to the ATM switch for a connection using those parameters. If support for different vendors' ATM switches was required, the flexibility of the switch control software means that this is the only interface requiring modification.

C.3 CALL/CONNECTION CONTROL

This section outlines the key features implemented on the broadband call control demonstrator, areas that caused problems, and where lessons were learnt.

Point-to-point connection control

The initial demonstrator phase was only required to provide a simple point-to-point switched virtual channel connection. This could be supported by Q.2931. It was decided from the outset of the project that there would be little achieved by implementing every feature of the protocol, especially those relating to timeout and release 'causes'. A sub-set was developed, covering the features below:

- single-sided call establishment,
- both ends able to release connections,
- simple bearer channel established across an ATM switch,
- Network release of the call request in the event of incorrect call set up.

Figure C.5 illustrates the basic call flows to establish a point-to-point connection. It represents the information exchange between the user-side signalling entity (Q.2931u) and the network-Side signalling entity (Q.2931n). The figure illustrates the particular implementation of the CALL PROCEEDING message, which is shown to originate at the called terminal. Generally, this message would be originated by the network.

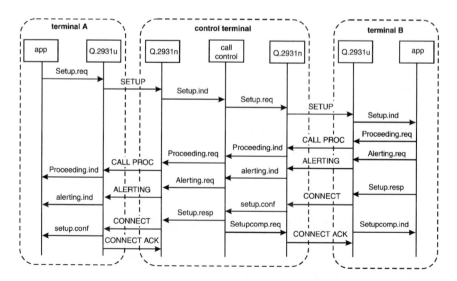

Figure C.5 Message Flows for Point-to-Point Connection

Point-to-multipoint connections

The next phase of development was the enhancement to support point-to-multipoint, unidirectional connections by implementing the additional functionality defined in ITU-T Q.2971. This configuration was used to model a 'video distribution' application, where a 'root' terminal was able to add 'leaf' terminals to an existing connection. This is represented by the information flows in Figure C.6.

Leaf Initiated Join (LIJ) capability

A natural extension of the point-to-multipoint connection was the implementation of the ATM Forum Leaf Initiated Join (LIJ) procedure, which allowed the leaves themselves to request connection to an existing call. This appears a simple addition, though certain problems emerge when this is migrated to a larger scale network:

- the leaf terminals need to identify which particular call they wish to join—a set of default values could be provided for 'known' or frequently used connections, but those formed dynamically would be more difficult to identify;

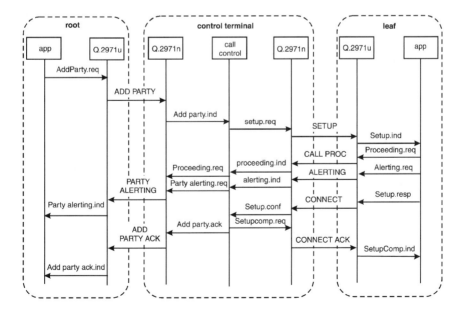

Figure C.6 Point-to-multipoint information flow—adding a party

- the point-to-multipoint and LIJ capabilities described so far only support unidirectional connections—when these later evolve to bidirectional connections, there needs to be some form of control to prevent the root terminal from becoming congested.

The role of proxy signalling

Proxy signalling (see Figure C.7) is a mechanism which allows a network element to handle signalling on behalf of a terminal that does not possess full signalling capabilities. There are several approaches for this, the one represented by this stage of the demonstrator allowed a network element (e.g. a video server) to be controlled by the root terminal initiating the point-to-multipoint connections. This is represented in Figure C.7, which shows a proxy UNI from the root terminal handling the signalling messages, while the data connections are made directly to the video server itself.

A more realistic scenario would be where a piece of a customer's equipment without full connection control capabilities (e.g. a Set Top Box (STB)) would use a Centralised Proxy Signalling Agent (PSA) to establish connections. This scenario is shown in Figure C.8.

The user's terminal or STB would either use a lightweight signalling protocol, or adopt a higher level view, and request the establishment of a 'session' (as described in Section C.4 and shown in Figure C.9).

The use of proxy signalling highlights the requirement for information to be exchanged between the session (S3) and the connection (S4) layers. In particular, the session layer must be informed of the lower layer connections that have been established to instruct the end user which VPI/VCI values to use for the S1 and S2 application flows. This information is returned in a resource descriptor information element as part of the 'client session set up confirm message', as shown in Figure C.10.

Figure C.10 identifies the stages, labelled (a) to (d), where additional processing is required:

1. Upon initialisation, the user's terminal needs to inform the network that proxy signalling is required. This will enable the network to route the connection-layer signalling messages to the proxy agent, and not back to the terminal itself.
2. The network needs to insert the VPI/VCI values used for the S1 and S2 connections into the 'connection level' message returned to the proxy agent.
3. The proxy agent needs to extract the VPI/VCI values from the connection layer message and insert them into a resource descriptor information element within the 'client session set up confirm' message.
4. The user extracts the VPI/VCI values from the session layer message to identify which channels the S1 and S2 information will use.

C.3 CALL/CONNECTION CONTROL

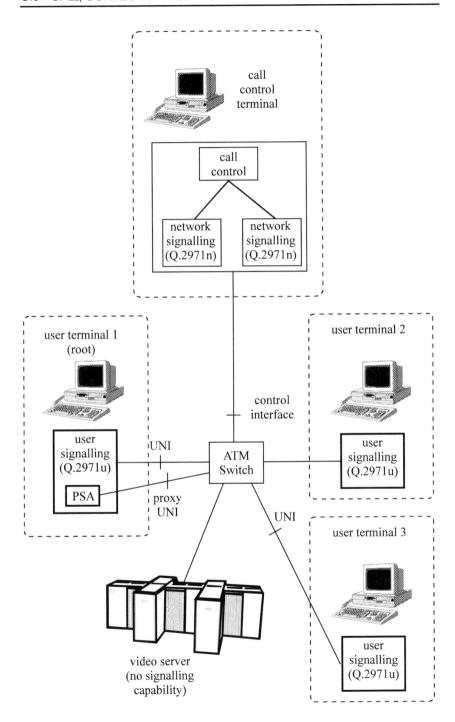

Figure C.7 Proxy Signalling on Behalf of a Network Element (Video Server)

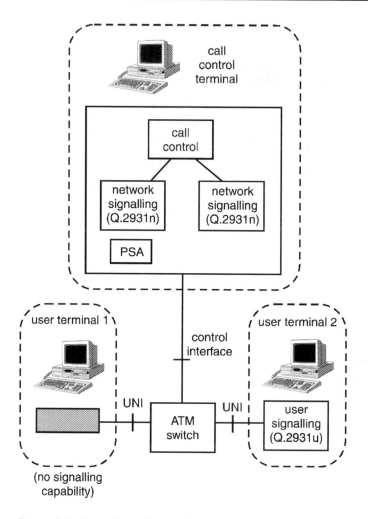

Figure C.8 Proxy Signalling on Behalf of a User's Terminal

Connection control Application Programming Interface (API)

A simple connection control API was developed to allow higher (application) layers to easily access the connection capabilities offered by the demonstrator. It comprised of a set of primitives, based on the Q.2931 message format, which allowed the application to request specific parameters for its connections.

C.3 CALL/CONNECTION CONTROL

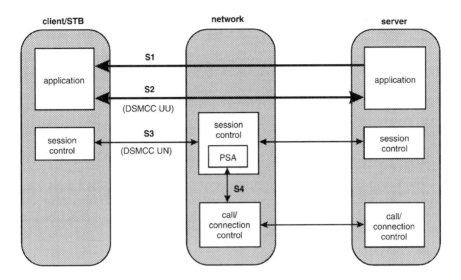

Figure C.9 Proxy Signalling Performed by Session Controller

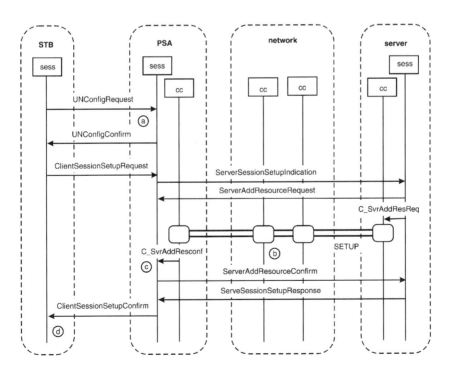

Figure C.10 Information Flows for PSA Operation

C.4 SESSION CONTROL

After the call/connection control and switch control functionality had been implemented, the next key area to consider was the little understood area of session control. This provides continuity of interactions between users, service providers and network providers when the delivery of a particular service is implemented through more than one call/connection. From the network's point of view, session control consists of a number of related call/connections which occur consecutively to provide some service to a user. It should be noted that a particular application may use its own session controller to control a number of (application) related calls (call control terminal entities or agents), but there is no requirement for the network to relate these concurrent calls. The session control described in the DAVIC specification was based on ISO DSM-CC, as explained in Appendix B.

'Methods' of session control

There are several methods of using DSM-CC within an ATM network. The methods differ in the way in which the session (as identified by the session identifier and the resource number) is tied to the call/connection (as identified by the call Reference). The four methods are discussed below:

- *Session Method*: the server informs the network (Session and Resource Manager, SRM) of the required session using the ServerAddResourceRequest message, and the network initiates the connection. The Q.2931 messages initiated by the network include the resourceId (session Id + Resource Number).

- *Network Method* (with AddResource messages): the server allocates resources in the network prior to sending an AddResource message to the SRM. The Q.2931 messages initiated by the Server include the resourceId (session Id + Resource Number).

- *Network Method* (without AddResource message): this allows any server to initiate Q.2931 signalling to the network. The SETUP message sent by the server includes a resourceId (session Id + Resource Number) and the ResourceGroupTag. The SRM is in the signalling path to allow it to perform resource management within a session.

- *Integrated Method*: here the DSM-CC U-N messages are mapped into the Q.2931 signalling messages. DSM-CC U-N messages that are not naturally coupled with Q.2931 signalling messages can be handled in Q.2932 facility messages. The SRM is in the signalling path for the Q.2931 signalling messages, and thus manages the session and resources.

C.4 SESSION CONTROL

DAVIC does not explicitly use one of the methods, rather the DAVIC session control concept is based loosely on the Network Method (with AddResource messages). To ensure evolution of the session control protocol, the team chose to follow the DSM-CC standard as opposed the DAVIC proprietary implementation of DSM-CC U-N.

Roles

DAVIC identify three key roles in their DAVIC V1.0 specifications, service consumer, delivery system and service provider. Through consideration of long-term business model work within TINA-C and ITU-T, it was felt that the DAVIC delivery system could in fact be separated into three roles: network connectivity provider, retailer and broker. Definitions for each of these roles are given below:

- service Consumer: enrols for and uses services.
- network Connectivity provider: provides the communication services that consist of calls and connections.
- Retailer: provides the contact point for the Consumer to arrange for, and the contact point for the service provider to offer, services which employ communication services as the delivery mechanism.
- Broker: provides location information that the business domains use to locate other entities for the purpose of establishing communications.
- service provider: provides a variety of services which employ communication services as the delivery mechanism.

Information flows

The information flows in Figure C.11 concentrate on the session control information flows. Other information flows are appropriate to the presentation, application and call/connection control levels.

Session set up

The main information flows are those between the session layer entities (Sess), carrying DSMCC UN messages to establish the session. The primitives passed from the End-to-End (E-E) control layer down to the session layer do not form part of the DAVIC/DSMCC specification, but are proprietary,

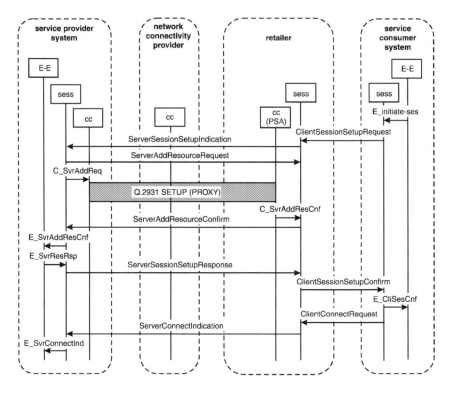

Figure C.11 Session Set-up

based on the DSMCC UN messages they trigger, e.g. the E_initiate_sess primitive is based upon the ClientSessionSetupRequest message as defined within the DSMCC UN specification.

Session release

The release phase is initiated by the service Consumer System generating an E_initiate-rel primitive which is passed to the session layer control element, which in turn generates a ClientReleaseRequest message, as shown in Figure C.12.

Session transfer

The session Transfer capability is supported via a session set up and a session Release.

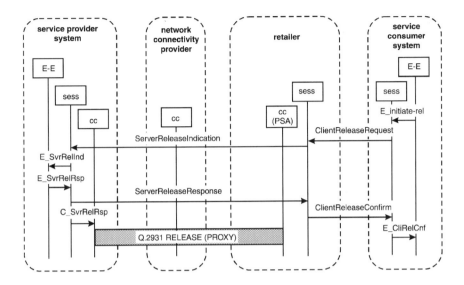

Figure C.12 Session Release

Session control API

As had been implemented earlier with the call/connection control, a high-level API to the session control level was defined for use by applications that would use this level. This API allows others to independently develop applications using the session control level.

C.5 THE TELECOMMUNICATIONS INFORMATION NETWORK ARCHITECTURE CONSORTIUM (TINA-C)

Following the work described above, the broadband call control demonstrator provided a complete implementation of a DAVIC-compliant system for supporting a Video on Demand (VoD) service using proxy signalling. The architecture to that point had evolved through a traditional, standards-based approach, building on low level connection capabilities to offer more complex services. It was decided to progress with a more revolutionary approach by considering the demonstrator in the light of work done within the TINA consortium.

Introduction to TINA-C

With the current speed of technical innovation, service operators have the challenge of incorporating new technologies within their networks and providing new services:

- with minimum development time,
- at minimum costs, and
- in a flexible and manageable way.

In 1992 the world's leading network operators, telecommunication and computer manufacturers formed a consortium to develop a common software architecture for multimedia services to support these aims. This is the Telecommunications Information networking Architecture Consortium (TINA-C).

The initial work of TINA-C centred on the development of a software architecture, encompassing all parts of telecommunication and information systems. This architecture is based on the principles of Object-Oriented (OO) design, the decoupling of software components and the development of an 'enterprise model' showing how different stakeholders (companies) can interact to develop a service.

TINA specifies three sub architectures.

- *Computing architecture*: this defines a Distributed Processing Environment (DPE) to separate service applications from the computing platforms on which they run. With the DPE, computational objects can interact with each other without needing to know the physical location or computational platform of their peer.

- *Service architecture*: this defines a set of principles for providing services, and is based on the concept of a session which represents the information required for the provision of a service for a given duration, and the separation of objects into generic objects (common to all services) and service specific objects.

- *Network architecture*: defines a technology-independent model for establishing connections and managing telecommunication networks.

Latterly, TINA-C has concentrated on the specification of a set of reference points between the different parties in a service; the consumer, connectivity provider, service provider, etc.

The architecture defined by TINA has been implemented in a number of validation projects in member companies, and TINA-C interacts with other standards bodies and industry consortia, including the ATM Forum, DAVIC, ITU-T and Object Management Group (OMG) to provide harmony and avoid duplication of effort

By the end 1997, TINA had produced a full set of coherent, validated architecture specifications and reference points. The TINA-based services have been implemented in trials by member companies, and already some TINA compliant products are on the market.

TINA and the broadband call control demonstrator

Given that TINA might move into a commercial phase, it becomes necessary to consider how one might interwork TINA compliant products with legacy (non-TINA) systems. TINA customers may wish to access non-TINA services, TINA service suppliers may wish to support non-TINA customers, and network operators may have to link networks based on TINA and non-TINA architectures. As a means of exploring these issues, it was decided to develop the broadband call control demonstrator to support one or more TINA components. The problems exposed in interworking DAVIC and TINA will be representative of generic interworking issues.

An initial problem faced by the development team was their lack of practical familiarity with TINA concepts and architectures. To address this issue, work was undertaken to implement one of the demonstrator components (the retailer) on a distributed computing environment. This would serve to:

- introduce the team members to DPE technology,
- allow the issue of supporting legacy (non OO) code on a DPE, and
- be a significant first step to developing expertise in TINA.

The DPE chosen for the work was ORBIX, a CORBA compliant Object Request Broker (ORB) supplied by the company IONA. The Q.2931 state machines supporting call control and the DSM-CC U-N state machines supporting the session control were reworked so that they communicated using Interface Definition Language (IDL) interfaces. These identify the set of procedure calls that the processes can support, and define them in a way that allows the procedures to interwork over the DPE. A software interworking unit (IWU) was developed which supported sockets and an interface to an ATM card to enable the retailer to communicate to the rest of the network and, in addition, supported an IDL interface. This provided the connection between the legacy network and the DPE on which the modified software was supported.

Finally, the single workstation of the retailer was replaced with a workstation and a PC to explore the issues of running the Retail functionality over a number of dissimilar hardware platforms. The completed TINA retailer is shown in Figure C.13.

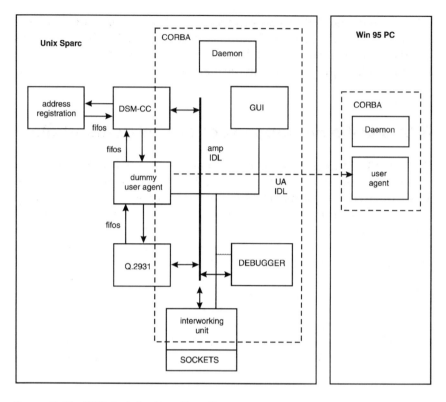

Figure C.13 TINA Retailer Functionality

The PC has a CORBA daemon and runs a single process, the user agent, which send information to the monitor showing the state of messages being processed in the retailer. The UNIX workstation runs the IWU to link the CORBA DPE to the non-CORBA environment, a GUI to allow the operator to view and control the retailer and various functions (dummy user agent, address registration) which are required by the legacy (non-CORBA) DSM-CC and Q.2931 state machines to run in the CORBA environment. Finally, a debugging function was developed to allow testing of the retailer.

This modified retailer successfully interworked with the call control demonstrator and provided a test bed for the exploration of CORBA related issues. However, it did not at this stage support any new functionality.

C.6 FUTURE WORK

A range of new topics are under investigation using the broadband call control demonstrator. They form the basis of an ongoing development of expertise within the team.

IP resource reservation protocol

Work to-date within the call control team, and specifically on the broadband call control demonstrator, has concentrated on a 'native ATM' approach. Increasingly, IP is being proposed as the networking approach of the future—even in support of real-time services, previously thought to be impossible or impractical. The advent of new protocols, such as the resource reservation protocol (RSVP), now makes it possible to attempt to reserve bandwidth through a routed network. Further work is needed to fully evaluate these claims under real networking conditions, and thereby understand the respective roles to be taken by ATM and IP in a future broadband network.

The broadband call control demonstrator platform could be used to investigate the control requirements of the underlying ATM network when supporting IP, which itself is operating an RSVP session. The exact scope of this stage is for further study.

Open control

The connection manager functionality within the switch control has been implemented such that the team can support other vendors' ATM switches with minimal change to the connection manager functionality only. The next phase in this is to progress towards a truly open interface that will mean signalling and management software will be fully interchangeable between vendors ATM' switches. This has the benefit of reducing the cost of this software development (through open competition), and reducing the time to upgrade software versions.

Network signalling

The evolution of the broadband call control demonstrator platform has always been directed at making it as representative of a future BT broadband network as possible. One of the key steps in doing this is to support a multiple switch environment, with an implementation of network signalling. The demonstrator is currently based around a single ATM switch with only access signalling implemented.

Earlier work has provided an implementation of ITU-T B-ISUP network signalling, and this will be integrated with the current ITU-T Q.2931 access signalling using the principles in ITU-T's Q.2650 interworking recommendation.

Implementation of a TINA ConS interface

The next phase in the exploration of interworking between TINA and DAVIC is to enhance the broadband call control demonstrator so that it supports one of the standard TINA reference points. The TINA ConS reference point exists between a connectivity provider (the network) and a TINA retailer. The TINA retailer can user this interface to instruct a network to establish a connection between a service provider and a service consumer. This would allow TINA compliant service consumers and third-party service suppliers to be supported by the ATM network of the demonstrator.

Completion of such an interface would:

- verify the ConS interface,
- substantially enhance the functionality of the demonstrator,
- provide a tutorial platform for the understanding of TINA concepts, and
- support non TINA-architectures that also work to a ConS interface.

Strictly, the broadband call control demonstrator should support a ConS interface on the workstation controlling the ATM switch. However, it is likely to be more convenient to implement the functionality on the retailer, as:

- this platform already had a CORBA platform and OO code,
- non-DAVIC compliant developments are limited to a single component of the enterprise model, and
- the retailer is likely to be a point of flexibility where DAVIC/TINA interworking issues could be addressed.

An overview of the proposed development is provided in Figure C.14.

It is hoped this stage of the development will allow the demonstrator to be interworked to other demonstrators within BT which support the other side of a ConS interface.

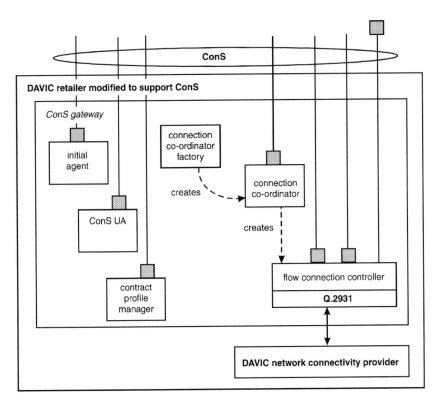

Figure C.14 ConS Interface Development

C.7 CONCLUSIONS

This appendix has provided an overview of some of the work undertaken at BT Laboratories to understand networking architectures and signalling systems. This is essential to provide a future broadband network capable of supporting on-demand ATM connections with guaranteed bandwidth and quality of service. It shows how the complexities of such a network can be better understood with the use of a demonstrator platform to model the information flows and processing requirements. By adopting a steady evolution of capabilities, the team was able to develop the platform, in a bottom-up approach, from simple connectivity into the support of complex services in a DAVIC-compliant manner.

In addition to providing a valuable learning tool for use within BT Laboratories, the platform has been used as a means of promoting the BT's expertise to its external customers. The platform has been used to illustrate future capabilities of BT's Cellstream service, where connections formed dynamically to a range of service providers give increased choice to the customer and more efficient use of network resources. More recently, at the International Broadcast Conference in Amsterdam, the demonstrator was positioned as a 'validation platform' for DAVIC/DVB architectures to allow closer working with vendors entering the consumer market with early prototypes of terminal implementations such as set-top boxes.

REFERENCES

King, T J (ed), Advanced data networking, *BT Tecnol. J.* Vol. 16, No. 1, January 1998.

Appendix D: Index of International Standards

Anselm Martin and Dick Knight

This book has been about the international standards that define broadband signalling, and even though standards are still being developed and new ones published, it is useful to have an index into the standards world.

D.1 ITU-T RECOMMENDATIONS

While the ITU-T publications need to be purchased (e.g. through the ITU's electronic bookshop on the world wide web), the current status can be determined from the list of Q recommendations at
http://www.itu.int/itudoc/itu-t/approved/q/
It is left to the reader to find the lists of the other series!

Networks and ATM: Definitions and concepts

I.113	*Vocabulary of terms for broadband aspects of ISDN's*, ITU-T (06/97)
I.211	*B-ISDN service aspects*, ITU-T, (03/93)
I.121	*Broadband aspects of ISDN*, ITU-T (04/91)
I.150	*B-ISDN ATM Functional Characteristics*, ITU-T (11/95)
I.310	*ISDN—Network Functional Principles*, ITU-T, (08/96)
I.311	*B-ISDN general networking aspects*, ITU-T (03/93)
I.313	*B-ISDN networking requirements*, ITU-T (09/97)
I.321	*B-ISDN protocol reference model and its application*, ITU-T (04/91)
I.325	*Reference configurations for ISDN connection types*, ITU-T (03/93)
I.326	*Functional architecture of transport networks based on ATM*, ITU-T (11/95)

I.327   *B-ISDN functional architecture*, ITU-T (03/93)
I.356   *B-ISDN ATM layer cell transfer performance*, ITU-T (revised, 10/96)
I.357   *B-ISDN semi-permanent connection availability*, ITU-T (08/96)
I.361   *B-ISDN ATM Layer Specification*, ITU-T (11/95)
I.363   *B-ISDN ATM adaptation Layer specifications*, ITU-T (08/96 onwards)
I.363.1   *AAL type 1*, ITU-T (08/96)
I.363.2   *AAL type 2*, ITU-T (09/97)
I.363.3   *AAL type 3/4*, ITU-T (08/96)
I.363.5   *AAL type 5*, ITU-T (08/96)
I.371   *Traffic control and congestion control in B-ISDN*, ITU-T (revised, 08/96)
I.375   *Network capabilities to support multimedia services: general aspects*, ITU-T, (06/98)
I.432.1   *B-ISDN user-network interface—Physical layer specification: General characteristics*, ITU-T (08/96)
I.610   *B-ISDN Operation and Maintenance Principles and Functions*, ITU-T (11/95)
Q.65   *The unified functional methodology for the characterization of services and network capabilities* , ITU-T, (06/97)

Signalling requirements, signalling concepts and layer 2 signalling

I.130   *Method for the characterization of telecommunication services supported by an ISDN and network capabilities of an ISDN*, ITU-T (11/98)
I.140   *Attribute technique for the characterization of telecommunication services supported by ISDN and network capabilities of an ISDN*, ITU-T, (03/93)
I.210   *Principles of telecommunication services supported by an ISDN and the means to describe them*, ITU-T (03/93)
Q.700   *Introduction to CCITT Signalling System No. 7*, ITU-T (03/93)
Q.701   *Functional description of the message transfer part of Signalling System No. 7*, ITU-T (03/93)
Q.702   *Signalling Data Link*, ITU-T (11/88)
Q.703   *Signalling Link*, ITU-T (07/96)
Q.704   *Signalling network functions and messages*, ITU-T (07/96)
Q.705   *Signalling network structure*, ITU-T (03/93).
Q.708   *Numbering of international signalling point codes*, ITU-T (03/93)
Q.709   *Hypothetical signalling reference connection*, ITU-T (03/93)
Q.767   *Application of the ISDN User Part of CCITT Signalling System No. 7 for international ISDN interconnections*, ITU-T (02/91)
Q.1400   *Architecture framework for the development of signalling and O&AM protocols using OSI concepts*, ITU-T (03/93)

D.1 ITU-T RECOMMENDATIONS

Q.2100   *B-ISDN signalling ATM adaptation layer (SAAL) overview*, ITU-T (07/94).

Q.2110   *B-ISDN—ATM adaptation layer—Service specific connection oriented protocol (SSCOP)*, ITU-T (07/94).

Q.2120   *B-ISDN meta-signalling protocol*, ITU-T (02/95)

Q.2130   *B-ISDN—signalling ATM adaptation layer—Service specific coordination function for support of signalling at the user-network interface (SSCF at UNI)*, ITU-T (07/94)

Q.2140   *B-ISDN—signalling ATM adaptation layer—Service specific coordination function for support of signalling at the network node interface (SSCF at NNI)*, ITU-T (02/95)

Q.2144   *B-ISDN—signalling ATM adaptation layer (SAAL)—Layer management function for the SAAL at the network node interface (NNI)*, ITU-T (10/95)

Q.2210   *Message transfer part level 3—functions and messages using the services of ITU-T recommendation Q.2140*, ITU-T (07/96). Table 20-1, Layer 2 User-Network Interface Standards

Q Series Supp. 7   *Technical report TRQ.2001—General aspects for the development of unified signalling requirements*, ITU-T (03/99)

Q Series Supp. 8   *Signalling requirements for AAL type 2 Capability Set 1 (CS1)*, ITU-T (03/99)

X.25   *Interface between Data Terminal Equipment and Data Circuit-terminating Equipment for terminals operating in the packet mode and connected to public data networks by dedicated circuit*, ITU-T (10/96)

X.200   *Information technology—Open Systems Interconnection—Basic reference model: The basic model* ITU-T (07/94)

Z.100   *CCITT specification and description language (SDL)* ITU-T, (03/93)

Access signalling: Layer 3 user-network interface standards—basic call

Q.920   *ISDN user-network interface data link layer—General aspects*, ITU-T (03/93)

Q.921   *ISDN user-network interface data link layer specification*, ITU-T (09/97)

Q.930   *ISDN user-network interface layer 3—General aspects*, ITU-T (03/93)

Q.931   *ISDN user-network interface layer 3 specification*, ITU-T (05/98)

Q.2931   *Digital Subscriber Signalling System No. 2 (DSS 2)—User-Network Interface (UNI) layer 3 specification for basic call/connection control*, ITU-T (02/95)

Q.2932.1   *Digital Subscriber Signalling System No. 2 (DSS 2) —Generic functional protocol: Core functions*, ITU-T (07/96).

Q.2933	*Digital Subscriber Signalling System No. 2—Signalling Specification for Frame Relay Service*, ITU-T (07/96)
Q.2934	*Digital Subscriber Signalling System No. 2—Switched Virtual Path capability*, ITU-T (05/98)
Q.2939.1	*Application of DSS2 service-related information elements by equipment supporting B-ISDN services*, ITU-T (09/97)
Q.2941.1	*DSS2 Generic Identifier Transport*, ITU-T (09/97)
Q.2961.1	*Additional signalling capabilities to support traffic parameters for the tagging option and the sustainable cell rate parameter set*, ITU-T (10/95)
Q.2961.2	*Support of ATM transfer capability in the broadband bearer capability information element*, ITU-T (06/97)
Q.2961.3	*DSS2 Additional traffic parameters: Signalling capabilities to support traffic parameters for the available bit rate (ABR) ATM transfer capability*, ITU-T (09/97)
Q.2961.4	*DSS2 Additional traffic parameters: Signalling capabilities to support the traffic parameters for the ATM block transfer (ABT) ATM transfer capability*, ITU-T (09/97)
Q.2961.5	*Additional traffic parameters: Signalling capabilities to support the traffic parameters for Cell Delay Variation Tolerance (CDVT) ATM transfer capability*, ITU-T (03/99)
Q.2961.6	*Additional signalling procedures for the support of the SBR2 and SBR3 ATM transfer capability*, ITU-T (05/98)
Q.2962	*Digital Subscriber Signalling System No. 2—Connection characteristics negotiation during call/connection phase*, ITU-T (05/98)
Q.2963.1	*Digital Subscriber Signalling System No. 2—Connection modification: Peak cell rate modification by the connection owner*, ITU-T (07/96)
Q.2963.2	*Digital Subscriber Signalling System No. 2—Connection modification: modification procedures for sustainable cell rate parameters*, ITU-T (07/96)
Q.2963.3	*DSS2 ATM traffic descriptor modification with negotiation by the connection owner*, ITU-T (05/98)
Q.2964.1	*Digital Subscriber Signalling System No. 2: Basic look ahead*, ITU-T (07/96)
Q.2965.1	*Digital Subscriber Signalling System No. 2: Support of Quality of Service*, ITU-T (03/99)
Q.2971	*Digital subscriber signalling system No. 2 (DSS 2)—user-network interface (UNI) layer 3 specification for point to multipoint call/connection control*, ITU-T (10/95)
Q.2981	*Rose based Call Control*, ITU-T (to be published)
Q.2982	*Call Control (Q.2931 based)*, ITU-T (to be published)
Q.2983	*Separated Bearer Control*, ITU-T (to be published)
Q.2984	*Pre-negotiation*, ITU-T (to be published)

Access signalling: Layer 3 network-node interface standards—basic call

Q.2721.1 B-ISDN User Part—*Overview of the B-ISDN Network Node Interface Signalling Capability Set 2, Step 1*, ITU-T (07/96)

Q.2722.1 B-ISDN User Part—*Network Node Interface specification for point-to-multipoint call/connection control*, ITU-T (07/96)

Q.2723.1 B-ISDN User Part—*Support of additional traffic parameters for Sustainable Cell Rate and Quality of Service*, ITU-T (07/96)

Q.2723.2 *Extensions to the B-ISDN User Part—Support of ATM transfer capability in the broadband bearer capability*, ITU-T (09/97)

Q.2723.3 *Extensions to the B-ISDN User Part—Signalling capabilities to support traffic parameters for the available bit rate (ABR) ATM transfer capability*, ITU-T (09/97)

Q.2723.4 *Extensions to the B-ISDN User Part—Signalling capabilities to support the traffic parameters for the ATM block transfer (ABT) ATM transfer capability*, ITU-T (09/97)

Q.2723.5 *Extensions to the B-ISDN User Part—Signalling capabilities to support the traffic parameters for Cell Delay Variation Tolerance (CDVT) ATM transfer capability*, ITU-T (03/99)

Q.2723.6 *Extensions to the B-ISDN User Part—Additional signalling procedures for the support of the SBR2 and SBR3 ATM transfer capability*, ITU-T (05/98)

Q.2724.1 B-ISDN User Part—*Look-ahead without state change for for the Network Node Interface (NNI)*, ITU-T (07/96)

Q.2725.1 B-ISDN User Part—*Support of negotiation during connection setup*, ITU-T (05/98)

Q.2725.2 B-ISDN User Part—*Modification procedures*, ITU-T (07/96)

Q.2725.3 *Extensions to the B-ISDN User Part—Modification procedures for sustainable cell rate parameters*, ITU-T (09/97)

Q.2725.4 *Extensions to the B-ISDN User Part—Modification procedures with negotiation*, ITU-T (05/98)

Q.2726.1 B-ISDN User Part—*ATM end system address*, ITU-T (07/96)

Q.2727 B-ISDN User Part—*Support of frame relay*, ITU-T (07/96)

Q.2761 *Functional description of the B-ISDN user part (B-ISUP) of Signalling System No. 7*, ITU-T (02/95)

Q.2762 *General functions of messages and signals of the B-ISDN user part (B-ISUP) of Signalling System No. 7*, ITU-T (02/95)

Q.2763 *Signalling system No. 7 B-ISDN user part (B-ISUP)—formats and codes*, ITU-T (02/95)

Q.2764 *Signalling system No. 7 B-ISDN user part (B-ISUP)—basic call procedures*, ITU-T (02/95)

The inter-exchange standards, B-ISDN User Part (B-ISUP), contained in ITU-T Recommendations Q.2761–2764 were proposed to be re-published as a single recommendation, B-ISUP 2000 (publication date 12/99.)

Q.2766.1 *B-ISDN User Part—Switched Virtual Path capability*, ITU-T (05/98)
Q.2767.1 *B-ISDN User Part—soft PVC capability*, ITU-T (05/98)

Common aspects of UNI and NNI signalling and interworking

Q.2610 *Use of cause and location in B-ISDN User Part and DSS2*, ITU-T (02/95)
Q.2650 *Interworking between signalling system No. 7—broadband ISDN user part (B-ISUP) and digital subscriber signalling system No. 2 (DSS 2)*, ITU-T (02/95)
Q.2660 *Interworking betwen signalling system No. 7—Broadband ISDN User Part (B-ISUP) and Narrowband ISDN User part (N-ISUP)*, ITU-T (02/95)
Q.2630.1 *Signalling for AAL2*, ITU-T (12/99)

Supplementary services

Q.2951.1 *DSS2—Direct-dialling-in (DDI)*, ITU-T (02/95)
Q.2951.2 *DSS2—Multiple subscriber number (MSN)*, ITU-T (02/95)
Q.2951.3 *DSS2—Calling Line identification presentation (CLIP)*, ITU-T (02/95)
Q.2951.4 *DSS2—Calling Line identification restriction (CLIR)*, ITU-T (02/95)
Q.2951.5 *DSS2—Connected Line identification presentation (COLP)*, ITU-T (02/95)
Q.2951.6 *DSS2—Connected Line identification restriction (COLR)*, ITU-T (02/95)
Q.2951.8 *DSS2—Sub-addressing (SUB)*, ITU-T (02/95)
Q.2955.1 *DSS2—Closed User Group (CUG)*, ITU-T (06/97)
Q.2735.1 *B-ISUP—Closed User Group (CUG)*, ITU-T (06/97)
Q.2957.1 *DSS2—User-to-user signalling implicit (UUS 1)*, ITU-T (02/95).

D.2 ETSI BROADBAND SIGNALLING STANDARDS

ETSI mainly perform endorsement of ITU-T standards (although endorsement of other bodies standards, such as ATM-F, has also occurred), adding to it the PICS, PIXIT and ATS, which provides a set of documents that may be procured against. ETSI have different approval levels, specifi-

D.2 ETSI BROADBAND SIGNALLING STANDARDS

cations that have begun the approval process, but not yet completed it, have the letters 'pr' pre-pended to indicate its draft status, while items that are being worked in the appropriate working groups are identified by a work item number. ETSI has recently changed its standards identification. The two document types (deliverables in ETSI), which used to be ETS (European Technical Standard) and ETR (European Technical Report), have been renamed EN (European Normative) and EG (European Guide). Since there are no plans to rename all the existing specifications, both types will be found.

Standards being drafted in the working groups have an identification format beginning with DEN/SPS and followed by five digits, a hyphen and one digit. The first two digits indicates the WG responsible for the work item, the next three digits identify the work item in the WG and the last digit identifies the document type (1 = Standard, 2 = Protocol Interface Conformance Specification, 3 = Test Specification Structure and Test Purposes; 4 = Abstract Test Suite). Work Items that are updating an existing standard will be identified by REN/SPS.

Thus, DEN/SPS-01054-1 indicates an SPS1 work item drafting a European Normative standard. PrETS 300 298-2 indicates a European Technical Standard Protocol Interface Conformance Statement (PICS) in the approvals process. ETS300 443-1 indicates a published specification. Wherever possible, the ETS/EN is cross-referenced to the appropriate ITU specification by the ITU recommendation number.

To update all published ETSI standards, use the ETSI 'tube'—a www indexing engine—at
http://www.etsi.org/ewpweb/queryform.asp

Basic call

ETSI Specifications	Equivalent ITU-T recommendation
ETS 300 298	I.150 B-ISDN ATM Functional Characteristics
ETS 300 298	I.361 B-ISDN ATM Layer specification
ETS 300 299	(B-ISDN Cell Based user network access, physical layer)
ETS 300 300	(B-ISDN SDH Based user network access, physical layer)
ETS 300 301	(Traffic and Congestion Control in B-ISDN)
ETS 300 337	Generic Frame Structures for transport of ATM and others)
ETS 300 349	(AAL Type 3/4)
ETS 300 353	(AAL Type 1)

ETS 300 354	(B-ISDN Protocol Reference Model)
ETS 300 404	(B-ISDN OAM Principles and Functions)
ETS 300 428	(AAL Type 5)
ETS 300 464	(ATM Cell Layer Performance for B-ISDN)
ETS 300 465	(Availability/Retainability of B-ISDN semi permanent connections)
ETS 300 486	Q.2120 (Meta Signalling)
ETS 301 064	Management B-ISDN ATM
ETR 092	(Framework for testing of B-ISDN lower layers)
ETR 117	(AAL Requirements)
ETR 168	(Interpretation of ETS 300 443-1)
ETR 155	(OAM Functions and Parameters)
ETR 180	(Optionally for AAL 1)
DEN/SPS-01054	Q.2722.1 (B-ISUP Pt-Mpt)
DEN/SPS-05131	DSS2/B-QSIG CS2 Pre-negotiation
DEN/SPS-05132	DSS2 Call Control (Call & Bearer Separation)
DEN/SPS-05133	DSS2 Bearer Control (Call & Bearer Separation)
EN 301 276	DSS2 UNI Modification SCR (Q.2963.2)
DEN/SPS-05148	DSS2 Call Control—Modification active call
DEN/SPS-05175	Switched Virtual Path (Q.2934)
DEN/SPS-05180	DSS2 GFP Additional Constructs (Q.2932.2)
EN 301 068	DSS2 UNI indication (Q.2961.1)
EN 301 067	DSS2 UNI negotiation (Q.2962.1)
EN 301 003	DSS2 UNI Modification PCR (Q.2963.1)
DEN/SPS-05202	Q.2963.3 DSS2 Modification with Negotiation by connection owner
DEN/SPS-05216	Q.2959.1 Support of ATM end system addressing format
REN/SPS-05217	Q.2971 amendments to include in revised ETS 300 771-1
ETS 300 443	Q.2931 (DSS2 Basic Call)
prEN 300 443	Q.2931 PICS (DSS2 Basic Call) to endorse ITU-T amendments to protocol
ETS 300 436	Q.2110 (SSCOP Protocol Spec)
EG 202 090	H.323/B-ISDN interoperability
ETS 300 437	Q.2130 (SAAL SSCF for UNI)
ETS 300 438	Q.2140 (SAAL SSCF for NNI)
ETS 300 486	Q.2120 (B-ISDN Metasignalling)
ETS 300 495	Q.2650 (B-ISUP/DSS2 Interworking)
ETS 300 496	Q.2660 (B-ISUP/ISUP Interworking)
ETS 300 647	Q.2144 (SAAL for NNI)
ETS 300 656	Q.2761 to Q.2764 inclusive (B-ISUP)
ETS 300 685	Q.2610 (SS7/DSS2 Cause values)
ETS 300 770	Q.2955.1 (DSS2 CUG)
ETS 300 771	Q.2971 (Point to Multipoint)
ETS 300 796	Q.2932 (Generic Functional Protocol)

D.2 ETSI BROADBAND SIGNALLING STANDARDS

EN 301 003	Q.2963.1 (DSS2 Connection Modification)
EN 301 004	Q.2210 (B-ISDN MTP-3)
EN 301 005	VB5.1
EN 301 217	VB5.2
EN 301 029	B-ISUP (10 parts)
EN 301 067	Q.2962 (DSS2 Connection Negotiation)
EN 301 068	Q.2961.1-.4 (DSS2 Connection Indication)
EN 301 174	Q.2933 (DSS2 Frame Relay)

Supplementary services

ETSI Specifications	Equivalent ITU-T recommendation
ETS 300 657	Q.2730 (B-ISUP Supp. Services)
ETS 300 661	Q.2951.1 (DDI Supp. Service)
ETS 300 662	Q.2951.2 (Multiple Subscriber No)
ETS 300 663	Q.2951.3 (CLIP SS)
ETS 300 664	Q.2951.4 (CLIR SS)
ETS 300 665	Q.2951.5 (COLP SS)
ETS 300 666	Q.2951.6 (COLR SS)
ETS 300 667	Q.2951.8 (Subaddressing)
ETS 300 668	Q.2957.1 (User to User signalling)
ETS 300 669	(Suppl. Service Interactions)

D.3 ATM FORUM SIGNALLING STANDARDS

Further information on the status of specifications and work in progress is available publicly from the ATM Forum's internet server at http://www.atmforum.com/atmforum/specs/approved.html; and http://www.atmforum.com/atmforum/specs/specwatch.html.

Readers are recommended to visit the site for the very latest information. The following tables were derived from the above public information pages.

Specification	Date	Description
af-bici-0013.000	Sep 1993	B-ICI 1.0 Signalling is not supported in this doc.
af-bici-0013.001	not known	B-ICI 1.1 Signalling is not supported in this doc.
af-bici-0013.002	Dec 1995	B-ICI 2.0 (delta document to B-ICI 1.1)
af-bici-0013.003	Dec 1995	B-ICI 2.0
af-bici-0068.000	Nov 1996	B-ICI 2.0 Addendum (or 2.1)
af-pnni-0026.000	Dec 1994	Interim inter-switch signalling protocol UNI 3.1 signalling plus hop by hop static routing
af-pnni-0055.000	Mar 1996	P-NNI V1.0 Consisting of two protocols—a dynamic source routing protocol and a signalling protocol based on ATMF UNI signalling
af-pnni-0066.000	Sep 1996	P-NNI V1.0 Addendum (soft PVC MIB)
af-pnni-0075.000	Jan 1997	P-NNI ABR Addendum
af-pnni-0081.000	Jul 1997	P-NNI V1.0 Errata and PICS
af-uni-0010.000	Jun 1992	ATM User-Network Interface Specification V2.0 Signalling is not supported in this document
af-uni-0010.001	Sep 1993	ATM User-Network Interface Specification V3.0 As 3.1 but with different S-AAL
af-uni-0010.002	1994	ATM User-Network Interface Specification V3.1
af-sig-0061.000	Jul 1996	ATM User-Network Interface Signalling Specification V4.0 (Note this doc is signalling only, unlike UNI3.1)
af-sig-0076.000	Jan 1997	ATM User-Network Interface Specification ABR Addendum
af-cs-0102.000	Oct 1998	PNNI Addendum on PNNI/B-QSIG Interworking and Generic Functional Protocol for the Support of Supplementary Services
af-cs-0107.000	Feb 1999	Addressing Addendum for UNI Signalling 4.0
af-vtoa-0113.000	Feb 1999	TM Trunking Using AAL2 for Narrowband Services
Draft	2000	ATM Inter-Network Interface (AINI)
Draft	2000	Network Call Correlation Identifier
Draft	2000	Generic Application Transport
Draft	2000	Soft PVC Extensions

D.4 DAVIC

http://www.davic.org/DOWN1.htm

DAVIC Specification 1.0	1995–1996
DAVIC Specification 1.1	September 1996
DAVIC Specification 1.2	December 1996
DAVIC Specification 1.3	September 1997
DAVIC Specification 1.3.1	March 1998
DAVIC Specification 1.4	1998

D.5 ECMA

ECMA specifications are available from http://www.ecma.ch/stand/standard.htm (a suggested search word is PISN).

ECMA-251 *PISN—Inter-Exchange Signalling Protocol—Common Information ANF (Q-SIG-CMN)*, Second Edition (December 1998)—equivalent to ISO/IEC 15772

ECMA-252 *B-PISN—Inter-Exchange Signalling Protocol—Transit Counter ANF (B-Q-SIG-TC)* (December 1996)—equivalent to ISO/IEC 15773

ECMA-254 *B-PISN—Inter-Exchange Signalling Protocol—Generic Functional Protocol (B-Q-SIG-GF)*, (December 1996)

ECMA-261 *B-PISN—Service Description—Broadband Connection Oriented Bearer Services (B-BCSD)*, (June 1997)—equivalent to ISO/IEC 15899

ECMA-265 *B-PISN—Inter-Exchange Signalling Protocol—Signalling ATM Adaptation Layer (B-Q-SIG-SAAL)*, (September 1997)—equivalent to ISO/IEC 13246

ECMA-266 *B-PISN—Inter-Exchange Signalling Protocol—Basic Call/Connection Control (B-Q-SIG-BC)*, (September 1997)—equivalent to ISO/IEC 13247

Appendix E:
List of Abbreviations

AAL	ATM Adaptation Layer
AAP	Atomic Action Protocol
ABR	Available Bit Rate
ABT	ATM block transfer
ACM	Address Complete Message
ADSL	Asymmetric Digital Subscribers Line
AE	Application Entity
AINI	ATM Inter-Network Interface
ALS	Application Layer Service
AN	Access Network
ANF	Access Network Function
ANS	Answer Message
ANSI	American National Standards Institute
API	Application Programming Interface
AS	Abstract Service
ASE	Application Service Element
ASG	Abstract Service Group
ASM	Abstract Service Module
ASO	Application Service Object
ATC	ATM transfer capability
ATM	Asynchronous Transfer Mode
ATM-F	ATM Forum
BBSM	Basic Bearer State Machine
B-ISPBX	Broadband Integrated Services Private Branch Exchange
B-ISUP	Broadband Integrated Services User Part
B-PISN	Broadband Private Integrated Services Network
B-TE	Broadband Terminal Equipment
BC	Bearer Control
BCAF	Bearer Control Agent Function

BCCMS	BUNI Command and Control Message Set
BCF	Bearer Control Function
BCSM	Basic Call State Model
BHCA	Busy Hour Call Attempts
B-HLI	Broadband High Layer Identifier
B-ICI	Broadband Inter-Carrier Interface
B-ISDN	Broadband Integrated Services Digital Network
BUNI	Broadband User Network Interface
C7	CCITT No 7 Signalling System
CAC	Connection Admission Control
CAMEL	Customised Applications for Mobile network Enhanced Logic
CBR	Constant Bit Rate
CC	Call Control (in ITU-T and ETSI)
CC	Connection control (in DAVIC)
CCA	Call Control Agent
CCAF	Call Control Agent Function
CCF	Call Control Function
CCITT	International Telephone and Telegraph Consultative Committee
CCITT SS No. 7	CCITT Signalling System No. 7
CCR	Commitment, Concurrency, and Recovery
CCSUA	Call Control Service User Application
CDMA	Code Division Multiple Access
CDME	Composite Downstream Mapping Element
CDV	Cell Delay Variation
CDVT	Cell Delay Variation Tolerance
CF	Control Function
CLIP	Calling Line Identification Presentation
CLP	Cell Loss Priority
CLR	Cell Loss Ratio
CM	Connection Manager
COBI	Connection Oriented Bearer Independent (Part of the Generic Functional Protocol used in access signalling)
Config	Configuration
CORBA	Common Object Request Broker Architecture
CP	Confirmed Party
CPCS	Common Part Convergence Sub-layer (Part of SAAL)
CPE	Customer Premises Equipment
CPH	Call Party Handling
CPN	Customer Premises Network
CPS	Content Provider System
CPSL	Confirmed Party Service Link
CRC	Cyclic Redundancy Check (Error Checking Field)
CS	Capability Set

LIST OF ABBREVIATIONS

CUG	Closed User Group
DAVIC	Digital Audio Visual Council
DBR	Deterministic Bit Rate
D-Channel	Data Channel (N-ISDN)
DB	Database
DCC	Data Country Code
DDI	Direct Dial In
DPC	Destination Point Code
DPE	Distributed Processing Environment
DS	Delivery system
DSM-CC	Digital Storage Media Command and Control
DSS	Digital Subscriber Signalling
DSS 1	Digital Subscriber Signalling System No. 1
DSS 2	Digital Subscriber Signalling System No. 2
DTL	Designated Transit List
DVB	Digital Video Broadcast
ECMA	(European) Computer Manufacturers Association
E-E	End-to-End control
EG	European Guide
EN	European Normative
ESI	End System Identifier
ETR	ETSI Technical Report
ETS	ETSI Technical Standard
ETSI	European Telecommunications Standards Institute
FDMA	Frequency Division Multiple Access
FE	Functional Entity
FTP	File Transfer Protocol
GCAC	Generic Connection Admission Control
GFC	Generic Flow Control
GFP	Generic Functional Protocol
GIT	Generic Identifier Transport Information element
GPRS	General Packet Radio Service
GSM	Global System for Mobile communication
HDTV	High Definition Television
HEC	Header Error Control
HLF	Higher Layer Functions
HMI	Human Machine Interface
HTML	Hypertext Mark-up Language
HTTP	Hypertext Transfer Protocol
IAA	IAM Acknowledge Message
IAM	Initial Address Message
ICD	International Code Designator
ID	Identifier
IDL	Interface Definition Language
IE	Information Element

IETF	Internet Engineering Task Force
IFAM	Initial and Final Address Message
IISP	Interim Interswitch Signalling Protocol
IMS	Interactive Multimedia Services
IN	Intelligent Network
INAP	Intelligent Network Application Protocol
Info	Information
Initial DP	Initial Detection Point
IP	Intelligent Peripheral (in network intelligence)
IP	Internetworking Protocol
ISAP	Internet Service Access Point
ISO	International Organisation for Standardisation
ISPBX	Integrated Services Private Branch Exchange
ISUP	ISDN User Part
IT	Information Technology
ITU-T	International Telecommunications Union - Telecommunications standardisation sector
IWU	Interworking Unit
J	Junction (exchange)
JTAPI	Java Telephony Application Programmers Interface
L1GW	Level 1 Gateway
LEX	Local Exchange
LGN	Logical Group Node
LIJ	Leaf Initiated Join
LLF	Lower Layer Functions
LME	Layer Management Entity
MAGIC	Multiservice Applications Governing Integrated Control
MBS	Maximum Burst Size
MC	Maintenance Control
MD	Management Data
MHEG	Multimedia and Hypermedia Experts Group
MPEG	Motion Pictures Expert Group
MSN	Multiple Subscriber Number
MSU	Message Signal Unit
MSVCI	Metasignalling Signalling Virtual Channel Indicator
MTP	Message Transfer Part
MTS	Message Transmission Subsystem
Mux	Multiplexer
NCCI	Network Call Correlation Identifier
N-ISDN	Narrowband Integrated Services Digital Network
NIP	Network Intelligence Platform
NISUP	Narrowband ISDN User Part
NIU	Network interface Unit
NNI	Network Network Interface (in the ATM Forum)
NNI	Network Node Interface (in ITU-T and ETSI)

LIST OF ABBREVIATIONS 419

NSAP	Network Service Access Point
OAM	Operations And Maintenance
OBCA	Originating Bearer Control Agent
OCCA	Originating Call Control Agent
ODP	Open Distributed Processing
OG	Outgoing
OMG	Object Management Group
OO	Object Oriented
OPC	Originating Point Code
ORB	Object Request Broker
OSI	Open Systems Interconnect
OSPF	Open Shortest Path First
PABX	Private Automatic Branch Exchange (private network switch)
PBX	Private Branch Exchange (private network switch)
PC	Point Code
PCM	Pulse Code Modulation
PCR	Peak Cell Rate
PDU	Protocol Data Unit
PEP	Party End Point
PG	Peer Group
PGID	Peer Group Identifier
PGL	Peer Group Leader
PICS	Protocol Implementation Conformance Statement
PIN	Personal Identification Number
PINX	Private Integrated service Network Exchange (private network switch)
PISN	Private Integrated Services Network
P-NNI	Private Network-Network Interface
PNO	Public Network Operator
POTS	Plain Old Telephony Service
PS	Party Set
PSA	Proxy Signalling Agent
PSPDN	Packet Switched Public Data Network
PSS1	Private Signalling System number 1
PSTN	Public Switched Telephone Network
PSVC	Point-to-point Signalling Virtual Channel
PSVCI	Point-to-point Signalling Virtual Channel Indicator
PTSP	PNNI Topology State Packet
PTT	Postal, Telegraph and Telephone
PVC	Permanent Virtual Circuit
QoS	Quality of Service
Q-SIG	Q reference point Signalling (private network signalling)
RACE	Research and Development into Advanced Communications in Europe

RASM	Remote Abstract Service Module
RBC	Relay Bearer Control
RC	Resource Control
REL	Release
RELIC	Recall Last Incoming Call
RLC	Release Complete
ROSE	Remote Operations Service Element
RSVP	Resource Reservation Protocol
SAAL	Signalling ATM Adaptation Layer
SAO	Single Association Object
SAP	Service Access Point
SAR	Segmentation And Reassembly sub-layer
SBR	Statistical Bit Rate
SCCP	Signalling Connection Control Part
SCP	Service Control Point
SCR	Sustainable Cell Rate
SCS	Service Consumer System
SDF	Service Data Function
SDH	Synchronous Digital Hierarchy
SDL	Specification Description Language
SDU	Signalling Data Unit (in the SAAL)
SDU	Service Data Unit
Sess	Session control
SG	Study Group (ITU-T)
SID	Signalling Identifier
SIF	Signalling Information Field
SLP	Service Logic Programs
SLS	Signalling Link Selector
SM	Session Manager
SMDS	Switched Multi-Megabit Data Service
SNF	Service Network Function
SNMP	Simple Network Management Protocol
SP	Signalling Point
SPAN	Services, Protocols And Networks (ETSI TC)
SPANS	Simple Protocol for ATM networks
S-PVC	Soft Permanent Virtual Circuit
SPS	Service Provider system
SRF	Special Resource Function
SRM	Session and Resource Manager
SSCF	Service Specific Co-ordination Function
SSCOP	Service Specific Connection Oriented Protocol
SSP	Service Switching Point
STB	Set-Top Box (equivalent to DAVIC SCS)
STU	Set-Top Unit
STUI	Service To User Information

LIST OF ABBREVIATIONS 421

SVC	Signalling Virtual Circuit
SVC	Switched Virtual Circuit
SVCI	Signalling Virtual Channel Indicator
SVP	Switched Virtual Path
SVPI	Switched Virtual Path Identifier
SW	Switch
TC	Transaction Capabilities
TCAP	Transaction Control Application Part
TCP	Transmission control Protocol
TCP/IP	Transmission Control Protocol / Internet Protocol
TDM	Time Division Multiplex
TDMA	Time Division Multiple Access
TE	Terminal Equipment
TEX	Transit Exchange
TINA-C	Telecommunications Information Networking Architecture Consortium
TRIBUNE	Testing Ratification and Interoperability of the Broadband User-Network interfacE
TV	Television
UBCF	User Bearer Control Function
UCCF	User Call Control Function
UD	Unnumbered Data
UDP	User Datagram Protocol
UFM	Unified Functional Model
UI	Unnumbered Information
UI	Unrecognised Information
UME	Upstream Mapping Element
UMTS	Universal Mobile Telecommunications System
U-N	User to network
UNI	User Network Interface
UPC	Usage Parameter Control
UPT	Universal Personal Telecommunications
UTRA	UMTS Terrestrial Radio Access
UTSI	User To Service Information
U-U	User to User
VADSL / VDSL	Very high speed Asymmetric Digital Subscriber Line
VAP	VADSL Access Point
VBR	Variable Bit Rate
VC	Virtual Channel
VCC	Virtual Channel Connection
VCI	Virtual Circuit Identifier
VCR	Video Cassette Recorder
VHE	Virtual Home Environment
VoD	Video-on-Demand
VP	Virtual Path (ATM transmission and switching)

VP	Virtual Party (in RACE MAGIC)
VPC	Virtual Path Connection
VPI	Virtual Path Identifier
VPN	Virtual Private Network
VPSL	Virtual Party Service Link
VTR	Video Tape Recorder
WTSC	World Telecommunications Standards Committee

Index

Addressing 48, 71, 109, 111, 115, 119–121, 126, 127, 129, 130, 132, 133, 136, 137, 141, 142, 153–155, 161, 163, 184, 202, 214, 216, 228–230, 233, 235–239, 245, 273, 281, 288, 309, 312, 319
Application entity 150–152, 162, 163, 196, 274, 290
Application programmers interface 281, 282, 309–311, 317, 319, 326
Application service element 150, 151, 195, 290, 298
Associated mode 156
Asynchronous transfer mode 7–15, 17–19, 33, 39, 44, 57, 61, 79, 81, 84, 91–93, 96, 111, 114, 115, 120–122, 125, 129, 130, 133, 137, 147, 148, 153–156, 158, 161–166, 170, 172, 173, 175, 176, 183, 184, 186, 194, 198, 199, 204, 210, 217–221, 227, 230–234, 236, 238, 241, 249–251, 253, 255, 259, 261, 262, 268, 273, 278, 287, 298, 306, 313, 314, 317
ATM adaptation layer 14–19, 33, 37, 39, 68, 69, 84, 90, 92, 102, 108, 138, 139, 149, 153, 156, 164, 176, 177, 182, 198, 286, 322
ATM end system address 114, 120, 129, 136, 321
ATM forum 46, 53, 57–59, 79, 107–122, 125–129, 135, 141–144, 155, 165, 166, 189, 235, 237, 239, 291, 310
ATM inter-network interface 53, 113, 116, 119, 143, 166
ATM layer 10, 11, 14, 18, 19, 33, 69, 92, 93, 138, 139, 149, 164, 171, 182, 198
ATM traffic characteristics 74, 84, 91–95, 154–156, 184, 278, 321
Atomic action 267, 269–273, 275, 292

Bearer branch 135, 280, 285, 287
Bearer capability 84–86, 153, 156, 162, 184, 321
Bearer control 150, 151, 172, 177, 181–183, 185, 186, 215–220, 256, 259, 264, 265, 273, 278–280, 283, 285–287, 290, 291, 293–295, 297, 307, 308, 317–319, 322, 323

B-ISUP 112, 113, 115, 117, 119, 120, 136, 138–143, 147–166, 179, 182, 185, 186, 261, 264, 273, 289, 291, 293, 321
Blocking 151, 158, 160, 169, 181, 187, 188
BQSIG 56, 57, 144, 227, 236–239, 298
Broadband virtual private network 227–240
Browser, browsing 196, 204, 312, 313, 321

Call/connection admission control 130, 131, 138, 140, 162, 185
Call/connection establish/set up 82, 83, 86, 87, 94, 104, 128, 133, 137, 141, 184, 185, 203, 205, 246, 291, 293, 295, 307, 321
Call/connection modification 74–76, 91, 96–99, 251, 255, 258, 269
Call control 85, 103, 104, 117, 150, 151, 205, 208, 215–220, 236, 248, 250–252, 256, 259, 264–266, 273, 275, 277–299, 302, 307, 308, 310, 312, 317–319, 322, 323
Call control agent 265, 279
Call-unrelated signalling 164, 305, 319, 321
Called party 86, 98, 112, 132–134, 152–155, 162–164, 237, 254, 263, 267, 274, 281, 319
Calling party 71, 82, 86, 101, 112, 133, 134, 136, 141, 163, 237, 262, 263, 267, 272, 273, 294, 302, 303, 305, 311, 321, 322
Cell delay variation 92, 141, 154
Cell loss priority 12, 93, 96, 97, 141
Closed user group 71, 120, 136, 137, 164, 237, 309
Codec 267, 274
Communication 28, 48, 65, 68, 72, 74, 75, 170, 177, 179, 187, 209, 210, 227, 228, 232, 237, 241–244, 247, 248, 253–256, 258, 259, 261, 262, 265, 268, 273, 277–281, 284, 288, 289, 292, 295, 298, 301–303, 305, 308, 315, 317, 321
Compatibility 85, 100–102, 110, 119, 131, 153, 156, 164, 196, 234, 239, 261, 287, 305
Conference 15–17, 67, 69, 203, 261, 264, 270–273, 317, 319
Connection 13–17, 20, 36, 72, 74, 75, 80, 82, 84–91, 94–99, 102–104, 111, 113, 114, 116, 120, 121, 125, 126, 130, 134, 135, 141, 144,

424 INDEX

connection (*cont.*)
  147–151, 153–156, 158, 161, 163–166, 171, 172, 174–177, 181–188, 193, 194, 196–198, 201–204, 206–208, 214, 227, 233–235, 238, 243, 244, 247, 248, 250, 252, 255, 256, 259, 261–275, 277–281, 283, 285–289, 291, 292, 294, 295, 299, 302, 304, 306–311, 313–315, 319–322
Connection oriented bearer independent 99, 290, 298
Connectionless 158, 243–245, 248–251, 254, 256, 259, 290
constant bit rate 18, 20, 155, 156, 278, 306
Control 23, 36, 92, 93, 104, 112, 113, 133, 138, 141, 147, 148, 158, 160, 161, 164, 170, 171, 181, 191, 194, 196–200, 202–204, 210, 342, 344, 248–250, 252, 256, 257, 259, 261–264, 270, 274, 277, 279, 281, 286, 289, 291, 299, 302, 304, 305, 308, 309, 315, 317–322
CORBA 197, 198
Correlation identifier 114, 120, 137, 143, 144
Crankback 112, 133, 141

DAVIC 46, 54, 58, 66, 189, 191–210, 249, 310, 312–314
DECT 242, 245
Digital storage media
  - command and control 192, 195, 198–200, 202, 203, 205, 207, 210, 313
Digital subscriber signalling system 56, 57, 79, 80, 88, 90, 91, 93, 95–97, 99, 101–103, 175, 178, 182, 184–186, 229–231, 237, 262, 293, 305, 321
Domain 115, 120, 132, 134, 135, 196, 201, 203, 204, 234–236, 242, 244, 246, 259, 310, 311, 315–317
Dynamic description 63

ECMA 52, 56–58, 65, 103, 165, 229, 235–237, 239
End point 72, 73, 89, 90, 111, 134, 157, 179, 183, 186–188, 234, 268, 278, 279, 281, 283–286, 292, 321–323
Entity 37, 83, 130, 144, 150–152, 169, 174, 177, 179, 192–197, 204, 207, 244, 251, 252, 265, 267, 269, 284, 289, 292–294, 297–299, 315, 316, 323

Facility 100–101
Facsimile 48, 278
Frame 20, 25, 48, 112, 131, 175, 233
Functional entity 63, 64, 195, 208, 214, 275, 278, 315–317, 319

Gateway 195, 196, 199–202, 214, 216–219, 222, 224, 225, 256, 257, 279, 309, 322
Generic call admittance control (*see* call/connection admission control)
Generic flow control 9, 12, 172

Generic functional protocol/transport 99–104, 290, 298
Generic identifier transport 131, 165
Global system for mobile communications (GSM) 55, 241–245, 249, 258, 311
Global tagging 92, 93
GPRS 243–245, 249

Handover 243, 244, 250, 255, 256, 259
Header error control 11, 13
Hierarchical 115, 125, 126, 128–130, 132, 133, 139, 142, 171, 235
High definition television 209

IMS 214, 217–219, 225
INAP 164, 217, 230, 231, 303, 304, 313, 322, 323
Information element 84, 88–90, 92–95, 97–101, 103, 130, 132–134, 141, 162, 175, 184, 204–206, 270, 272, 273, 287, 293, 298, 305, 320
Intelligent network 48, 49, 51, 58, 136, 142, 217, 222, 225, 228, 230, 233, 235, 236, 239, 241, 242, 245, 248, 249, 251, 253–257, 274, 301–324
Intercarrier 112
Interconnect 113–117, 119, 122, 135, 136, 143, 165, 172, 230, 232
Internet (IP) 46, 53, 76, 122, 124, 144, 166, 209, 225, 243–245, 248–251, 253–257, 259, 274, 302, 309, 310, 312–314, 317, 320
Internetworking 109, 115
Internodal 138, 139, 147–166, 291
Interoperable 55, 110, 136, 186, 191, 192, 196
Interswitch 112
Interwork 107–111, 113–116, 120–122, 136, 143, 144, 165, 166, 182, 186, 238, 239, 251, 258, 263–267, 273, 274, 280, 320, 324
Intranet 210, 243
ISDN 43, 44, 47, 49, 50, 56, 62, 63, 71, 80–85, 104, 121, 166, 170, 171, 227–235, 238, 243, 249, 250, 261, 294, 305, 320
ISO 26, 46, 52, 57, 59, 68, 120, 129, 136, 198, 203, 289

Java 196, 209, 313

LIJ 112, 135
Link 27, 29, 31, 112, 116–118, 126, 127, 131, 133, 135, 140, 153, 156–160, 173, 177, 178, 186, 247, 250, 263, 267, 268, 289, 293, 294
Load sharing 213, 221, 226, 247, 309
Look ahead 99–102, 164, 307

Maximum burst size 92, 131
Message transfer part 29, 30–33, 38, 139–142, 148, 149, 152, 158–161, 163–165
Meta signalling 80

INDEX   **425**

Mobile   48, 49, 121, 142, 164, 166, 241–259,
   301, 308, 309, 311, 314, 317–320
Mode   139, 177, 242, 245
Moving pictures expert group (MPEG)   192,
   196–198, 200, 203, 204, 263, 266, 286, 309,
   313
Muticast   14, 112, 165, 289, 291
Multi-connection   102–104, 164, 261, 262, 275,
   289, 291, 292, 297, 304, 307
Multi-media   19, 49, 70, 191, 196, 214, 216,
   244, 246, 255, 264, 304, 320, 321
Multimedia and hypermedia expert group
   (MHEG)   196
Multi-party   261, 262, 264, 265, 274, 275, 282,
   284, 289, 292, 296, 298, 299, 304
Multi-point   14–17, 85–90, 104, 111, 113, 121,
   131, 135, 165, 177, 178, 227, 280, 281,
   289–291, 307, 319

Negotiate   91, 93–95, 104, 110, 131, 184, 246,
   247, 255, 264, 268, 281, 291, 307, 310,
   321
Network node interface   9, 20, 28, 35, 38, 39,
   53, 65, 109, 112, 113, 117, 119, 137, 161,
   162, 165, 172, 174, 234, 235, 262, 264

Object-oriented   186–188, 193, 195–202,
   261–275, 281–285, 314, 315
Open network provision (ONP)   171, 173
Open systems interconnect   26, 27, 39, 68,
   149, 289
Operations and maintenance   156, 171, 172,
   178, 184
OSI seven layer model   26–28, 79

Packetisation   126, 127, 133, 135, 139, 164,
   243, 244, 248–252, 254–256
Payload   12
PCM   7
Peak cell rate   92, 94, 96, 97, 131, 278
Peer   115, 126–129, 138, 144, 179, 193, 200
Permanent virtual circuit   122, 125, 131, 133,
   141, 147, 165, 234, 244, 250
Point Code   29, 31, 140, 142, 153, 159, 160,
   162, 163
Policing   163
Primitive   36, 39, 40, 152, 200, 290
Proxy signalling   112, 165, 197, 214, 219, 284,
   313, 314
Proxy signalling agent   214, 216–220, 222,
   224
Public switched telephone network, (PSTN)
   43, 44, 48, 230, 249, 261, 320

QSIG   229–231, 236
Quality of service   48, 92, 111, 121, 124, 130,
   141, 148, 153–156, 162, 165, 183, 184, 206,
   207, 210, 228, 230, 232, 244, 246, 247, 249,
   283, 310, 312, 321

Quasi-associated   139, 140, 157
Remote operations service element   103, 298
Reservation (of resource), resource
   allocation   24, 71, 83, 104, 153, 162, 177,
   205, 244, 247, 251, 258, 278, 280, 310
Reset   151, 177, 181, 183
Resource reservation protocol   310
Restart   90, 160, 204, 289, 290
Restrictions   135, 200, 234, 252, 274, 278, 304,
   308
Roaming   241–243, 245, 246, 311

Screening   137, 237, 238, 309, 320
Service access point   34, 39
Service control point/function   222, 225,
   302–305, 311, 314, 316, 319–324
Service element   71, 150, 195, 196, 289
Session   27, 191, 192, 194–204, 206–208, 210,
   215–217, 220, 250, 251, 254, 255, 258, 281,
   283, 284, 310–317, 320
Set-top box/unit   194, 195, 206, 214, 215,
   217–219, 221, 225, 305, 312, 313
Signalling ATM adaptation layer   33, 81, 90,
   111, 138, 139, 148, 149, 158–160, 163, 164,
   178, 182, 188, 236, 289, 290
Signalling identifier   152, 163
Signalling message   29, 38, 82–90, 94, 95, 97,
   100, 101, 111, 132–134, 138–141, 151–154,
   158, 160, 161, 163–165, 175, 182–185, 199,
   203, 207, 208, 214, 217, 219, 221, 224, 235,
   248, 271–273, 280, 281, 295, 297, 298, 304,
   305, 307, 309
Soft PVC   133, 134, 141, 166, 250
Supplementary services   71, 76, 91, 95, 99,
   112, 120, 121, 136, 236–238, 241, 245, 287,
   288, 290, 298, 320
Sustainable cell rate   92, 93, 131, 278
Switched virtual circuit   81, 122, 124, 130,
   133, 143, 147, 148, 166, 233, 234, 237, 238,
   244, 250, 314

TCAP   142, 148, 149, 164, 298

UMTS   241–259
Uplink   126, 129, 244, 247, 267, 268
User-network interface   9, 20, 35, 38, 39, 53,
   56, 65, 79, 81, 90, 108, 111, 113, 119, 129,
   132, 133, 136, 141, 144, 161, 162, 169, 170,
   172–174, 176, 177, 181, 186–188, 234, 235,
   237, 262, 264, 271–273, 288, 314

Variable bit rate   155, 156, 246, 278
VB interface   65, 169–189, 320
Video on demand   68, 75, 191, 194, 196, 210,
   310
Videoconference   69, 227, 270–274, 306,
   310
Videotex   69

Virtual circuit   9–12, 20, 80, 81, 83, 84, 122, 126, 131, 133, 134, 147, 153, 156, 161, 165, 172, 175–177, 182, 185–188, 233, 244, 258, 273
Virtual home environment   242–245, 252, 253, 257, 259
Virtual path   9–12, 68, 81, 83, 84, 112, 126, 131, 133, 134, 147, 153, 156, 161, 165, 171–182, 185–188, 233, 272, 273
Virtual private network   227–231, 238
Voice   227, 228, 230–233, 239, 278, 321

Andersonian Library

Gifted through the
University of Strathclyde Alumni Fund Endowment

ANDERSONIAN LIBRARY
★
WITHDRAWN FROM LIBRARY STOCK
UNIVERSITY OF STRATHCLYDE

Books are to be returned on or before the last date below.

1 5 MAR 2001

3 0 JUN 2003

3 1 MAR 2004

LIBREX—